Arsenic Contamination in the Environment

Dharmendra Kumar Gupta • Soumya Chatterjee
Editors

Arsenic Contamination in the Environment

The Issues and Solutions

Springer

Editors
Dharmendra Kumar Gupta
Institute for Radioecology and Radiation Protection (IRS)
University of Hannover
Hannover, Germany

Soumya Chatterjee
Defence Research Laboratory
Defence Research and Development Organization (DRDO)
Tezpur, Assam, India

ISBN 978-3-319-54354-3 ISBN 978-3-319-54356-7 (eBook)
DOI 10.1007/978-3-319-54356-7

Library of Congress Control Number: 2017933683

Printed on acid-free paper

This Springer imprint is published by Springer Nature
The registered company is Springer International Publishing AG
The registered company address is: Gewerbestrasse 11, 6330 Cham, Switzerland

Dedicated to

Dr. APJ Abdul Kalam
(Born: October 15, 1931, Rameswaram, India;
Died: July 27, 2015, Shillong, India)

This book is dedicated to the memory of Dr. Avul Pakir Jainulabdeen Abdul Kalam (commonly known by the name Dr. A.P.J. Kalam also the "Missile man"), Bharat Ratna and the 11th President of India from 2002 to 2007. An esteemed scientist by profession, his highly inspirational attitude and words motivate all of us.

Foreword

It is a great honor for me to be invited to introduce this timely new multiauthored monograph on arsenic contamination of the environment, edited by Drs. D.K. Gupta and S. Chatterjee. There is no doubt that the issue of widespread environmental arsenic contamination in the human environment and its movement through ecosystems has become one of the critical issues of the decade. Indeed, arsenic is considered a most problematic element for control and remediation. While the impacts of this toxic metalloid have become global, it is of especial concern in Asia where contamination of potable water sources and staple foodstuffs, notably rice, is extremely serious. It is therefore timely to review and update the major challenges presented and possible solutions for environmental arsenic remediation. The 10 chapters presented in this book discuss the origin and sources and extent of arsenic contamination, its toxicology and analysis in environmental samples, regulatory issues and their effectiveness, and bio- and phytoremediation possibilities. I congratulate the authors in bringing together such a comprehensive coverage of this important topic.

Prof. Alan J.M. Baker, DSc
Honorary Professorial Fellow, School of BioSciences,
The University of Melbourne, Australia

Honorary Professor, Centre for Mined Land Rehabilitation, Sustainable Minerals Institute, The University of Queensland, Australia

Honorary Professor, Department of Animal and Plant Sciences, The University of Sheffield, UK

Visiting Professor, Laboratoire Sols et Environnement, ENSAIA, UMR 1120 UL – INRA, LABEX Ressources 21, Université de Lorraine, Nancy, France

Visiting Professor, School of Life Sciences, Sun Yat-sen University, Guangzhou, China

Preface

In the earth's crust, metalloid arsenic (As) is widely distributed with an average concentration of 2 mg kg^{-1}. Arsenic has been known as a poison since ages. It is commonly used in several industries like paints, dyes, metals, soaps, insecticides, and semi-conductors. Various anthropogenic activities such as mining, burning fossil fuels, paper production, and cement manufacturing lead to the release of As into the environment. Arsenic may exist in several forms, with As(III) and As(V) being the most prevalent toxic form of inorganic arsenic, which naturally occurs in the ground and surface water of many parts of the world including soils. Millions of people worldwide are exposed directly to this Group A carcinogen through drinking water and other uses. Even ingestion of low dose of As via food or water is the main pathway of this particular metalloid into the organism, where absorption takes place in the stomach and intestines, followed by release into the bloodstream. At chronic poisoning, As is converted by the liver to a less toxic form; however, the poisoning is also associated with a range of cancers in different organs prevalently in bladder, skin, and lung, with the highest relative risks found for cancer in the bladder. Arsenic toxicity depends on the chemical species [As(III) > As (V)] > organic species.

Generally As contamination affects plant growth and crop yield and its accumulation in food may severely affect animals as well as humans. Arsenate, chemically analogous to phosphate, is taken up by plants mainly through root phosphate absorption systems. Once it enters inside the plant cell, As(V) interferes with essential cellular processes, such as oxidative phosphorylation and ATP synthesis in mitochondria by replacing phosphate moiety, while the toxicity of As(III) is due to its propensity to bind to sulfhydryl groups, with subsequent damaging effects on general protein functioning. Only a few plants are well known for which they may suppress phosphate transport systems and there are even few candidates, where As methylation and volatilization would occur to a significant level. The efflux of As is also recently discovered and it is still ambiguous to what extent it is prevalent among plants. However, the synthesis of metalloid (As) binding peptides like glutathione (GSH) and phytochelatins (PCs) is observed in almost all plants studied to date and leads to instant detoxification (in terms of loss of reactivity in chelated form) of a specific ion. Complexation of arsenite [As(III)] with phytochelatins (PCs) is an important mechanism employed by plants to detoxify arsenic; how this complexation affects As mobility is yet to be fully understood. In plant tissue, solubility of As increased through chelation with cyclohexylenedinitrotetraacetic acid, nitrilotriacetic acid, or As-sulfur complexes and thus facilitates mobilization.

Natural biodegradation/bioremediation processes can clean up wide-ranging environmental contaminants. Organisms like bacteria and fungi or plants have the capability to degrade or detoxify hazardous environmental pollutants into less toxic forms. Understanding the mechanisms of As homeostasis may enable the selective breeding of more tolerant varieties or varieties with increased concentrations of essential minerals or decreased concentrations of As in the edible parts of the crops. Moreover, such knowledge could be used for developing plants for the purpose of phytoremediation.

The main purpose of the book is to provide detailed and comprehensive knowledge to the academicians, scientists, and students (undergraduate, graduate, and master) who are working in the field of arsenic contamination, sources, and their impact on environment. This book will encompass a range of topics like historical perspectives to recent advancement in the field of arsenic contamination and associated theories, effects of As on human health, available and prospective biological and physico-chemical options for combating arsenic contamination issue.

Dr. Dharmendra K. Gupta and Dr. Soumya Chatterjee personally convey their sincere thanks to Springer publisher as well as contributing authors for their valuable time and enthusiasm to write their perspective chapters on this very important issue to bring this book into the present shape.

Hannover, Germany — Dharmendra Kumar Gupta
Tezpur, Assam, India — Soumya Chatterjee

Contents

Arsenic Contamination from Historical Aspects to the Present 1
Dharmendra K. Gupta, Sarita Tiwari, B.H.N. Razafindrabe, and Soumya Chatterjee

Arsenic: Source, Occurrence, Cycle, and Detection 13
Soumya Chatterjee, Roxana Moogoui, and Dharmendra K. Gupta

Studies on Arsenic and Human Health 37
Soumya Chatterjee, Sibnarayan Datta, and Dharmendra K. Gupta

Epigenetics in Arsenic Toxicity: Mechanistic Response, Alterations, and Regulations 67
Dibyendu Talukdar

Prospects of Combating Arsenic: Physico-chemical Aspects 103
Soumya Chatterjee, Mridul Chetia, Anna Voronina, and Dharmendra K. Gupta

Arsenic and Its Effect on Major Crop Plants: Stationary Awareness to Paradigm with Special Reference to Rice Crop 123
Soumya Chatterjee, Sonika Sharma, and Dharmendra K. Gupta

Uptake, Transport, and Remediation of Arsenic by Algae and Higher Plants 145
Anindita Mitra, Soumya Chatterjee, and Dharmendra K. Gupta

Genomics and Genetic Engineering in Phytoremediation of Arsenic 171
Sarma Rajeev Kumar, Gowtham Iyappan, Hema Jagadeesan, and Sathishkumar Ramalingam

Potential of Plant Tissue Culture Research Contributing to Combating Arsenic Pollution 187
David W.M. Leung

Potential Role of Microbes in Bioremediation of Arsenic 195
Anindita Mitra, Soumya Chatterjee, and Dharmendra K. Gupta

Index 215

Contributors

Soumya Chatterjee Defence Research Laboratory, Defence Research and Development Organization (DRDO), Ministry of Defence, Tezpur, Assam, India

Mridul Chetia Department of Chemistry, D.R. College, Golghat, Assam, India

Sibnarayan Datta Defence Research Laboratory, Defence Research and Development Organization (DRDO), Ministry of Defence, Tezpur, Assam, India

Dharmendra K. Gupta Institut für Radioökologie und Strahlenschutz (IRS), Gottfried Wilhelm Leibniz Universität Hannover, Hannover, Germany

Gowtham Iyappan DRDO—BU Centre for Life Sciences, Bharathiar University, Coimbatore, India

Hema Jagadeesan Department of Biotechnology, PSG College of Technology, Coimbatore, India

David W.M. Leung School of Biological Sciences, University of Canterbury, Christchurch, New Zealand

Anindita Mitra Bankura Christian College, Bankura, West Bengal, India

Roxana Moogoui Department of Environmental Planning, Management and Education, Islamic Azad University, Tehran, North Branch, Tehran, Iran

Sathishkumar Ramalingam DRDO—BU Centre for Life Sciences, Bharathiar University, Coimbatore, India

Plant Genetic Engineering Laboratory, Department of Biotechnology, Bharathiar University, Coimbatore, India

B.H.N. Razafindrabe Faculty of Agriculture, University of the Ryukyus, Nishihara, Okinawa, Japan

Sarma Rajeev Kumar Molecular Plant Biology and Biotechnology Laboratory, CSIR—Central Institute of Medicinal and Aromatic Plants, Research Centre, Bangalore, India

Sonika Sharma Defence Research Laboratory, Defence Research and Development Organization (DRDO), Ministry of Defence, Tezpur, Assam, India

Dibyendu Talukdar Department of Botany, R.P.M. College, Hooghly, India

Sarita Tiwari Environmental Biotechnology Division, National Environmental Engineering Research Institute (CSIR-NEERI), Nagpur, India

Anna Voronina Department of Radiochemistry and Applied Ecology, Physical Technology Institute, Ural Federal University, Ekaterinburg, Russia

About the Editors

Dharmendra K. Gupta is Senior Scientist of Environmental Biotechnology at Institut für Radioökologie und Strahlenschutz, Gottfried Wilhelm Leibniz Universität Hannover, Hannover, Germany, and has published more than 80 refereed research papers/review articles in peer-reviewed journals and edited nine books. His field of research includes abiotic stress by heavy metals/radionuclides and xenobiotics in plants; antioxidative system in plants; and environmental pollution (heavy metal/radionuclide) remediation through plants (phytoremediation).

Soumya Chatterjee is Senior Scientist and head of the Department of Biodegradation Technology at Defence Research Laboratory (DRDO) at Tezpur, Assam, India. His area of research includes microbial biodegradation, abiotic stress in plants, bioremediation and phytoremediation, wastewater bacteriophages, sanitation, and metagenomics. He has already published more than 60 refereed research papers/review articles and book chapters in peer-reviewed journals/books (including as editor of book and journal special issue).

Arsenic Contamination from Historical Aspects to the Present

Dharmendra K. Gupta, Sarita Tiwari, B.H.N. Razafindrabe, and Soumya Chatterjee

Contents

1 Introduction ... 1
2 Historical Uses of Arsenic ... 3
3 Different Sources of Arsenic in the Environment ... 5
4 Some Epidemiological Studies of Arsenic Exposure from Water/Food Sources ... 6
5 Toxic Effects of Arsenic ... 7
6 Metabolic Pathway of Arsenic ... 8
7 Remediation Measures and Recommendations for Arsenic Decontamination ... 8
8 Conclusion ... 9
References ... 10

1 Introduction

Arsenic (As) is well known as "the king of poisons." The word "arsenic" elicits an appalling reaction for its mutagenic, carcinogenic, and teratogenic effects. Arsenic gains entry into the human body through various means, and chronic exposure even

D.K. Gupta (✉)
Institut für Radioökologie und Strahlenschutz (IRS), Gottfried Wilhelm Leibniz Universität Hannover, Herrenhäuser Str. 2, Hannover 30419, Germany
e-mail: guptadk1971@gmail.com

S. Tiwari
Environmental Biotechnology Division, National Environmental Engineering Research Institute (CSIR-NEERI), Nehru Marg, Nagpur 440020, India

B.H.N. Razafindrabe
Faculty of Agriculture, University of the Ryukyus, 1 Senbaru, Nishihara, Okinawa 903-0213, Japan

S. Chatterjee
Defence Research Laboratory, Defence Research and Development Organization (DRDO), Ministry of Defence, 2, Tezpur 784001, Assam, India

D.K. Gupta, S. Chatterjee (eds.), *Arsenic Contamination in the Environment*, DOI 10.1007/978-3-319-54356-7_1

at low levels are enough for the onset of visible toxic symptoms due to poisoning. The safe concentration of arsenic in drinking water is 10 μg L^{-1} (Ly 2012), according to the World Health Organization (WHO), with a maximum limit of 100 μg L^{-1} in untreated water prior to being treated for consumption. The maximum safe limit of arsenic ingestion for an average middle-aged person is approximately 220 μg day^{-1} (Ahuja 2008).

As per the elemental abundance, arsenic ranks twentieth in nature, fourteenth in seawater, and twelfth in the human body (Mandal and Suzuki 2002). According to one estimate the average concentration of As in the continental crust is 1–2 mg kg^{-1} (Taylor and McLennan 1985). The mean concentration of As in igneous rocks is about 1.5–3.0 mg kg^{-1} and in sedimentary rocks between 1.7 and 400 mg kg^{-1} (Smith et al. 1998). Arsenic is a major component in more than 245 minerals, ranking 52nd in crustal abundance (O'Neill 1995). Small amounts of biogenic As also enter in the water and soil through various biological sources. Arsenic is listed by the US Environmental Protection Agency (USEPA) as one of the priority pollutants and is listed among the most hazardous substances having significant potential threat to human health (Irwin et al. 1997). Arsenic was isolated in 1250 AD, and since then has been used in various fields such as medicine, agriculture, livestock, electronics, and metallurgy (Nriagu and Azcue 1990). Arsenic contamination causes severe health problem worldwide, because of its toxicity and elevated unhealthful presence in drinking water and other important water supplies, particularly groundwater, As contamination affects more than 70 countries on six continents (Ravenscroft et al. 2009). Arsenic is widely distributed in the environment, especially in soils, water, air, and marine biota. In soil, it is part in various minerals in the form of arsenates, arsenides, arsenites, sulfides, sulfosalts, oxides, silicates, and elemental As (Yan 1994). Several factors regulate As transport and accumulation in groundwater, including adsorption/desorption, precipitation/dissolution, redox potential (Eh), pH, concentration of competing ions, biological transformation, and methylation.

Arsenic contamination in groundwater in India was first studied in West Bengal in the year of 1983. Several other states, namely, Uttar Pradesh, Bihar, and Jharkhand in the flood plain of the Ganga River basin; Assam and Manipur in the flood plain of the Brahmaputra and its tributaries; and Rajnandgaon village in Chhattisgarh state, have been exposed to As-contaminated drinking water from hand pumps, with levels above the permissible limit of 50 μg L^{-1}. In West Bengal, As-contaminated groundwater has increased from 33 affected villages in four districts in 1983 to 3417 in 111 blocks in nine other districts in 2008. (Information downloaded from http://www.cgwb.gov.in/documents/papers/incidpapers/Paper%208%20-%20Ghosh.pdf; Shanmugapriya et al. 2015). In 1999, groundwater contamination by As and its health effects in the Rajnandgaon district of Chhattisgarh state were also detected. In 2002, Bhojpur district, located in the western part of the Bihar state, were reported to have contamination exceeding 50 μg L^{-1}. As of 2008, among 38 districts of Bihar, 57 blocks from 15 districts, with a total population of about 10 million, were marked as affected by As, especially groundwater contamination above 50 μg L^{-1}. (Information downloaded from http://www.cgwb.gov.in/documents/papers/incidpapers/Paper%20

8%20-%20Ghosh.pdf;http://www.who.int/bulletin/volumes/90/11/11-101253/en/; http://wrmin.nic.in/writereaddata/16_Estimates_on_arsenic.pdf.)

2 Historical Uses of Arsenic

Historically As dates back to the fourth century BC, when the Greek philosopher Aristotle referred to a material that perhaps was the arsenic sulfide mineral, realgar (AsS). Arsenic was infamous too in different dynasties, especially in imperial Rome. It seems certain that As compounds were used for murdering a person by poisoning (Vaughan 2006). Naturally, As can associate with various metals including copper, cobalt, nickel, silver, lead, and tin. These metals are usually exploited for various industrial purposes and As is produced as a by-product of their mining. A worthy historical example is the mines in Cornwall in Southwest England, famous for mining mainly copper and tin. This minefield was first developed by the Romans following their conquest of Britain in 43 AD. Until the late twentieth century, for about 2000 years, mining continued uninterrupted.

Arsenic, through ages, is being persistently regarded as a high-profile poison and was related to several conspicuous murder cases, among them the infamous death of Napoleon Bonaparte in 1851, which was claimed to be a political murder by some conspiracy theorists (Cullen 2008). Arsenic contamination is undetectable in food or beverages as it is odorless and tasteless and readily available (Bartrip 1992). In 1832, James Marsh invented a method to detect As by analytical methods to provide confirmation of "visible arsenic" (Cullen 2008). In 1888, Hutchinson (1888) was the first to associate arsenic as a reason for the skin carcinomas seen in patients taking arsenic for psoriasis. Pye-Smith (1913) published the first large clinical case series in 1913, reviewing 31 cases of skin cancer in patients who had been treated with arsenicals for many years mostly for psoriasis. Bradford-Hill and Faning (1948) reviewed the causes of mortality in a factory in Britain where sodium arsenite-containing powder was made and packaged.

Arsenic was first used for tick control in South Africa in 1893. Subsequently, the use of sodium salt of arsenous acid became popular worldwide for application to cattle, using cattle dipping vats (Baker 1982). The highly toxic nature of As has prompted an enormous amount of studies on its behavior in grassland, marine, and other related ecosystems (Jackson et al. 1979; Leah et al. 1992), soils at cattle tick dipping vats (Kimber et al. 2002), and the areas neighboring various metal smelters (Pilgrim and Hughes 1994; Temple et al. 1977), among others to enumerate its routes of contamination and introduction via potable water (Valentine et al. 1979; Sheppard 1992). The cattle-dipping program was banned in late 1960s, however, disused and dismantled vats containing As were eventually blended into the surrounding areas and caused major accidental poisonings time to time (Kimber et al. 2002).

In the first half of the nineteenth century, inorganic pesticides were normally used in agriculture and were found to be stable in the environment having an affinity to water. Sulfur, one of the common pesticides, was used as a fungicide against powdery mildew and was proved to be potent in controlling mites of all species. The discovery of Bordeaux mixture, a combination of copper sulfate and lime, as a fungicidal discovered accidentally in 1885 in France when a farmer applied it to the grapes of his garden to keep away children from eating the fruits (Schooley et al. 2008). In the early twentieth century, countries such as the United States, the United Kingdom, and France started finding major amounts of resources in agricultural research. Since the industrial revolution, the use of As as an insecticide, fungicide, and herbicide gradually peaked in 1950s when it was one of the most common pesticides in use (Peters et al. 1996), because it was an inexpensive by-product of the smelting of copper, iron, silver, cobalt, nickel, lead, gold, zinc, manganese, and tin. Agricultural workers favored using As over lead because of its accumulative poison characteristics over insects and pests (Schooley et al. 2008). Lead arsenate was extensively used in the United States. In Massachusetts, lead arsenate ($PbHAsO_4$) was first used as an insecticidal spray in 1892 against the gypsy moth, *Lymantria dispar* (Linnaeus). Furthermore, cultivators began using it to fight against the codling moth, *Cydia pomonella* (Linnaeus), a destructive insect pest of apples. Application of lead arsenate is very well known among farmers because of its instant action, cost effectiveness, and portable handling (Schooley et al. 2008). A report published in the USDA's *Agricultural Statistics* yearbook states that the US consumption rate of lead arsenate was 29.1 million pounds in 1929 (Murphy and Aucott 1998). The rate peaked with an estimated consumption of 86.4 million pounds in 1944 and then dropped to 3.9 million pounds in 1973. Lead arsenate uses declined from 1970s onwards because of the availability of more effective pesticides. Earlier in 1907, the German Imperial Health Commission opposed the utilization of lead arsenate on grapes as traces of arsenic and lead were found in the wines (Schooley et al. 2008). However, it is still used widely as a popular component having wide range of applications. Such long-term anthropogenic input of As into the environment is also a grave concern.

In 1955, it was reported that nearly 12,131 Japanese infants were poisoned by dried milk adulterated with arsenic and 130 died (Gorby 1988). Again in 1956, more than 400 people in Japan were poisoned by soy sauce accidentally contaminated with inorganic arsenic (Mizuta et al. 1956). Paracelsus, first described a lung disease predominant in miners in the Schneeberg mines of central Europe which caused early death (Paracelsus 1925).

Since the evolution of chemistry, arsenic has been used both as medicine and poison. For centuries humans have had some level of knowledge about the toxicity of arsenic. The medicinal properties of As were known to the ancient Greek physicians, and in recent times, possibly the most well known is Fowler's solution (1% potassium arsenite), which was used for many years as a tonic and a treatment for psoriasis and asthma (Graeme and Pollack 1998). In 1878, it was found that Fowler's solution could be effective in lowering the white blood cell count in leukemia patients (Antman 2001), thus, As has a history in cancer chemotherapeutic. In 1880, pharmacology texts reported the use of arsenical pastes for the treatment of skin and breast cancer (Antman 2001). On the other hand, As along with opium was used as a fatal poisoning in England, Wales, and France (Polson and Tattersall 1969). In the Middle Ages, As was in the limelight as a result of its notoriety as an effective homicidal and suicidal agent, both because of the frequency of its use and because of its involvement in many high-profile murders. In fact, As is often termed the "king of toxins" and the "toxic king" due to its potency and the discretion by which it could be directed, mainly with the aim of eliminating members of the governing class during the Middle Ages and Renaissance (Vahidnia et al. 2007), although known to society, As poisoning still occurs. In 2003, As poisoning made headlines when As was found in the coffee served at a church meeting in Maine (Maine Rural Health 2008; Zernike 2003). Thus, understanding of exposure to As may affect human health, and its countermeasures, are still important area of research.

3 Different Sources of Arsenic in the Environment

Under natural conditions, groundwater may contain different concentrations and ranges of arsenic, mainly as the result of the resilient effect of water–rock interfaces. Furthermore, favorable physical and geochemical conditions in aquifers help in As mobility and accumulation. This geogenic origin of As is intricately connected to the groundwater flow regime and aquifer geometry. There is no single accepted mechanism of the release of As in groundwater; rather it has been mainly accepted as being of natural geological origin, especially considering the conditions closely associated with the oxidation-reduction process of pyrite and iron oxide that caused As contamination in groundwater in Indian states including the northeastern part (Kumar 2015).

Major sources of As discharged in the environment are from varied anthropogenic activities and industries, including generation of commercial wastes (40%), coal ash (22%), mining industry (16%), and the atmospheric consequence from the steel industry (13%; Eisler 2004). Apart from the major sources as mentioned above, other industrial and related processes such as electrolytic processes, combustion of fossil fuels, wood preservation, urban wastes, medicinal use, fertilizers, sewage sludge, crop desiccants, pigments, biocides, glass, alloys, electronics are also significantly contributing towards arsenic-related pollution affecting large unconsolidated aquifers along numerous alluvial and deltaic plains around the world (Mulligan et al. 2001; Smedley and Kinniburgh 2002). Elevated As content in groundwater is one of the gravest concerns predominantly in southern, southeastern, and eastern parts of Asia, which is again due to the huge withdrawal of groundwater for various purposes such as drinking water supply, agriculture, industry, and so on (Kumar 2015).

Arsenic is analogous to phosphorus as they belong to the same chemical group and both have analogous dissociation constants for their acids and solubility products for their salts. Therefore, in soils, $H_2AsO_4^-$ and $H_2PO_4^-$ ions compete for the same sorption sites, although some sites are preferentially available for the sorption of either $H_2PO_4^-$ or $H_2AsO_4^-$ ions. A number of reports suggest that among the competing anions, the $H_2PO_4^-$ suppresses As(V) sorption in soil more significantly than chloride (Cl^-), nitrate (NO^-_3), and sulfate (SO_4^{2-}; O'Neill 1995; Matera and Le 2001). Figure 1 depicts some of the anthropogenic sources of arsenic and different forms of As in the environment.

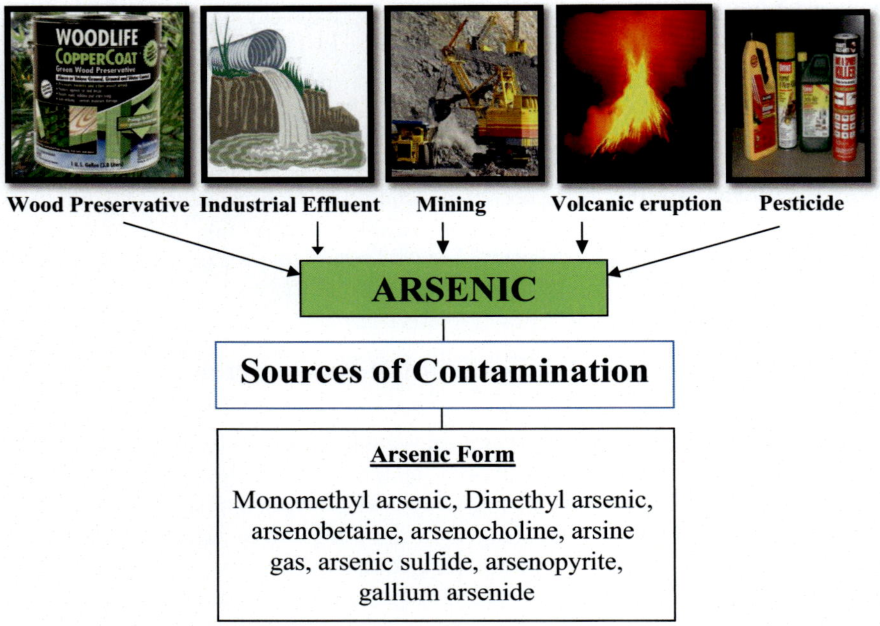

Fig. 1 Different sources and forms of arsenic in the environment

4 Some Epidemiological Studies of Arsenic Exposure from Water/Food Sources

In 1990s, there were only five major incidents of As contamination reported in groundwater in three Asian countries: Bangladesh, India (West Bengal state), and China. However, within the next five years, the reports of arsenic-related groundwater problems have arisen in different new sites on the Asian continent, including China, Mongolia, Nepal, Cambodia, Myanmar, Afghanistan, Iran, Vietnam, Korea, and Pakistan affecting several million people surviving with an extensive risk of chronic arsenic poisoning (Mukherjee et al. 2006). In most cases, human interference is the basic cause of As contamination. For example, it was reported that an anthropogenic source was primarily responsible for groundwater contamination in Vietnam. Since the late 1900s, Vietnamese farmers have been using arsenicals including monosodium methane arsenate (MSMA), disodium methane arsenate (DSMA), and cacodylic acid as pest control for crops in rural areas. These additions of As into food sources given rise to an inevitable uptake of As in plants, animals, and eventually, humans (Ly 2012).

Similarly, on the other side of the planet Earth, in many South American states such as Argentina, Uruguay, Ecuador, Cuba, Honduras, Dominican Republic, Colombia, Venezuela, Paraguay, Peru, and Chile, the prolonged problem of both geogenic and anthropogenic As contamination affects millions of people living in both urban and rural populations lacking safe drinking water facilities (Castro de Esparza 2003; Litter 2006; Bundschuh et al. 2009; Litter et al. 2010).

The indiscriminate use of pesticide such as lead arsenatein Australia before 1970 has been reported to cause chronic arsenic poisoning among workers (Tallis 1989). In China, the first area to be identified as a problematic area because of health effects arising from chronic arsenic exposure was the southwest coastal zone of Taiwan. Arsenic problems have also been documented in the northeastern parts of this country (Hsu et al. 1997). Concentrations of As above 50 μg L^{-1} have been recognized in groundwater from alluvial sediments associated with the River Danube in southern part of the Great Hungarian Plain, and also concentrations up to 150 μg L^{-1} (average 32 μg L^{-1}, 85 samples) have been reported by Varsányi et al. (1991). High As concentrations have also been identified in groundwater in the state of Sonora in northwest Mexico. Wyatt et al. (1998) found concentrations in the range of 2–305 μg L^{-1} (76 samples) with the highest concentration in groundwater from the towns of Hermosillo, Etchojoa, Magdalena, and Caborca.

Adverse health effects of As are strongly governed by the dose and duration of exposure, specific dermatological illness are caused due to chronic exposure towards arsenic. Again, localized groundwater arsenic-related problems are now being reported in an increasing number of countries and many new cases are being followed up or are likely to be discovered. Till lately, arsenic was not conventionally in the list of elements regularly tested by water quality testing laboratories and therefore many arsenic-rich sources will undeniably continue to be identified.

5 Toxic Effects of Arsenic

The Asian region is much more affected by As as compared to other regions of the world (Brinkel et al. 2009). Globally, aquifers of Bangladesh and West Bengal, India, represent the most severe As contamination and related health problems in the people residing in those areas. Concentrations in groundwater from the affected areas have a very huge range from <0.5 μg L^{-1} to about 3200 μg L^{-1} (DPHE/BGS/MML 1999- downloaded from: https://www.bgs.ac.uk/downloads/start.cfm?id=2223; CGWB, 1999- downloaded from: cgwb.gov.in/NEW/WQ/Geogenic%20Final.pdf). In late 1980s, resultant health complications were first recognized in West Bengal but the first diagnosis in Bangladesh was not made until 1993. Between 30 and 36 million peoples in Bangladesh are assessed as being exposed to As in drinking water at concentrations above 50 μg L^{-1} (BGS and DPHE 2001-downloaded from: https://www.bgs.ac.uk/downloads/start.cfm?id=2223) and up to 6 million in West Bengal, India. The affected aquifers are from the Holocene age and comprise a mixed sequence of micaceous sands, silts, and clays dumped by the Ganges, Brahmaputra, and Meghna river systems and their precursors and are usually shallow (less than 100–150 m deep; Chetia et al. 2011).

In many cases, due to the ease of As solubility in water, its concentration is elevated in groundwater above its recommended level of 0.05 mg L^{-1}. The trivalent oxidation state of arsenic compounds is the most toxicologically potent. This compound readily reacts with sulfur-containing compounds and generates reactive oxygen species (ROS). However, humans are exposed to both trivalent and pentavalent arsenicals (Hughes et al. 2011). The entry of As in humans is responsible for chromosomal irregularities, oxidative stress, altered DNA repair, altered DNA methylation, altered growth factors, cell proliferation and so on (Kapaz et al. 2006). Chronic exposure of As to humans exerts detrimental effects such as lesions to the skin, and mucous membrane of the digestive tract that result in a disease called arsenicosis. Furthermore, sufferers of arsenicosis develop cardiovascular, hepatic, renal, gastrointestinal, neurological, and reproductive problems and malignancies (Khan et al. 2009). About 38,000 cases of arsenicosis patients have been reported to date, but there is no specific treatment. Thus far, there are no effective medicinal drugs available for arsenicosis but some preventive symptomatic measures to fight with the disease such as use of safe drinking water, nutritious food, vitamins, and physical exercise are recommended (Mitra et al. 2004). Use of antioxidant multivitamins, various skin lotions, and drinking arsenic-free water have been shown to be beneficial to some extent for people who are in the initial stages of arsenicosis.

6 Metabolic Pathway of Arsenic

In mammals, fungi, and algae, detoxification of As usually involves methylation and other biotransformations such as incorporation of As into organic molecules by the formation of arsenocholine, arsenobetaine, or arsenosugars (Schmoger et al. 2000). Among the urinary arsenic species, dimethyl arsenic (DMA), inorganic arsenic (in the form of trivalent As), and monomethyl arsenic (MMA) are most common. In mammals, As^{III} and As^{V} are taken up by aquaglyceroporins and phosphate transporter inside cells. The key step in As detoxification is through conversion of As^{V} to As^{III} (Tiwari et al. 2015). In tolerant prokaryotes and some eukaryotes, the conversion of the arsenic form is enzyme-mediated in the presence of the arsenate reductase enzyme. However, there has been no report on the enzyme responsible for arsenate reduction in mammals. The As^{III} thus formed complex with glutathione and was transported out of the cell through Mrp isoforms (Sarangi et al. 2009). An alternative fate of As is methylation of As in the liver which is then removed from the body through excretion.

7 Remediation Measures and Recommendations for Arsenic Decontamination

Arsenic elimination from water for human consumption seems to be a very challenging task. Not a single universal method exists, and also the selection of method is reliant on the composition of water to be treated. Socioeconomic structures should

Table 1 Summary of selected conventional methods for arsenic remediation

Remediation Technology	Description
Ion exchange	• Reversible interchange of ions between solid and liquid phases • Resins have been used, • for example, Amberlite IRA 743
Ultrafiltration	• Semipermeable membrane is used • Suspended solids and solutes of high molecular weight are separated
Excavation	• Ex situ method that removes soil from site • Contaminated soil stored in designated landfill
Chemical precipitation	• pH adjustment to basic conditions (pH −11), for example, • alkali, sulfide, coagulant
Reverse osmosis	• Semipermeable membrane
Adsorption	• Semipermeable ion selective membranes, for example, • activated alumina
Solidification and stabilization	• Reduces the mobility of arsenic in soil • Contaminated soil is mixed with stabilizers in situ
Electrocoagulation	• Flocculating agent is generated by electro-oxidation of a sacrificial anode • Adsorbed onto activated carbon

be prudently considered to select the technology. Any method suitable for a specific area may not be generalized for other affected regions due to (a) geographical and geomorphological variations and (b) differing socioeconomic and literacy conditions of the people. As(V) is more efficiently removed from source waters than As(III) by iron coagulants, by precipitation of natural iron, and by adsorptive media. To remove As(III), a previous oxidation to As(V) is needed. Conventional technologies (coagulation cum precipitation, adsorption, reverse osmosis, etc.) can be applied at medium or large scale (Litter et al. 2010). Information regarding some of the conventional techniques used for As cleanup from different matrices is given in Table 1. Biological methods such as bioremediation and phytoremediation are also effective alternative techniques to that of the conventional method which is mostly preferable because of its cost-effective and ecofriendly approach.

8 Conclusion

Huge concern over the As contamination in groundwater is a challenging task. Aside from its primary role as part of a public health program, onsite village-scale testing of wells for As is required which may lead to a better understanding of the As problem and help in guiding future development of mitigating the contamination problem of groundwater resources. Therefore, an important task is to generate database of water quality in wells and tube-wells of contaminated regions meticulously.

Ideally the geographic coordinates using common GPS devices for water source should be recorded to prepare reliable large-scale maps of the locality. In areas having considerable arsenic problems, detailed consolidated studies should be undertaken to provide a comprehensive basis for understanding the problem and its variations in space and time. These could include more wide-range of water quality parameters including the entire major, and/or a range of minor, constituents. Again, problems can be tested for their environmental impact and sustenance-related parameters using isotopes such as ^3H and ^{14}C. Local geology and hydrogeology including water-level monitoring, pumping tests, and studies of the mineralogy, sedimentology, chemical composition, and adsorption behavior of the aquifer materials should also be considered. Either purpose-built piezometers or appropriate production wells should be examined regularly for changes in As and other water quality parameters. Appropriate user-friendly and cost-effective sustained effort is compulsory to mitigate mass poisoning by the element called arsenic.

References

Ahuja S (2008) Arsenic contamination of groundwater. Wiley & Sons Inc., Hoboken, NJ

Antman KH (2001) Introduction: the history of arsenic trioxide in cancer therapy. Oncologist 6:1–2

Baker JAF (1982) Some thoughts on resistance to ixodicides by ticks in South Africa, Symposium on ectoparasites of cattle, 15–16 March 1982. South African Bureau of Standards, Pretoria, South Africa, pp 53–67

Bartrip P (1992) A pennurth of arsenic for rat poison: the arsenic act, 1851 and the prevention of secret poisoning. Med Hist 36:53–69

Bradford-Hill BA, Faning EL (1948) Studies in the incidence of cancer in a factory handling inorganic compounds of arsenic. Br J Ind Med 5:1–15

Brinkel J, Khan MH, Kraemer A (2009) A systematic review of arsenic exposure and its social and mental health effects with special reference to Bangladesh. Int J Environ Res Public Health 6:1609–1619

Bundschuh J, Armienta MA, Bhattacharya P, Matschullat J, Birkle P, Mukherjee AB (2009) Natural arsenic in groundwater of Latin America–occurrence, health impact and remediation. Balkema Publisher, Lisse

Castro de Esparza ML (2003) The problem of arsenic in drinking water in Latin America. In: Murphy T, Guo J (eds) Aquatic arsenic toxicity and treatment. Backhuys Publisher, Leiden, pp 67–76

Chetia M, Chatterjee S, Banerjee S, Nath MJ, Singh L, Srivastava RB, Sarma HP (2011) Ground water arsenic contamination in Brahmaputra river basin: a water quality assessment in Golaghat (Assam), India. Environ Monit Assess 173:371–385

Cullen WR (2008) Is arsenic an aphrodisiac? The sociochemistry of an element. Royal Society of Chemistry, Cambridge

Eisler R (2004) Arsenic hazards to humans, plants, and animals from gold mining. Rev Environ Contam Toxicol 180:133–165

Gorby MS (1988) Arsenic poisoning. West J Med 149:308–315

Graeme KA, Pollack CV (1998) Heavy metal toxicity, part 1: arsenic and mercury. J Emerg Med 16:45–56

Hsu KH, Froines JR, Chen CJ (1997) Studies of arsenic ingestion from drinking-water in northeastern Taiwan: chemical speciation and urinary metabolites. In: Chappell WR, Abernathy CO,

Calderon RL (eds) Arsenic: exposure and health effects. Chapman & Hall, London, pp 190–209

Hughes MF, Beck BD, Chen Y, Lewis AS, Thomas DJ (2011) Arsenic exposure and toxicology: a historical perspective. Toxicol Sci 123:305–332

Hutchinson J (1888) On some examples of arsenic keratoses of the skin and of arsenic cancer. Trans Path Soc (London) 39:352–363

Irwin RJ, Van Mouwerik M, Stevens L, Seese MD, Basham W (1997) Environmental contaminants encyclopedia. National Park Service, Water Resources Division, Fort Collins, CO

Jackson DR, Ausmus BS, Levin M (1979) Effects of arsenic on nutrient dynamics of grassland microcosms and field plots. Water Air Soil Pollut 11:13–21

Kapaz S, Peterson H, Liber K, Bhattacharya P (2006) Human health effects from chronic arsenic poisoning–a review. J Environ Sci Health A Tox Hazard Subst Environ Eng 41:2399–2428

Khan NI, Owens G, Bruce D, Naidu R (2009) Human arsenic exposure and risk assessment at the landscape level: a review. Environ Geochem Health 31:143–166

Kimber SWL, Sizemore DI, Slavich PG (2002) Is there evidence of arsenic movement at cattle tick dip sites? Aust J Soil Res 40:1103–1114

Kumar C (2015) Status and mitigation of arsenic contamination in groundwater in India. Int J Earth Environ Sci 1:1–10

Leah RT, Evans SJ, Johnson MS (1992) Arsenic in place (*Pleuronectes platessa*) and whiting (*Merlangius merlangus*) from the north east Irish Sea. Mar Pollut Bull 24:544–549

Litter MI (2006) Final results of the OAS/AE/141 project: Research, development, validation and application of solar technologies for water potabilization in isolated rural zones of Latin America and the Caribbean. OAS Project AE141. OEA, Buenos Aires, Argentina. http://www.cnea.gov.ar/xxi/ambiental/agua-pura/default.htm

Litter MI, Morgada ME, Bundschuh J (2010) Possible treatments for arsenic removal in Latin American waters for human consumption. Environ Pollut 158:1105–1118

Ly TM (2012) Arsenic contamination in groundwater in Vietnam: An overview and analysis of the historical, cultural, economic, and political parameters in the success of various mitigation options. Pomona Senior Theses. Paper 41. http://scholarship.claremont.edu/pomona_theses/41

Maine Rural Health (2008) Maine Rural Health Association 2003 Outstanding service award. http://www.maineruralhealth.org/award.htm. Accessed 3 Dec 2010

Mandal BK, Suzuki KT (2002) Arsenic around the world: a review. Talanta 58:201–235

Matera V, Le HI (2001) Arsenic behaviour in contaminated soils: mobility and speciation. In: Selim HM, Sparks DL (eds) Heavy metals release in soils. Lewis Publishers, Washington, DC, pp 207–235

Mitra SR, Guha Mazumder DN, Basu A, Block G, Haque R, Samanta S, Ghosh N, Smith MM, von Ehrenstein O, Smith AH (2004) Nutritional factors and susceptibility to arsenic caused skin lesions in West Bengal, India. Environ Health Perspect 112:1104–1109

Mizuta N, Mizuta M, Ito F, Ito T (1956) An outbreak of acute arsenic poisoning caused by arsenic contaminated soy sauce. Bull Yamaguchi Med School 4:131–149

Mukherjee A, Sengupta MK, Hossain MA, Ahamed S, Das B, Nayak B, Lodh D, Rahman MM, Chakraborti D (2006) Arsenic As contamination in groundwater: a global perspective with emphasis on the Asian scenario. J Health Popul Nutr 24:142–163

Mulligan CN, Yong RN, Gibbs BF (2001) Remediation technologies for metal contaminated soils and groundwater: an evaluation. Eng Geol 60:193–207

Murphy EA, Aucott M (1998) An assessment of the amounts of arsenical pesticides used historically in a geographical area. Sci Total Environ 218:89–101

Nriagu JO, Azcue JM (1990) In: Nriagu JO (ed) Arsenic in the environment. Part I: cycling and characterization. John Wiley & Sons, Inc., New York, pp 1–15

O'Neill P (1995) Arsenic. In: Alloway BJ (ed) Heavy metals in soil. Blackie Academic & Professional, Glasgow

Paracelsus (1925) Von der Bergsucht und anderen Krankheiten. In: Koelsch F (ed) Schriften aus dem Gesamtgebiet der Gewerbehygiene. Neue Folge, Heft 12, J Springer, Berlin

Peters GR, McCurdy RF, Hindmarsh TJ (1996) Environmental aspects of arsenic toxicity. Crit Rev Clin Lab Sci 33:457–493

Pilgrim W, Hughes RN (1994) Lead, cadmium, arsenic and zinc in the ecosystem surrounding a lead smelter. Environ Monit Assess 32:1–2

Polson CJ, Tattersall RN (1969) Clinical toxicology. Pitman, London, p 181

Pye-Smith RJ (1913) Arsenic cancer with description of a case. Proc Roy Soc Med 5:229

Ravenscroft P, Brammer H, Richards K (2009) Arsenic pollution: a global synthesis. Wiley-Blackwell, Hoboken, NJ, p 618

Sarangi BK, Kalve S, Pandey RA, Chakrabarti T (2009) Transgenic plants for phytoremediation of arsenic and chromium to enhance tolerance and hyperaccumulation. Transgen Plant J 3:57–86

Schmoger MEV, Matjaz O, Grill E (2000) Detoxification of arsenic by phytochelatins in plants. Plant Physiol 122:793–801

Schooley T, Weaver MJ, Mullins D, Eick M (2008) The history of lead arsenate use in apple production: comparison of its impact in Virginia with other states. J Pestic Saf Educ 10:22–53

Shanmugapriya SP, Rohan J, Alagiyameenal D (2015) Arsenic pollution in India–an overview. J Chem Pharm Res 7:174–177

Sheppard SC (1992) Summary of phytotoxic levels of soil arsenic. Water Air Soil Pollut 64:539–550

Smedley PL, Kinniburgh DG (2002) A review of the source, behaviour and distribution of arsenic in natural waters. Appl Geochem 17:517–568

Smith E, Naidu R, Alston AM (1998) Arsenic in the soil environment. A review. Adv Agron 64:149–195

Tallis GA (1989) Acute lead arsenate poisoning. Aust NZ J Med 19:730–732

Taylor SR, McLennan SM (1985) The continental crust: its composition and evolution. Blackwell Scientific, London

Temple SN, Linzon L, Chai BL (1977) Contamination of vegetation and soil by arsenic emissions from secondary lead smelters. Environ Pollut 12:311–320

Tiwari S, Sarangi BK, Nasim J, Yadav D (2015) In silico arsenate reductase gene evolution. Online J Bioinform 16:303–317

Vahidnia A, van der Voet GB, de Wolff FA (2007) Arsenic neurotoxicity–a review. Hum Exp Toxicol 26:823–832

Valentine HK, Kang OT, Spivey G (1979) Arsenic level in human blood, urine and hair in response to exposure via drinking water. Environ Res 20:24–32

Varsányi I, Fodre Z, Bartha A (1991) Arsenic in drinking water and mortality in the southern great plain, Hungary. Environ Geochem Health 13:14–22

Vaughan DJ (2006) Arsenic. Elements 2:2. doi:10.2113/gselements.2.2.71

Wyatt CJ, Fimbers C, Romo L, Mendez RO, Grijalva M (1998) Incidence of heavy metal contamination in water supplies in northern Mexico. Environ Res 76:114–119

Yan CH (1994) Arsenic distribution in soils. In: Nriagu J (ed) Arsenic in the environment, part I: cycling and characterization. John Wiley & Sons, Inc., New York, pp 17–49

Zernike K (2003) Arsenic case is considered homicide, main police say. The New York Times, New York

Arsenic: Source, Occurrence, Cycle, and Detection

Soumya Chatterjee, Roxana Moogoui, and Dharmendra K. Gupta

Contents

1 Introduction to Arsenic: Its Sources and Occurrence in the Environment 14
2 Properties of Arsenic and Its Different Species 15
3 Organic Arsenic Compounds 15
4 Arsenic Sinks and Iron Oxides 16
5 Volatile Arsenicals 18
6 Arsenic Biogeochemical Cycle 19
7 Environmental Transport and Distribution 19
8 Arsenic Kinetics and Metabolism 21
9 Arsenic Volatilization by Microorganisms 22
 9.1 Fungi 23
 9.2 Bacteria 23
 9.3 Methanoarchaea 25
 9.4 Other Eukaryotic Microorganisms 25
10 Analytical Methods for Arsenic Detection 26
 10.1 Sample Preparation and Treatment 26
 10.2 Atomic Spectrometry-Based Methods 27
 10.3 Electrochemical Detection 28
 10.4 Radiochemical Methods 28
 10.5 X-Ray Spectroscopy 29
11 Conclusion 29
References 29

S. Chatterjee
Defence Research Laboratory, Defence Research and Development Organization (DRDO), Ministry of Defence, 2, Tezpur 784001, Assam, India

R. Moogoui
Department of Environmental Planning, Management and Education, Islamic Azad University, Tehran, North Branch, Tehran, Iran

D.K. Gupta (✉)
Institut für Radioökologie und Strahlenschutz (IRS), Gottfried Wilhelm Leibniz Universität Hannover, Herrenhäuser Str. 2, Hannover 30419, Germany
e-mail: guptadk1971@gmail.com

D.K. Gupta, S. Chatterjee (eds.), *Arsenic Contamination in the Environment*,
DOI 10.1007/978-3-319-54356-7_2

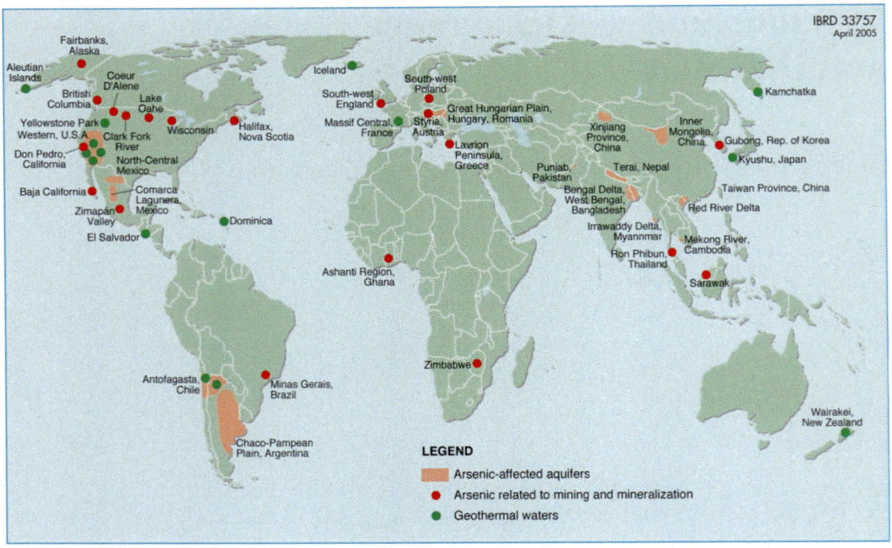

Fig. 1 Arsenic distribution in groundwater and in the environment worldwide. [Photo downloaded from http://web.worldbank.org/WBSITE/EXTERNAL/COUNTRIES/SOUTHASIAEXT/0,,contentMDK:22392781~pagePK:146736~piPK:146830~theSitePK:223547,00.html (Downloaded on 27.04.2016)]

1 Introduction to Arsenic: Its Sources and Occurrence in the Environment

Worldwide, arsenic exposure is one of the most dreaded public health crises with millions of people exposed to drinking water with arsenic concentrations that surpass the recommended limit of 10 μg L^{-1} (George et al. 2014). Most affected individuals live in southern Asian countries such as Bangladesh, India, Cambodia, Nepal, and Vietnam, and many countries in the world are affected by arsenic, including the United States, Argentina, Bolivia, Chile, Peru, and Mexico (George et al. 2014; Shibata et al. 2016; Yunus et al. 2016) (Fig. 1). The extent of the difficulty is very severe in Bangladesh, with nearly 85 million people (of a total population of ca. 125 million people) are affected by this curse. The adjoining state of West Bengal (India), with more than 6 million people, is also affected. Therefore, arsenic poisoning of groundwater affects about one third of the people in the Bengal basin (Sen 2013). In India, the Ministry report (2014; downloaded from: http://wrmin.nic.in/writereaddata/16_Estimates_on_arsenic.pdf) suggests 10 affected states: West Bengal, Bihar, Assam, Chhattisgarh, Manipur, Punjab, Haryana, Karnataka, Jharkhand, Uttar Pradesh.

Arsenic (As, with atomic number 33) is an element that naturally occurs in many minerals and can exist in various allotropes. More than 200 species of minerals contain arsenic, among which arsenopyrite (FeS) is most common. It is predicted that about one third of the atmospheric flux of arsenic is naturally derived, with volcanic eruption the most significant source. The geological origin of inorganic arsenic (iAs) has also become a considerably important source as it is found to be associated with groundwater. Arsenates present in the soil can dissolve easily in groundwater, which becomes the source of contaminated water flowing in rivers to the sea (Wang et al. 2014; Kim et al. 2015). The typical richness of As in the Earth's crust is between approximately 2 and 5 mg kg^{-1}. However, an enriched amount may be found in shale and coal deposits of sedimentary and igneous rocks (Smedley and Kinniburgh 2002). The occurrence of iAs with other metals such as iron (Fe), cobalt (Co), copper (Cu), nickel (Ni), silver (Ag), and gold (Au) is common (Tamaki and Frankenberger 1992). Direct mixing of arsenic into the aquatic environment may also be occurring through geothermal water, for example, the hot springs in Hot Creek, Nevada (Wilkie and Hering 1998). Arsenic adsorption to mineral surfaces (including Fe-, Mn-, and Al-rich soil and sediments) act as an important sink. This kind of mineral association is thought to be one of the important reasons for arsenic toxicity in groundwater, especially in the case of the Bengal delta aquifer, where Fe(III) oxide coatings on weathered alluvial sediments helping in release of arsenic into groundwater upon reductive dissolution of the Fe(III) oxide coating (Nickson et al. 1998; Nickson et al. 2000; Chowdhury et al. 2000; Acharya 2002; McArthur et al. 2004; Khalequzzaman et al. 2005; Ahuja 2008).

2 Properties of Arsenic and Its Different Species

Arsenic and its compounds occur naturally in trace quantities in all rock, soil, water, and air in various forms including crystalline, amorphous, powder, and vitreous. However, concentrations and the specific nature of species may vary due to a number of factors such as weathering, pH and Eh of the ambience, physical, chemical, biological (microbial), and anthropogenic activities, among others. Typically, arsenite compounds predominantly present under reducing or anoxic and waterlogged conditions (<200 mV) whereas arsenate species dominate in oxidizing and aerated conditions. The identity of chemical arsenic is represented in Table 1. In addition, naturally occurring and environmentally important arsenic species are presented in Table 2.

3 Organic Arsenic Compounds

There is a range of organic arsenic (oAs) compounds present, according to the metabolism and transformation of iAs into a less toxic form within the body of organisms. In marine creatures these oAs include arsenobetaine, tetramethylarsonium salts, arsenocholine, lipids (arsenolipids), and arsenic containing sugars (arsenosugars), even though some of these compounds have also been found in

Table 1 Identity of elemental arsenic

Periodic table:	Group 15 element
Atomic number:	33
Atomic mass:	74.91
Valence states:	–3, 0, +3, and +5
Chemical Abstract Service (CAS):	7440-38-2
National Institute for Occupational Safety	HSB 509
Health Registry of Toxic Effects of Chemicals (RTECS):	CG 05235 000
Hazardous Substances Data Bank (HSDB):	03300100X
UN transport class numbers:	UN 1558

terrestrial species (Table 3). The major sources of anthropogenic arsenic contamination of water, soil, and air are mining and smelting of nonferrous metals, burning of fossil fuels, coal-fired power generation plants, and indiscriminate use of arsenic-containing products (Carlin et al. 2016). Approximately 70% of the world arsenic production is used in manufacturing copper chrome arsenate (CCA) for wood/timber treatment and 22% in making a variety of herbicides, pesticide-like agricultural chemicals. Arsenic (As_2O_3) in the atmosphere mainly exists as adsorbed on particulate matter, which circulates and is returned to the Earth by wet or dry deposition and simultaneous oxidation and reconversion of arsenic to nonvolatile forms. In anoxic/reducing conditions, As(III) is thermodynamically stable, whereas As(V) is stable under more aerobic conditions, but they often co-occur in both anoxic and oxic soils and water (Anderson and Bruland 1991; Dowdle et al. 1996). Depending on the pH, redox potential (Eh), oxygen enrichment status, organic and other dissolved matter contents, clay particle composition, and biological processes (bacterial action), oxidation interchange may be effective between arsenate and arsenite species and can lead to potential arsenic release.

4 Arsenic Sinks and Iron Oxides

Abundantly present natural Fe(III) oxides (Fe(III) oxyhydroxides, Fe(III) hydroxides, and Fe(III) oxides) also play an important role as an arsenic sink (Hering and Kneebone 2001; Aide et al. 2016). Again, predominant biotransformations of arsenic species include interconversion between arsenite and arsenate, reduction and arsenic methylation, and organoarsenic biosynthesis. Because of relatively neutral pH, mobility of organoarsenicalsis is greater in sediment environments, generally contributing approximately 10% of the total arsenic in water (NRC 2001; Newman 2000). A schematic of configurations for the arsenic adsorbed on iron oxide surfaces follows (adopted from Cornell and Schwertmann 1996).

Table 2 Naturally occurring and environmentally important inorganic arsenic species (after Gomez-Caminero et al. 2001; WHO 2011)

	Name	CAS No.	Structure
Naturally occurring arsenic species	Arsenate		AsO_4^{3-}
	Arsenite		(different arsenite anions are known) AsO_3^{3-} orthorsenite, $[AsO^{-2}]^n$ metarsenite, $As_2O_5^{4-}$ pyrorsenite, $As_3O_7^{5-}$ polyarsenite, $As_4O_9^{6-}$ polyarsenite, $[As_6O_{11}^{4-}]^n$, polymeric anion
	Methylarsonic acid/monomethylarsonic acid/MMA	124-58-3	CH_5AsO_3
	Dimethylarsinic acid/cacodylic acid/DMA	75-60-5	$C_2H_7AsO_2$
	Trimethylarsine oxide	4964-14-1	C_3H_9AsO
	Tetramethylarsonium ion	27742-38-7	$C_4H_{12}As^+$
	Arsenobetaine	64436-13-1	$C_5H_{11}AsO_2$
	Arsenocholine	39895-81-3	$C_5H_{14}AsO^+$
	Lead arsenate	10102-48-4	$PbHAsO_4$
	Potassium arsenate	7784-41-0	KH_2AsO_4
	Potassium arsenite	10124-50-2	$KAsO_2HAsO_2$
	Dimethylarsinoylribosides		
	Trialkylarsonioribosides		
	Dimethylarsinoylribitolsulfate		
Inorganic As, trivalent	As(III) oxide/As trioxide/arsenous oxide/white As	1327-53-3	As_2O_3 (or As_4O_6)
	Arsenenous acid/arsenious acid	13768-07-5	$HAsO_2$
	As(III) chloride/As trichloride/arsenoustrichloride	7784-34-1	$AsCl_3$
	As(III) sulfide/As trisulfide orpiment/Auripigment	1303-33-9	As_2S_3
Inorganic As, pentavalent	As(V) oxide/As pentoxide	1303-28-2	As_2O_5
	Arsenic acid/*ortho*arsenic acid		H_3AsO_4
	Arsenenic acid/*meta*arsenic acid	10102-53-1	$HAsO_3$
	Arsenates/salts of *ortho*arsenic acid		$H_2AsO_4^-$, $HAsO_4^{2-}$, AsO_4^{3-}

Mononuclear monodentate:	Fe	–O–	As
Mononuclear bidentate:	Fe<	O O>	As
Binonuclear bidentate:	Fe Fe	–O –O>	As

Table 3 Naturally occurring and environmentally important organic arsenic species (after Gomez-Caminero et al. 2001)

	Name	CAS No.	Structure
Organic arsenic	Methylarsine	593-52-2	CH_3AsH_2
	Dimethylarsine	593-57-7	$(CH_3)_2AsH$
	Trimethylarsine	593-88-4	$(CH_3)_3As$
	(4aminophenyl) arsonic Acid/arsanilic acid/*p*aminobenzenearsonic acid	98-50-0	
	4,4arsenobis(2aminophenol) Dihydrochloride/arsphenamine/salvarsan	139-93-5	
	[4[aminocarbonylamino]Phenyl] arsonic acid/ Carbarsone/ *n*carbamoylarsanilic acid	121-59-5	
	[4[2amino2oxoethyl)amino]phenyl]arsonic acid/ tryparsamide	554-72-3	
	3nitro4hydroxyphenylarsonic acid	121-19-7	
	4nitrophenylarsonic acid/*p*nitrophenylarsonic acid	98-72-6	
	Dialkylchloroarsine		R_2AsCl
	Alkyldichloroarsine		$RasCl_2$

5 Volatile Arsenicals

Approximately 20 As species are arsines (AsH_3, monomethylarsine-($(CH_3)AsH_2$), dimethylarsine-($CH_3)AsH_2$), trimethylarsine-$(CH_3)_3As$, and diarsine (As_2H_4) (bp:) with their boiling points −62.5, −2, +36, +52, and +100°C, respectively (Planer-Friedrich et al. 2006). The complete methylated arsines constitute a volatile group of trivalent arsenic compounds that are partitioned into the atmosphere from aqueous solutions due to low (below 150°C) boiling points (Gong et al. 2002; Mestrot et al. 2013). Interestingly, arsine formation could well alleviate As poisoning. Global arsenic volatilization from land surfaces to the atmosphere has been estimated about 2.1×10^7 kg (Srivastava et al. 2011). Several microorganisms (bacteria, fungi, and algae) are capable of reducing intercellular As to form arsines (Mestrot et al. 2013; Yin et al. 2011a, b; Jia et al. 2012, 2013). In contact with HCl vapor, arsine can form volatile chloroarsine ($AsCl_3$), monomethylchloroarsine (CH_3) $AsCl_2$, and dimethylchloroarsine $(CH_3)_2AsCl$ (Mester and Sturgeon 2001). Furthermore, volatile As-S and As-Cl species ($(CH_3)_2AsCl$, $(CH_3)_2AsSCH_3$, and CH_3AsCl_2) have also been identified as some degree of microbial interaction for volatilization (Planer-Friedrich et al. 2006). Studies have shown that the toxicity of soluble inorganic and organic arsenic species in the order dimethylarsenite (DMAs(III)) and monomethylarsenite (MMAs(III)) > As(III) > As(V) > dimethylarsenate, (DMAs(V)), monomethylarsenate (MMAs(V)) > trimethylarsine (TMAs), trimethylarsine oxide (TMAsO) (Akter et al. 2005; Wang et al. 2014).

6 Arsenic Biogeochemical Cycle

Between organic and inorganic arsenic, iAs is of more interest due to its contribution to the biogeochemical cycle (Fig. 2), although involvement of organoarsenicals is insignificant. The iAs present in four oxidation states: As(III), As(0), As(III), and As(V). Microorganisms play a key role in the arsenic geocycle (Mukhopadhyay et al. 2002) by the processes of oxidation, reduction, methylation, and demethylation of arsenic species. However, microbial transformations between 3 and 5 oxidation states affect its mobility and speciation in the environment. To date various species belongs to different genera have been reported to be involved in arsenic chemistry. Some bacteria such as sulphate reducers and iron oxidizers are reported to precipitate As with their metabolic product out of the cell resulting in an insoluble form (Keimowitz et al. 2007; Battaglia-Brunet et al. 2012). Bacteria mediating Fe(II) oxidation through nitrate reduction in anoxic conditions have been shown to play a key role in arsenic cycling by forming solid hydrous ferric oxide on which As(V) (Senn and Hemond 2002; Hohmann et al. 2009) or As(III) (Gibney and Nusslein 2007; Hohmann et al. 2009) sorbs. As(III) can also be trapped with Fe(III), tightly bound to extracellular polymeric substances. Prokaryote metabolisms actively participated in mobilization of arsenic from the solid phase into the aqueous phase in a subsurface drinking water aquifer. Some fungi biotranform arsenic into volatile As gas (Pakulska and Czerczak 2006).

7 Environmental Transport and Distribution

The emission of arsenic into the atmosphere is either related to high-temperature processes (such as volcanic eruptions, coal-fired power generation plants, burning contaminated vegetation, etc.) or normal release through reduction and biomethylation to arsines (Mukhopadhyay et al. 2002). Arsenic (as As_2O_3) is primarily released into the atmosphere, adsorbed on particulate matter, circulated by the wind, and released to the Earth by dry or wet deposition. Similarly, the oxidation process in the air reconverts volatile arsines into the nonvolatile forms that settle back to the Earth. Arsenic species including arsenate, arsenite, monomethylarsonic acid (MMA), and dimethylarsinic acid (DMA) are usually reported in water, with the thermodynamically more stable pentavalent state (arsenate) predominating in oxygenated water and sediments (Gomez-Caminero et al. 2001; David and Hemond 2002; Chen G et al. 2015; Chen Y et al. 2015). The interchange in oxidation and solubility of arsenic species depend upon various factors such as Eh, pH, soluble arsenic concentration, organic content, and biological activities. These factors further affect the environmental behavior of many arsenic species and subsequent transportation either by wind or by water (Gomez-Caminero et al. 2001). As discussed, the concentration of the arsenic level in the environment depends upon various factors. Mean total arsenic concentrations in air, water, soil, sediments, and marine water are represented in Table 4.

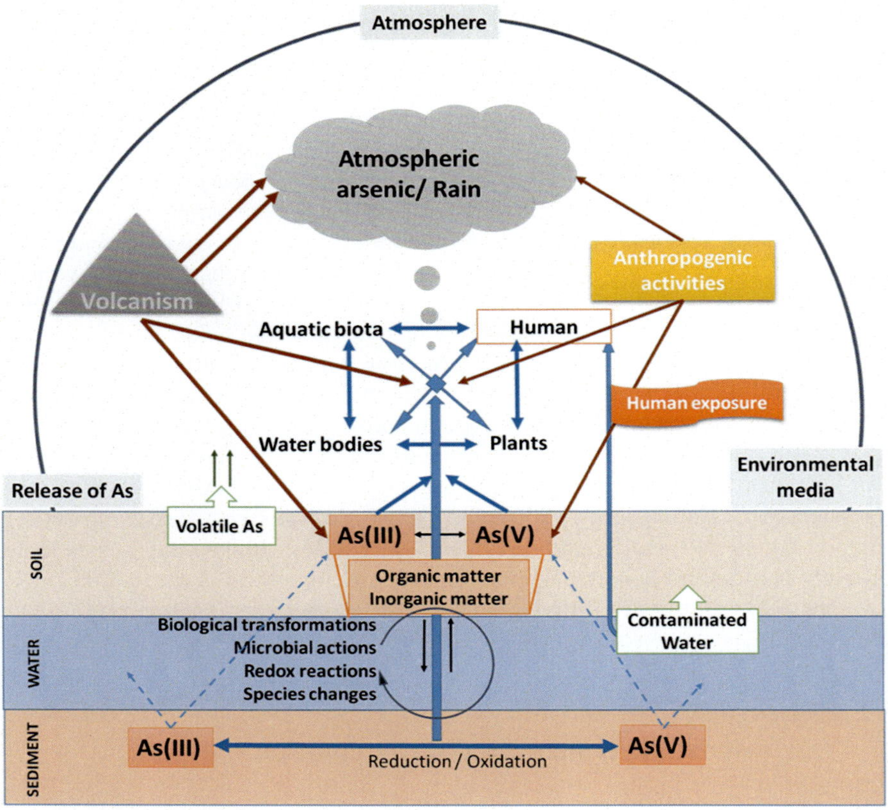

Fig. 2 Schematic representation of the biogeochemical cycle of arsenic

Arsenic concentrations in biota vary widely and depend upon exposure (higher in places with geothermal or anthropogenic sources) to arsenic. Bioconcentration factors (BCFs) in freshwater fish and invertebrates are less than for marine organisms regarding arsenic compounds (Chen G et al. 2015; Chen Y et al. 2015). Background arsenic concentrations in freshwater and terrestrial biota are usually less than 1 mg kg^{-1} (fresh weight). Root uptake or leaf-deposited arsenic and its subsequent adsorption leads to accumulations in terrestrial plants. Less than 1 mg kg^{-1} to more than 100 mg kg^{-1} arsenic compounds are reported to be present in marine biota with common compounds: arsenosugars (macroalgae) and arsenobetaine (invertebrates and fish; Gomez-Caminero et al. 2001; López-Serrano Oliver et al. 2011; Duan et al. 2015).

Human exposure to arsenic can either be occupational (industrial and other activities) or nonoccupational (contaminated water and food ingestion). There is hardly any limit for occupational exposure in small-scale unorganized industrial sectors, however, in workplaces with advanced occupational hygiene practices,

Table 4 Mean total arsenic concentrations in different components of environment

Components	Areas	Mean total arsenic concentrations
Air	Remote and rural areas	0.02–4 ng/m^3
	Urban areas	3–200 ng/m^3
	Vicinity of industrial sources	> 1000 ng/m^3
Freshwater	Rivers and lakes	<10 μg L^{-1} (vary widely)
	Groundwater	1–2 μg L^{-1}
	Groundwater (areas with volcanic rock and sulfide mineral deposits)	3 mg L^{-1}
Sediment	Natural	5–3000 mg kg^{-1}
Soil	Natural	1–40 mg kg^{-1}, (mean ~ 5 mg kg^{-1})
	Contaminated due to human activities	Up to several grams per 100 ml
Ocean	Seawater	1–2 μg L^{-1}

Adopted from, WHO 1987, 2011; USNRC 1999, 2001; Gomez-Caminero et al. 2001

exposure generally does not exceed 10 μg/m^3 8 h TWA (time-weighted average; Gomez-Caminero et al. 2001; Daria and Czerczak 2006; Rahman et al. 2013; Meliker et al. 2010). In nonoccupational exposures, ingestion and inhalation are the primary route of exposure. Although varying according to food stuff, approximately between 20 and 300 μg per person of total arsenic are being ingested daily (with 25% inorganic arsenic). Inorganic arsenic in cereals, poultry, meat, and dairy products is usually high, whereas in fish and shellfish is low. A smoker may be exposed to arsenic of approximately 10 μg day^{-1} and a nonsmoker about 1 μg day^{-1}. The pulmonary exposure of arsenic may also vary according to the air quality of the ambience. Inorganic arsenic and its metabolite (MMA and DMA) concentration in individual urine usually ranges from 5 to 20 μg As L^{-1}, but may even exceed 1000 μg L^{-1} (Gomez-Caminero et al. 2001; Rahman et al. 2013; Meliker et al. 2010).

8 Arsenic Kinetics and Metabolism

In general, the exposure route of arsenic in humans is ingestion, that is, oral intake of contaminated drinking water and food. Individuals who are occupationally exposed to arsenic are more prone to arsenic-related diseases. Typically less than 10 μg of arsenic is the standard for mean daily intake from drinking water. However, elevated arsenic concentrations in drinking water become a considerable source of inorganic arsenic within the body. Similarly, food stuffs are also important sources. In some circumstances where the drinking water requirement for preparation of food (such as rice, soups, or similar dishes) is obligatory, total arsenic ingestion occurs in even greater amounts (WHO 2011).

Ingested elemental arsenic is largely eliminated unchanged as it is poorly absorbed in the gut; however, soluble arsenic compounds are rapidly absorbed (Hindmarsh and McCurdy 1986), enter into the blood, and are rapidly and almost completely eliminated via kidney (especially arsenic(V) and organic arsenic; Buchet et al. 1981b; Luten et al. 1982; WHO 2011). Excess iAs is accumulated in the skin, liver, kidney, bone, and muscle (WHO 2011). Both the trivalent and pentavalent forms of arsenic are subjected to urinary excretion (half-life in humans is between 2 and 40 days; Pomroy et al. 1980). Sequential methylation to MMA and DMA in both As(III) and As(V) also takes place, but if daily intake exceeding more than 0.5 mg, saturates the process (Buchet et al. 1981b). Measuring the arsenic species in urine is the common technique to estimate the inner quantity of inorganic arsenic in individuals. Nonexposed individuals are reported generally having lower than 10 μg L^{-1} arsenic in urine, whereas in highly exposed individuals from areas such as West Bengal, India, and Bangladesh, urinary arsenic concentrations above 1 mg-L^{-1} have been observed (IPCS 2001; WHO 2011)

9 Arsenic Volatilization by Microorganisms

Several microorganisms including bacteria, archaea, fungi, and a couple of eukaryotes are capable of volatilizing arsenic (Table 5).

9.1 Fungi

A number of fungi are known to metabolize arsenicals into its volatile forms. The involvement of fungi and arsenical poisoning came into the picture after severe cases were reported in Germany and other parts of Europe as early as 1815 (Challenger 1945; Bartrip 1994). At that time, arsenical pigments (including Scheele's green (copper arsenite), Schweinfurt green (Paris green), Emerald green (copper acetoarsenite), and Vienna green contained in paints and wallpapers (because arsenic has an insecticidal property) were widely used in houses (Meharg 2003). In damp conditions, these arsenicals usually produce toxic compounds as fungi develop. The Italian physician, Bartolomeo Gosio, first reported the involvement of certain fungi (*Mucormucedo* [sic] and subsequent report of *Penicillium brevicaule*) in converting iAs into garlic-smelling volatile arsenicals (Gosio 1892a, b, 1893). This volatile garlic-odor toxic gas was called Gosio gas and identified as TMA (Challenger et al. 1933). Different studies have identified a long list of fungal species, including *Aspergillus glaucus*, *A. virens*, *A. niger*, *A. clavatus*, *Neosartorya fischeri*, *Mucor ramosus*, *Gliocladium roseum*, *Penicillium* sp., *Fusarium oxysporum*, *Trichoderma asperellum*, *Cephalothecium roseum*, *Trichoderma* sp., *Neocosmospora* sp., *Rhizopus* sp., and *Sterigmatocystis ochracea*, having the capability to volatilize arsenic (Bentley and Chasteen 2002; Urik et al. 2007; Cernansky et al. 2009; Su et al. 2010; Srivastava et al. 2011; Wang et al. 2014). The efficiency of arsenic volatilization of these fungi may vary with species and the conditions of the ambience, with a mean from 3 to 30% (Srivastava et al. 2011; Wang et al. 2014).

9.2 *Bacteria*

Woolson and Kearney (1973) first reported arsenic volatilization from soil under anaerobic flooded conditions. With the decrease in redox potential (Eh), soil arsenic methylation increases, signifying efficient arsenic volatilization in anaerobic conditions (Frohne et al. 2011). Reports indicate a number of intestinal microbiota, dominated by anaerobic bacteria and archaea thriving in different anaerobic conditions have the capability to volatilize arsenicals (Eckburg et al. 2005; Meyer et al. 2007; Sun et al. 2016). Michalke et al. (2008) reported that the arsines are produced in the human intestine ranging from a rate of 0.7–5 pmol h^{-1} kg^{-1} (dry weight), and the total As content of the samples was about 1.3 μmol kg^{-1} (dry weight) (Michalke et al. 2008; Wang et al. 2014). In another study on a simulated in vitro gastrointestinal model SHIME (simulator of the human intestinal microbial ecosystem), it has been reported that the human gut microbiome has the capacity of volatilization of metal(loid)s including arsenic (Van de Wiele et al. 2010).

Furthermore, several aerobic bacteria were also reported to volatilize arsenic. Bacterium of the Flavobacterium–Cytophaga group have been shown to generate TMA in the gas phase from a medium containing As(III; Honschopp et al. 1996). Majumder et al. (2013) reported arsenic volatilizing native soil bacteria (aerobic, anaerobic, and facultative), having the capacity to volatilize more than 30% of iAs in three days. Cyanobacteria, the lifeforms especially abundant in freshwaters and soils play an important role in the biogeochemical cycling of arsenic in aquatic systems (Yin et al. 2011a). The reported bacterial species (Table 5) include *Corynebacterium* sp., *E. coli*, *Proteus* sp., *Achromobacter* sp., *Aeromonas* sp., *Alcaligenes* sp., *Flavobacterium* sp., *Pseudomonas* sp., *Nocardia* sp., *Clostridium collagenovorans*, *Desulfovibrio gigas*, *D. Vulgaris*, and *Staphylococcus aureus* (Shariatpanahi et al. 1981; Wickenheiser et al. 1998; Michalke et al. 2000; Wang et al. 2014).

9.3 *Methanoarchaea*

In strict anaerobic conditions (such as in human and animal gut, wetland mud, etc.), methanogens (archaea) and sulphate-reducing bacteria (SRB) are other groups that can volatilize arsenic (Michalke et al. 2000). Methanoarchaea have the innate characteristic of volatilizing arsenic, which is very similar to the production of methane, where methylcobalamin acts as a methyl donor (Thomas et al. 2011). Among other methanobacteria, *Methanobacterium formicicum*, *Methanosphaera stadtmanae*,

Table 5 Different microbes and their activity for production of volatile arsenic species

Arsenic species (volatile)	Species	Substrate
Bacteria		
Dimethylarsine	*Corynebacterium* sp.	Arsenate
	Escherichia coli	
	Proteus sp.	
Mono-, di-, and trimethylarsine; arsine	*Pseudomonas* sp.	Methylarsonate, arsenate
	Nocardia sp.	Methylarsonate
Mono- and dimethylarsine	*Achromobacter* sp	
	Aeromonas sp.	
	Alcaligenes sp.	Arsenate, arsenite, methylarsonate
	Flavobacterium sp.	Methylarsonate
	Enterobacter sp.	
Trimethylarsine	*Veillonella alcalescens*	Trimethylarsine oxide
	Streptococcus sanguis	
	Fusobacterium nucleatum	
	Bacillus subtilis	
	Staphylococcus aureus	
	Flavobacterium cytophaga	Arsenite
	Clostridium collagenovorans	
	Desulfovibrio gigas	
	Desulfovibrio vulgaris	
	Rhodopseudomonas palustris	Arsenate or arsenite
Methanoarchaea		
Arsine	*Methanothermobacter thermautotrophicus*	Arsenate
	Methanobacterium barkeri	
Di- and trimethylarsine	*Methanosphaera stadtmanae*	
Mono-, di- and trimethylarsine	*Methanococcus vannielii*	
	Methanobacterium formicicum	
	Methanobrevibacter smithii	
Trimethylarsine	*Methanoplanus limicola*	
	Methanosarcina mazei	
Eukaryotic microorganisms		
Trimethylarsine	*Cyanidioschyzon* sp.	Arsenite
	Microcystis sp.	
	Nostoc sp.	
	Synechocystis sp.	
Total volatile arsenic	*Tetrahymena pyriformis*	
Fungi		

(continued)

Table 5 (continued)

Arsenic species (volatile)	Species	Substrate
Trimethylarsine	*Penicillium brevicaule*	Arsenic acid, monomethylarsonic acid
	Mucormucedo	
	Cephalothecium	
	Sterigmata cystis	
	Paecilomyces	Methylarsonate, dimethylarsinate
	Gliocladium roseum	
	Candida humicola	Trimethylarsine arsenate, arsenite and methylarsonate, dimethylarsinate
	Rhodotorula rubra	Arsenic trioxide
	Aspergillus glaucus	
	Aspergillus versicolor	Monomethylarsonic acid
	Penicillium chrysogenum	
	Penicillium notatum	
	Aspergillus fischeri	
Total volatile arsenic	*Ulocladium* sp.	Arsenate
	Neocosmospora sp.	
	Rhizopus sp.	
	Aspergillus clavatus	Arsenic acid
	Aspergillus niger	
	Trichoderma viride	
	Penicillium glabrum	
	Neosartorya fischeri	Arsenate or arsenite
	Fusarium oxysporum	

Adopted from: Wang et al. 2014

Methanobrevibacter smithii, and *Methanococcus vannielii* are reported to produce volatile arsenic compounds, $MeAsH_2$, Me_2AsH, and TMAs; *Methanobacterium barkeri*, *M. thermoautotrophicum*, and *M. formicicum* produce AsH_3 (Meyer et al. 2008; Wang et al. 2014).

9.4 Other Eukaryotic Microorganisms

The involvement of eukaryotic microorganisms, including aquatic alga and protozoans also has the capacity to volatilize arsenic. Qin et al. (2009) reported thermoacidophilic eukaryotic alga, *Cyanidioschyzon* sp., can reduce and methylate iAs to TMAs and DMAs at an optimal temperature of 60–70°C. There are other species such as marine green microalgae, *Ostreococcus tauri*, *Cyanidioschyzon* sp., that have the capacity through unique algal methyltransferases to methylate iAs

(Qin et al. 2009; Zhang et al. 2013). *Tetrahymena pyriformis*, a freshwater eukaryotic protozoan, is also capable of methylating iAs (Yin et al. 2011a, b; Zhang et al. 2012; Wang et al. 2014).

10 Analytical Methods for Arsenic Detection

Analysis for arsenic content of water and other environmental samples as well as food stuffs is an important issue as it is directly correlated with key decision making regarding the maximum contaminant level (MCL). Recently, the World Health Organization (WHO) recommended a guideline value of 10 $\mu g\ L^{-1}$ (10 ppb) arsenic in drinking water and considers all dissimilar arsenic species irrespective of their toxicity. However, guidelines for other environmental samples such as soil, sludge, and foodstuffs vary widely. In the United Kingdom, the value of arsenic content in domestic gardens, allotments, and play areas is 10 mg kg^{-1}, whereas landscapes and building hardcovers have a threshold of 40 mg kg^{-1}. But in Canada, the minimum value for contaminated soil is 12 mg kg^{-1}, soil for agriculture use 20 mg kg^{-1}, parkland 30 mg kg−1, and commercial and industrial land 50 mg kg^{-1} (CCME 2003; Francesconi 2007). Australia established a guideline value of foodstuffs as 1 mg kg^{-1} for total arsenic. In addition, some countries, such as China have incorporated import recommendations (for food items including rice) based on inorganic arsenic (Williams et al. 2005; Hossain et al. 2009; Garnier et al. 2010). Therefore, it is important to note that for any analytical instrument or analysis methodologies, performance parameters will vary according to the requirements. The qualitative and quantitative determination of arsenic in environmental samples (Table 6) should concentrate on maximum sensitivity, accuracy, user friendly, cost-effective technology with high throughput (Feldmann and Salaun 2008; Flora 2014).

10.1 Sample Preparation and Treatment

The collection of samples is the very first important step for appropriate measurement of arsenic present in the sample. Collection bottles must be cleaned properly before use to avoid contamination and avert speciation changes of the sample during transportation to the laboratory and storage before examination. (Freezing at −80°C is recommended.) Concentrated HCl may be added to biological fluid samples (such as urine) to prevent bacterial growth. Several types of filters (including polytetrafluoroethylene, glass microfiber, and cellulose ester) are used in sampling aerosol particles (Crecelius 1986; Tripathi et al. 1997; Gomez-Caminero et al. 2001; Hughes et al. 2011; Tripathi et al. 2012). Oxidative digestion (acid digestion, dry ashing, and microwave digestion) of samples is required before analysis (George and Roscoe 1951; George et al. 1973; Thomas et al. 1997).

A solvent extraction process may be followed for speciation study on arsenic (arsenite and arsenate). Several researchers have followed different techniques and

Table 6 Representing major methods used for analysis of different arsenic-containing samples

Method/instrument used	Samples for arsenic analysis
HGAAS	Blood, hair, urine, nails, food, water, air
NAA	Serum, urine, water
Colorimetric photometry	Urine, water/soil
XRF	Urine, soil, rock, sediments
GFAAS	Soft tissues, food, air
IEC-HGAAS	Urine
AES (direct current plasma)	Urine
HGAAS/TLC/ HRMS	Urine
GLC/ECD	Blood, tissue
HPLC/ICP-MS	Blood, tissue, urine, organism, food
HG-HCT/GC-MID	Water/soil
AAS	Water/soil
ICP-AES/ ICP-MS	Air/water/soil/ sediments/ solid wastes

Adopted from Gomez-Caminero et al. 2001; Feldmann and Salaun 2008
AAS atomic absorption spectrophotometry, *AES* atomic emission spectroscopy, *GC-MID* gas chromatography-multiple ion detection, *GFAAS* graphite furnace atomic absorption spectrometry, *HGAAS* hydride generation-atomic absorption spectroscopy, *HG-HCT* hydride generation-heptane cold trap, *HPLC* high-performance liquid chromatography, *ICP-AES* inductively coupled plasma-atomic emission spectrometry, *ICP-MS* inductively coupled plasma-mass spectrometry, *IEC* ion exchange chromatography, *NAA* neutron activation analysis, *XRF* X-ray fluorescence

chemicals to extract the arsenic compounds, which include chloroform, methanol-based extraction, extraction by ultrasonic treatment, selective chelation by sodium bis(trifluoroethyl) dithiocarbamate (NaFDDC), gas chromatograph (GC) detection, or supercritical fluid chromatography (SFC) detection (Hansen et al. 1992; Kuehnelt et al. 1997; Ng et al. 1998a, b).

10.2 Atomic Spectrometry-Based Methods

Laboratory detection of total arsenic can be accomplished using atomic spectrometric techniques such as atomic absorption spectrometry (AAS), atomic fluorescence spectrometry (AFS), inductively coupled plasma atomic emission spectrometry (ICP-AES), or inductively coupled plasma mass spectrometry (ICP-MS). These techniques are highly sensitive and accurate and can be used in a quality assurance/quality control (QA/QC) system (detection limit: sub-μg L^{-1} with ±10% accuracy) for detection of total arsenic, but are extremely expensive and require controlled laboratory practices including a skilled operator and stable power supplies (thus they are not suitable for field applications). Again, qualitative determination of contamination at the species level cannot be done unless coupled with other expensive instruments such as high-performance liquid chromatography (HPLC) or gas chromatography (GC; Feldmann and Salaun 2008; Duflou et al. 1987).

10.3 Electrochemical Detection

Stripping analysis is the basis of electroanalysis, where arsenic in samples accumulates on the surface of the working electrode (WE) by generating deposition potential (*E*dep*)*. The accumulated arsenic is then measured using potentiometric or voltammetry methods. This laboratory-based technique is a highly sensitive one (detection limit: sub-μg L^{-1}); cathodic stripping voltammetry (CSV) at a hanging mercury drop electrode (HMDE) is commonly used for arsenic detection (Holak 1980; Rasul et al. 2002; He et al. 2004; Muñoz and Palmero 2005). However, this technique consumes a high amount of mercury and is relatively complex and therefore not very acceptable for field conditions (Feldmann and Salaun 2008; Luong et al. 2014).

However, the use of solid electrodes is better for field conditions if anodic stripping voltammetry (ASV) or potentiometric stripping analysis (PSA; Muñoz and Palmero 2005) is utilized.

$$\underbrace{\mathrm{As(III)/As(V)} + 3/5e^- \rightarrow \mathrm{As(0)}}_{\text{Deposition}} \underbrace{\rightarrow \mathrm{As(III)} + 3e^-}_{\text{Stripping}}$$

In both cases, arsenic is deposited at a negative potential (As 0) and stripped back into solution either by sweeping the potential anodically (ASV) or by enforcing a continuous anodic current (CCPSA). Stripping is generally done under high acidic conditions (HCl is the electrolyte; Muñoz and Palmero 2005; Salaün et al. 2007; Feldmann and Salaun 2008). But the deterioration of the electrode is common due to strong hydrogen and chlorine generation and the formation of gold–chloride complexes (Feldmann and Salaun 2008). A few studies reported arsenite detection only under neutral or alkaline conditions with comparatively low deposition potential (Salaün et al. 2007).

Interference by metals such as Fe, Bi, Cu, Ni, Se, Sb, Hg, and Zn is common, affecting the analysis of arsenic in natural samples. Electrodes are also affected by dissolved organic matter (humic acid, fulvic acid); surfactants form intermetallic compounds and interfere strongly by adsorbing on the electrode surface and obstructing the electron transfer reaction (Feldmann and Salaun 2008). Proper sample preparation to protect the electrode against adsorption is necessary for the accurate measurement of arsenic (Majid et al. 2006). There are different brands of electrochemical-based measurement systems available on the market that are potentially applicable for field measurement (Feldmann and Salaun 2008).

10.4 Radiochemical Methods

Neutron activation analysis (NAA) is a radiochemical method (first reported by Orvini et al. 1974) for determination of arsenic, zinc, selenium, cadmium, and mercury. A variety of samples and standard reference materials was analyzed by the technique with a recovery of 98–100% (Chutke et al. 1994; Landsberger and Wu 1995; Gomez-Caminero et al. 2001; Calderon et al. 2013).

10.5 X-Ray Spectroscopy

PIXE (particle-induced X-ray emission) spectrometry and XRF (X-ray fluorescence) spectroscopy are important multielemental detection techniques. In PIXE, the sample (target) is bombarded with charged particles resulting in the emission of characteristic X-rays of the elements present with a detection limit of approximately 0.1 μg As g^{-1}. PIXE is a nondestructive technique that requires small samples (1 mg or less) and can be used for a wide range of environmental samples (Maenhaut 1987). Similarly, XRF is also widely used for the determination of arsenite and arsenate with a detection limit of 3.1 ng g^{-1} (Hagiwara et al. 2015; Sánchez-Rodas et al. 2015). X-ray absorption fine structure (XAFS) spectroscopy is another technique capable of detecting speciation data at realistic concentrations of 10–100 mg kg^{-1} (Huffman et al. 1993; Gomez-Caminero et al. 2001; Wang et al. 2011; Rouff et al. 2016).

11 Conclusion

The problem of arsenic in drinking water, food stuffs, and so on is growing day by day. Chronic exposure and/or ingestion of arsenic can cause a range of diseases such as skin lesions, diseases of the respiratory system, nervous system, reproductive system, cancer, and painful death. Development of appropriate, cost-effective, user-friendly field technology for assessing arsenic in environmental samples, water, and biological samples such as urine and blood is urgently required. Only concerted efforts from different sectors, including engineers, scientists, social scientists, government, and nongovernmental organizations along with the political willingness to curb the grave problem by developing, implementing, and monitoring better technologies can save humanity from this mass poisoning.

References

Acharya SK (2002) Arsenic contamination in groundwater affecting major parts of southern West Bengal and parts of western Chhattisgarh: source and mobilization processes. Curr Sci 82:740–744

Ahuja S (2008) Arsenic contamination of groundwater: mechanism, analysis, and remediation. John Wiley & Sons, Inc., Hoboken, NJ

Aide M, Beighley D, Dunn D (2016) Arsenic in the soil environment: a soil chemistry review. Intl J Appl Agri Res 11:1–28. ISSN 0973-2683

Akter KF, Owens G, Davey DE, Naidu R (2005) Arsenic speciation and toxicity in biological systems. Rev Environ Contam Toxicol 184:97–149

Anderson LC, Bruland KW (1991) Biogeochemistry of arsenic in natural waters: the importance of methylated species. Environ Sci Technol 25:420–424

Battaglia-Brunet F, Crouzet C, Burnol A, Coulon S, Morin D, Joulian C (2012) Precipitation of arsenic sulphide from acidic water in a fixed-film bioreactor. Water Res 46:3923–3933

Bartrip PWJ (1994) How green was my valance?: environmental arsenic poisoning and the Victorian domestic ideal. Eng Histor Rev 109:891–913

Bentley R, Chasteen TG (2002) Microbial methylation of metalloids: arsenic, antimony, and bismuth. Microbiol Mol Biol Rev 66:250–271

Buchet JP, Lauwerys R, Roels H (1981b) Urinary excretion of inorganic arsenic and its metabolites after repeated ingestion of sodium metaarsenite by volunteers. Int Arch Occup Environ Health 48:111–118

Calderon RL, Hudgens EE, Carty C, He B, Le XC, Rogers J, Thomas DJ (2013) Biological and behavioural factors modify biomarkers of arsenic exposure in a U.S. population. Environ Res 126:134–144

Carlin DJ, Naujokas MF, Bradham KD, Cowden J, Heacock M, Henry HF, Lee JS, Thomas DJ, Thompson C, Tokar EJ, Waalkes MP, Birnbaum LS, Suk WA (2016) Arsenic and environmental health: state of the science and future research opportunities. Environ Health Persp 124:890–899

CCME (Canadian Council of Ministers of the Environment)(2003) Canadian Environmental Quality Guidelines

Cernansky S, Kolenck M, Sevc J, Urik M, Hiller E (2009) Fungal volatilization of trivalent and pentavalent arsenic under laboratory conditions. Biores Technol 100:1037–1040

Challenger F (1945) Biological methylation. Chem Rev 36:315–361

Challenger F, Higginbottom C, Ellis L (1933) The formation of organo-metalloidal compounds by microorganisms. Part I Trimethylarsine and dimethylethylarsine. J Chem Soc 5:95–101

Chen G, Liu X, Brookes PC, Xu J (2015) Opportunities for phytoremediation and bioindication of arsenic contaminated water using a submerged aquatic plant: *Vallisneria natans* (lour.) Hara. Int J Phytorem 17:249–255

Chen Y, Moore KL, Miller AJ, McGrath SP, Ma JF, Zhao FJ (2015) The role of nodes in arsenic storage and distribution in rice. J Exp Bot 66:3717–3724

Chowdhury UK, Biswas BK, Chowdhury TR, Samantha G, Mandal B, Basu GC, Chanda CR, Lodh D, Saha KC, Mukherjee SK, Roy S, Kabir S, Quamruzzaman Q, Chakrabarti D (2000) Groundwater arsenic contamination in Bangladesh and West Bengal, India. Environ Health Perspect 108:393–397

Chutke NL, Ambulkar MN, Aggarwal AL, Garg AN (1994) Instrumental neutron activation analysis of ambient air dust particulates from metropolitan cities in India. Environ Pollut 85:67–76

Cornell RM, Schwertmann U (1996) The iron oxides-structure, properties, reactions, occurrence and uses. VCH Publishers, New York, p 573

Crecelius EA, Bloom NS, Cowan CE, Jenne EA (1986) Determination of arsenic species in limnological samples by hydride generation atomic absorption spectroscopy. In: Speciation of selenium and arsenic in natural waters and sediments, vol 2, Arsenic speciation, Electric Power Research Institute, Palo Alto, EA-4641, Project 2020-2, pp 1–28

Daria P, Czerczak S (2006) Hazardous effects of arsine: a short review. Int J Occup Med Environ Health 19:36–44

David SB, Hemond HF (2002) Nitrate controls on iron and arsenic in an urban lake. Science 296:2373–2376

Duan GL, Hu Y, Schneider S, McDermott J, Chen J, Sauer N, Rosen BP, Daus B, Liu Z, Zhu YG (2015) Inositol transporters AtINT2 and AtINT4 regulate arsenic accumulation in Arabidopsis seeds. Nat Plants 2:15202. doi:10.1038/nplants.2015.202

Dowdle PR, Laverman AM, Oremland RS (1996) Bacterial dissimilatory reduction of arsenic(V) to arsenic(III) in anoxic sediments. Appl Environ Microbiol 62:1664–1669

Duflou H, Maenhaut W, De Reuck J (1987) Application of PIXE analysis to the study of the regional distribution of trace elements in normal human brain. Biol Trace Elem Res 13:1–17
Eckburg PB, Bik EM, Bernstein CN, Purdom E, Dethlefsen L, Sargent M, Gill SR, Nelson KE, Relman DA (2005) Diversity of the human intestinal microbial flora. Science 308:1635–1638
Feldmann J, Salaun P (2008) Field test kits for arsenic: Evaluation in terms of sensitivity, reliability, applicability, and cost. In: Ahuja S (ed) Arsenic contamination of groundwater: Mechanism, analysis, and remediation. John Wiley & Sons, Inc., Hoboken, NJ
Flora SJS (2014) Handbook of arsenic toxicology. Academic Press, United Kingdom
Francesconi KA (2007) Toxic metal species and food regulations making a healthy choice. Analyst 132:17–20
Frohne T, Rinklebe J, Diaz-Bone RA, Laing GD (2011) Controlled variation of redox conditions in a floodplain soil: impact on metal mobilization and biomethylation of arsenic and antimony. Geoderma 160:414–424
Garnier JM, Travassac F, Lenoble V, Rose J, Zheng Y, Hossain MS, Chowdhury SH, Biswas AK, Ahmed KM, Cheng Z, van Geen A (2010) Temporal variations in arsenic uptake by rice plants in Bangladesh: the role of iron plaque in paddy fields irrigated with groundwater. Sci Total Environ 408:4185–4193
George CM, Sima L, Arias MHJ, Mihalic J, Cabrera LZ, Danz D, Checkley W, Gilman RH (2014) Arsenic exposure in drinking water: an unrecognized health threat in Peru. Bull WHO 92:565–572
George GM, Frahm LJ, McDonnell JP (1973) Dry ashing method for determination of total arsenic in animal tissues: collaborative study. J AOAC Int 56:793–797
George RK, Roscoe RS (1951) Microdetermination of arsenic and its application to biological material. Anal Chem 23:914–919
Gibney BP, Nüsslein K (2007) Arsenic sequestration by nitrate respiring microbial communities in urban lake sediments. Chemosphere 70:329–336
Gomez-Caminero A, Howe P, Hughes M, Kenyon E, Lewis DR, Moore M, Ng JC, Aitio A, Becking G (2001) Environmental health criteria 224 arsenic and arsenic compounds, 2nd edn. World Health Organization, Geneva
Gong ZL, Lu XF, Ma MS, Watt C, Chris LX (2002) Arsenic speciation analysis. Talanta 58:77–96
Gosio B (1892a) Action of microphytes on solid compounds of arsenic: a recapitulation. Science 19:104–106
Gosio B (1892b) S'ulriconoscimentodell' arsenico per mezzo di alcunemuffe. RivIg Sanit'a Pub 3:261–273
Gosio B (1893) Action de quelquesmoisissures Sur les compos'es fixesd' arsenic. Arch Ital Biol 18:253–265
Hagiwara K, Inui T, Koike Y, Aizawa M, Nakamura T (2015) Speciation of inorganic arsenic in drinking water by wavelength-dispersive X-ray fluorescence spectrometry after in situ preconcentration with miniature solid-phase extraction disks. Talanta 134:739–744
Hansen SH, Larsen EH, Pritzl G, Cornett C (1992) Speciation of seven arsenic compounds by high performance liquid chromatography with on-line detection by hydrogen-argon flame atomic absorption spectrometry and inductively coupled plasma mass spectrometry. J Anal Spectrom 7:629–634
He Y, Zheng Y, Ramnaraine M, Locke DC (2004) Differential pulse cathodic stripping voltammetric speciation of trace level inorganic compounds in natural water samples. Anal Chim Acta 511:55–61
Hering JG, Kneebone PE (2001) Biogeochemical controls on arsenic occurrence and mobility in water supplies. In: Frankenberger WT (ed) Environmental chemistry of arsenic. Marcel Dekker, New York, pp 155–181
Hindmarsh JT, McCurdy RF (1986) Clinical and environmental aspects of arsenic toxicity. Crit Rev Clin Lab Sci 23:315–347

Holak W (1980) Determination of arsenic by cathodic stripping voltammetry with a hanging mercury drop electrode. Anal Chem 52:2189

Hohmann C, Winkler E, Morin G, Kappler A (2009) Anaerobic Fe (II)-oxidizing bacteria show As resistance and immobilize As during Fe (III) mineral precipitation. Environ Sci Technol 44:94–101

Honschopp S, Brunken N, Nehrkorn A, Breunig HJ (1996) Isolation and characterization of a new arsenic methylating bacterium from soil. Microbiol Res 151:37–41

Hossain MB, Jahiruddin M, Loeppert RH, Panaullah GM, Islam MR, Duxbury JM (2009) The effects of iron plaque and phosphorus on yield and arsenic accumulation in rice. Plant Soil 317:167–176

Hughes MF, Beck BD, Chen Y, Lewis AS, Thomas DJ (2011) Arsenic exposure and toxicology: a historical perspective. Toxicol Sci 123:305–332

Huffman GP, Ganguly B, Zhao J, Rao KRPM, Shah N, Feng Z, Huggins FE, Taghiei MM, Lu F, Wender I, Pradhan VR, Tierney JW, Seehra MM, Ibrahim MM, Shabtai J, Eyring EM (1993) Structure and dispersion of iron-based catalyst for direct coal liquefaction. Energy Fuels 7:285–296

IPCS (2001) Arsenic and arsenic compounds. Geneva, World Health Organization, International Programme on Chemical Safety (Environmental Health Criteria 224)

Jia Y, Huang H, Sun GX, Zhao FJ, Zhu YG (2012) Pathways and relative contributions to arsenic volatilization from rice plants and paddy soil. Environ Sci Technol 46:8090–8096

Jia Y, Huang H, Zhong M, Wang FH, Zhang LM, Zhu YG (2013) Microbial arsenic methylation in soil and rice rhizosphere. Environ Sci Technol 47:3141–3148

Khalequzzaman M, Faruque FS, Mitra AK (2005) Assessment of arsenic contamination of groundwater and health problems in Bangladesh. Int J Environ Res Public Health 2:204–213

Keimowitz AR, Mailloux BJ, Cole P, Stute M, Simpson HJ, Chillrud SN (2007) Laboratory investigations of enhanced sulfate reduction as a groundwater arsenic remediation strategy. Environ Sci Technol 41:6718–6724

Kim HS, Kim YJ, Seo YR (2015) An overview of carcinogenic heavy metal: molecular toxicity mechanism and prevention. J Cancer Prev 20:232–240

Kuehnelt D, Goessler W, Irgolic KJ (1997) Arsenic compounds in terrestrial organisms II: arsenocholine in the mushroom amanita muscaria. Appl Organomet Chem 11:459–470

Landsberger S, Wu D (1995) The impact of heavy metals from environmental tobacco smoke on indoor air quality as determined by Compton suppression neutron activation analysis. Sci Total Environ 173-174:323–337

López-Serrano Oliver A, Sanz-Landaluze J, Muñoz-Olivas R, Guinea J, Cámara C (2011) Zebra fish larvae as a model for the evaluation of inorganic arsenic and tributyltin bioconcentration. Water Res 45:6515–6524

Luong JHT, Lamb E, Maleb KB (2014) Recent advances in electrochemical detection of arsenic in drinking and ground waters. Anal Methods 6:6157–6169

Luten JB, Riekwel-Booy G, Rauchbaar A (1982) Occurrence of arsenic in plaice (*Pleuronectes platessa*), nature of organo-arsenic compound present and its excretion by man. Environ Health Perspect 45:165–170

Maenhaut W (1987) Particle-induced X-ray emission spectrometry: an accurate technique in the analysis of biological environmental and geological samples. Anal Chim Acta 195:125–140

Majid E, Hrapovic S, Liu Y, Male KB, Luong JH (2006) Electrochemical determination of arsenite using a gold nanoparticle modified glassy carbon electrode and flow analysis. Anal Chem 78:762–769

Majumder A, Bhattacharyya K, Kole SC, Ghosh S (2013) Efficacy of indigenous soil microbes in arsenic mitigation from contaminated alluvial soil of India. Environ Sci Pollut Res 20:5645–5653

McArthur JM, Banerjee DM, Hudson-Edwards KA, Mishra R, Purohit R, Ravenscroft P, Cronin A, Howarth RJ, Chatterjee A, Talukder R, Lowry D, Houghton S, Chadha DK (2004) Natural

organic matter in sedimentary basins and its relation to arsenic in anoxic groundwater: the example of West Bengal and its worldwide implications. Appl Geochem 19:1255–1293
Meharg A (2003) The arsenic green. Nature 423:688
Meliker JR, Slotnick MJ, Av Ruskin GA, Schottenfeld D, Jacquez GM, Wilson ML, Goovaerts P, Franzblau A, Nriagu JO (2010) Lifetime exposure to arsenic in drinking water and bladder cancer: a population-based case-control study in Michigan, USA. Cancer Causes Control 21:745–757
Mester Z, Sturgeon RE (2001) Detection of volatile arsenic chloride species during hydride generation: a new prospectus. J Anal At Spectrom 16:470–474
Mestrot A, Planer-Friedrich B, Feldmann J (2013) Biovolatilization: a poorly studied pathway of the arsenic biogeochemical cycle. Environ Sci Processes Impacts 15:1639–1651
Meyer J, Schmidt A, Michalke K, Hensel R (2007) Volatilization of metals and metalloids by the microbial population of an alluvial soil. Syst Appl Microbiol 30:229–238
Meyer J, Michalke K, Kouril T, Hensel R (2008) Volatilisation of metals and metalloids: an inherent feature of methanoarchaea? Syst Appl Microbiol 31:81–87
Michalke K, Wickenheiser EB, Mehring M, Hirner AV, Hensel R (2000) Production of volatile derivatives of metal(loid)s by microflora involved in anaerobic digestion of sewage sludge. Appl Environ Microbiol 66:2791–2796
Michalke K, Schmidt A, Huber B, Meyer J, Sulkowski M, Hirner AV, Boertz J, Mosel F, Dammann P, Hilken G, Hedrich HJ, Dorsch M, Rettenmeier AW, Hensel R (2008) Role of intestinal microbiota in transformation of bismuth and other metals and metalloids into volatile methyl and hydride derivatives in humans and mice. Appl Environ Microbiol 74:3069–3075
Mukhopadhyay R, Rosen BP, Phung LT, Silver S (2002) Microbial arsenic: from geocycles to genes and enzymes. FEMS Microbiol Rev 26:311–325
Muñoz E, Palmero S (2005) Analysis and speciation of arsenic by stripping potentiometry: a review. Talanta 65:613–620
Newman DK (2000) Arsenic. In: encyclopedia of microbiology. Academic Press, San Diego, CA, pp 332–338
Ng JC, Johnson D, Imray P, Chiswell B, Moore M (1998a) Speciation of arsenic metabolites in the urine of occupational workers and experimental rats using an optimised hydride cold-trapping method. Analyst 123:929–933
Ng JC, Kratzmann SM, Qi L, Crawley H, Chiswell B, Moore MR (1998b) Speciation and absolute bioavailability: risk assessment of arsenic-contaminated sites in a residential suburb in Canberra. Analyst 123:889–892
Nickson RT, Mc Arthur JM, Burgess WG, Ahmed KM, Ravenscroft P, Rahman M (1998) Arsenic poisoning of Bangladesh groundwater. Nature 395:338
Nickson RT, Mc Arthur JM, Ravenscroft P, Burgess WG, Ahmed KM (2000) Mechanism of arsenic release to groundwater, Bangladesh and West Bengal. Appl Geochem 15:403–413
NRC (2001) Arsenic in drinking water update. National Academy Press, Washington, DC
Orvini E, Gillis TE, LaFleur PD (1974) Method for determination of selenium, arsenic, zinc, cadmium and mercury in environmental samples by neutron activation analysis. Anal Chem 46:1294–1299.
Pakulska D, Czerczak S (2006) Hazardous effects of arsine: a short review. Intl J Occup Med Environ Health 19:36–44
Planer-Friedrich B, Lehr C, Matschullat J, Merkel BJ, Nordstrom DK, Sandstrom MW (2006) Speciation of volatile arsenic at geothermal features in Yellowstone National Park. Geochim Cosmochim Acta 70:2480–2491
Pomroy C, Charbonneau SM, McCullough RS, Tam GK (1980) Human retention studies with 74As. Toxicol Appl Pharmacol 53:550–556
Qin J, Lehr CR, Yuan CG, Le XC, Mc Dermott TR, Rosen BP (2009) Biotransformation of arsenic by a Yellowstone thermoacidophilic eukaryotic alga. Proc Natl Acad Sci U S A 106:5213–5217

Rahman M, Sohel N, Yunus M, Chowdhury ME, Hore SK, Zaman K, Bhuiya A, Streatfield PK (2013) Increased childhood mortality and arsenic in drinking water in Matlab, Bangladesh: a population-based cohort study. PLoS One 8:e55014

Rasul SB, Munir AK, Hossain ZA, Khan AH, Alauddin M, Hussam A (2002) Electrochemical measurement and speciation of inorganic arsenic in groundwater of Bangladesh. Talanta 58:33–43

Rouff AA, Ma N, Kustka AB (2016) Adsorption of arsenic with struvite and hydroxylapatite in phosphate-bearing solutions. Chemosphere 146:574–581

Salaün P, Planer-Friedrich B, van den Berg CM (2007) Inorganic arsenic speciation in water and seawater by anodic stripping voltammetry with a gold microelectrode. Anal Chim Acta 585:312–322

Sánchez-Rodas D, de la Campa AM, Alsioufi L (2015) Analytical approaches for arsenic determination in air: a critical review. Anal Chim Acta 898:1–18

Sen G (2013) Cooperation between India and Bangladesh on control of arsenic poisoning: IDSA comment, Institute of Defence Studies and Analysis (Downloaded from: http://idsa.in/idsacomments/CooperationBetweenIndiaandBangladesh_GautamSen_270113)

Senn DB, Hemond HF (2002) Nitrate controls on iron and arsenic in an urban Lake. Science 296:2373–2376

Shariatpanahi M, Anderson AC, Abdelghani AA, Englande AJ, Hughes J, Wilkinson RF (1981) Biotransformation of the pesticide sodium arsenate. J Environ Sci Health Part B 16:35–41

Shibata T, Meng C, Umoren J, West H (2016) Risk Assessment of Arsenic in Rice Cereal and Other Dietary Sources for Infants and Toddlers in the U.S. Int J Environ Res Public Health 25:13

Smedley PL, Kinniburgh DG (2002) A review of the source, behaviour and distribution of arsenic in natural waters. Appl Geochem 17:517–568

Srivastava PK, Vaish A, Dwivedi S, Chakrabarty D, Singh N, Tripathi RD (2011) Biological removal of arsenic pollution by soil fungi. Sci Total Environ 409:2430–2442

Su SM, Zeng XB, Bai LY, Jiang XL, Li LF (2010) Bioaccumulation and biovolatilization of pentavalent arsenic by *Penicillin janthinellum*, *Fusarium oxysporum* and *Trichoderma asperellum* under laboratory conditions. Curr Microbiol 61:261–266

Sun J, Chillrud SN, Mailloux BJ, Stute M, Singh R, Dong H, Lepre CJ, Bostick BC (2016) Enhanced and stabilized arsenic retention in microcosms through the microbial oxidation of ferrous iron by nitrate. Chemosphere 144:1106–1115

Tamaki S, Frankenberger JWT (1992) Environmental biogeochemistry of arsenic. Rev Environ Contam Toxicol 124:79–110

Thomas F, Diaz-Bone RA, Wuerfel O, Huber B, Weidenbach K, Schmitz RA, Hensel R (2011) Connection between multimetal(loid) methylation in methanoarchaea and central intermediates of methanogenesis. Appl Environ Microbiol 77:8669–8675

Thomas P, Finnie JK, Williams JG (1997) Feasibility of identification and monitoring of arsenic species in soil and sediment samples by coupled high-performance liquid chromatography inductively coupled plasma mass spectrometry. J Anal Spectrom 12:1367–1372

Tripathi N, Kannan GM, Pant BP, Jaiswal DK, Malhotra PR, Flora SJ (1997) Arsenic-induced changes in certain neurotransmitter levels and their recoveries following chelation in rat whole brain. Toxicol Lett 92:201–208

Tripathi RD, Tripathi P, Dwivedi S, Dubey S, Chatterjee S, Chakrabarty D, Trivedi PK (2012) Arsenomics: omics of arsenic metabolism in plants. Front Physiol 3:275. doi:10.3389/fphys.2012.00275

Urik M, Cernansky S, Sevc J, Simonovicˇova A, Littera P (2007) Biovolatilization of arsenic by different fungal strains. Water Air Soil Pollut 186:337–342

USNRC-United States National Research Council (1999) Arsenic in drinking water. National Academy Press, Washington, DC

USNRC-United States National Research Council (2001) Arsenic in drinking water, 2001 update. National Academy Press, Washington, DC

Van De Wiele T, Gallawa CM, Kubachk KM, Creed JT, Basta N, Dayton EA, Whitacre S, Du Laing G, Bradham K (2010) Arsenic metabolism by human gut microbiota upon in vitro digestion of contaminated soils. Environ Health Perspect 118:1004–1009

Wang P, Sun G, Jia Y, Meharg AA, Zhu Y (2014) A review on completing arsenic biogeochemical cycle: microbial volatilization of arsines in environment. J Environ Sci 26:371–381

Wang Y, Morin G, Ona-Nguema G, Juillot F, Calas G, Brown GE Jr (2011) Distinctive arsenic(V) trapping modes by magnetite nanoparticles induced by different sorption processes. Environ Sci Technol 45:7258–7266

WHO (1987) Air quality guidelines for Europe. Copenhagen, WHO Regional Office for Europe (European Series No. 23)

WHO (World Health Organization) (2011) Arsenic in drinking-water: WHO/SDE/WSH/03.04/75/Rev/1 (Downloaded from: http://www.who.int/water_sanitation_health/dwq/chemicals/arsenic.pdf)

Wickenheiser EB, Michalke K, Drescher C, Hirner AV, Hensel R (1998) Development and application of liquid and gas-chromatographic speciation techniques with element specific (ICP-MS) detection to the study of anaerobic arsenic metabolism. Fresenius J Anal Chem 362:498–501

Wilkie J, Hering JG (1998) Rapid oxidation of geothermal arsenic(III) in stream waters of the eastern sierra Nevada. Environ Sci Technol 32:657–662

Williams PN, Price AH, Raab A, Hossain SA, Feldmann J, Meharg AA (2005) Variation in arsenic speciation and concentration in paddy rice related to dietary exposure. Environ Sci Technol 39:5531–5540

Woolson EA, Kearney PC (1973) Persistence and reactions of 14C cacodylic acid in soils. Environ Sci Technol 7:47–50

Yin XX, Chen J, Qin J, Sun GX, Rosen BP, Zhu YG (2011a) Biotransformation and volatilization of arsenic by three photosynthetic cyanobacteria. Plant Physiol 156:1631–1638

Yin XX, Zhang YY, Yang J, Zhu YG (2011b) Rapid biotransformation of arsenic by a model protozoan *Tetrahymena thermophila*. Environ Pollut 159:837–840

Yunus FM, Khan S, Chowdhury P, Milton AH, Hussain S, Rahman M (2016) A review of groundwater arsenic contamination in Bangladesh: the millennium development goal era and beyond. Int J Environ Res Public Health 13:215

Zhang SY, Sun GX, Yin XX, Rensing C, Zhu YG (2013) Biomethylation and volatilization of arsenic by the marine microalgae *Ostreococcus tauri*. Chemosphere 93:47–53

Zhang YY, Yang J, Yin XX, Yang SP, Zhu YG (2012) Arsenate toxicity and stress responses in the freshwater ciliate *Tetrahymena pyriformis*. Eur J Protistol 48:227–236

Studies on Arsenic and Human Health

Soumya Chatterjee, Sibnarayan Datta, and Dharmendra K. Gupta

Contents

1 Arsenic and Clinical Issues 38
2 Arsenic Poisoning—An Overview 39
3 Types of Arsenic Poisoning to Human 40
3.1 Acute Poisoning 40
3.2 Chronic Poisoning 40
4 Clinical Symptoms of Arsenic Poisoning 41
4.1 Asymptomatic (Preclinical) 41
4.2 Symptomatic (Clinical) 41
4.3 Stages of Other Health Issues and Malignancy 42
5 Arsenic Toxicity to Humans 42
6 Arsenic Metabolism in Humans 44
7 Arsenic Methylation and Toxicity—The Interlinkages 46
8 Biological Basis of Arsenic Methylation 47
9 Arsenic and Cancer 48
9.1 Arsenic Induces Chromosomal and Genomic Instability 50
9.2 Arsenic-Induced Epigenetic Alterations and Aberrant DNA Methylation 51
9.3 Arsenic and Micro-RNA Expression 52
10 Development of Cancer: Receptors and Signaling Pathways 52
10.1 PI3K/AKT Signaling Pathway 52
10.2 Nrf2-KEAP1 Signaling Pathway and Arsenic 53
11 Therapeutic Applications of Arsenic 53
11.1 Application on Acute Promyelocytic Leukemia (APML, APL) 54
11.2 Arsenic and Chronic Myelogenous Leukemia (CML) Treatment 54

S. Chatterjee (✉) • S. Datta
Defence Research Laboratory, Defence Research and Development Organization (DRDO), Ministry of Defence, Post Bag No. 2, Tezpur 784001, Assam, India
e-mail: drlsoumya@gmail.com

D.K. Gupta
Institut für Radioökologie und Strahlenschutz (IRS), Gottfried Wilhelm Leibniz Universität Hannover, Herrenhäuser Str. 2, Hannover 30419, Germany

D.K. Gupta, S. Chatterjee (eds.), *Arsenic Contamination in the Environment*,
DOI 10.1007/978-3-319-54356-7_3

11.3 Arsenic Therapy in Other Malignancies ... 55
11.4 Genetic Susceptibility ... 55
12 Conclusion ... 55
References ... 56

1 Arsenic and Clinical Issues

"The word 'arsenic' elicits a fearful response in most people" (Hughes et al. 2011), since it has a long history of being a poison to humans. Arsenic is an ever-present element in the environment. It cannot be destroyed but can change its form by reacting with oxygen or other molecules present in the ambience (Ng et al. 2003; ATSDR 2007). There are three forms of inorganic arsenic—red arsenic (As_4S_4, or "Realgar"), yellow arsenic (As_2S_3, or "Orpiment"), and white arsenic (or As_2O_3; "arsenic trioxide," "ATO"). Realgar or orpiment burning produces ATO (Chen et al. 2011). Arsenic compounds that contain carbon are termed as organic arsenic, and can be found in nature in natural gas, water, and shale oil. Various forms of organic arsenic are as follows: methylarsine [CH_3AsH_2], dimethylarsine [$(CH_3)_2AsH$], trimethylarsine [$(CH_3)_3As$], monomethylarsonic acid [$CH_3AsO(OH)_2$, MMAV], monomethylarsenous acid [$CH_3As(OH)_2$, MMAIII], dimethylarsinic acid [$(CH_3)_2AsO(OH)$,DMAV], dimethylarsenous acid [$(CH_3)_2AsOH$, DMAIII], trimethylarsinic oxide [$(CH_3)_3AsO$, TMAO], tetramethylarsonium ion[$(CH_3)_4As^+$, TMA^+], arsenobetaine [$(CH_3)_3As$ + CH_2COO^-, AB], arsenocholine [$(CH_3)As$ + CH_2 CH_2OH, AC], etc. Again, 5,7–9 darinaparsin is an organic arsenic composed of dimethylated arsenic linked to glutathione (Mann et al. 2009; Chen et al. 2011).

Regarding elemental toxicities, arsenic has become one of the major clinical concerns from both human health and environmental perspectives (Saha et al. 1999; Iyer et al. 2016). Because of its ubiquitous nature, humans may often come across arsenic through arsenic-contaminated water, food, dusts, fumes, mists or by eating soil (specially in the case of children) (Nriagu and Azcue 1990; Saha et al. 1999; ATSDR-Agency for Toxic Substances and Disease Registry 2007; Muenyi et al. 2015; Sanchez et al. 2016). The main food source of arsenic is seafood, fish and shellfish, rice/rice cereal, poultry, and mushrooms (ATSDR-Agency for Toxic Substances and Disease Registry 2007). Apart from normal level of arsenic in water, air, soil, or food, the higher levels of exposure may occur through various means. The most common route of exposure is living in an area where high natural levels of arsenic are present in either rock or soil (includes agricultural fields) or water. Hazardous waste sites are also unsafe if the material is not properly disposed of, as it could pollute nearby water, air, or soil. Occupational exposure is another concern those who are involved with arsenic production or use. Further, use of arsenic-treated products (like arsenic-treated woods—chromated copper arsenate (CCA treatment)) can also elevate the level of its exposure (ATSDR-Agency for Toxic Substances and Disease Registry 2007; Koedrith et al. 2013).

The toxicity of arsenic was familiar much earlier in the thirteenth century when Albertus Magnus prepared its elemental form (Buchanan 1962). During the course of history, arsenic has played many deadly and decisive roles in internal and imperial conspiracies, where inorganic arsenic (iAs) was used for poisoning, till detection methods were developed in the nineteenth century (Jolliffe 1993; Cullen 2008; Drobna et al. 2009). iAs as both arsenite (iAs^{III}) and arsenate (iAs^{V}) causes different cancers (Lubin et al. 2000; Kryeziu et al. 2016). Further, for long time, arsenicals have been used as medicines. For example, Fowler's solution (an alcoholic solution of potassium arsenite) was used in dermatological, respiratory, and hematological illnesses, while Salvarsan (arsphenamine, compound 606 synthesized by Paul Ehrlich) was used for the treatment for syphilis, initiating the foundation of contemporary chemotherapeutics (Schwarz 2004). Till 1940s organoarsenicals were extensively used as antibiotics. Currently, iAs as arsenic trioxide is again practiced as a treatment for acute promyelocytic leukemia (APML, APL) (Douer and Tallman 2005; Drobna et al. 2009).

2 Arsenic Poisoning—An Overview

Early accounts of chronic arsenic poisoning due to consumption of contaminated drinking water were reported by WP Tseng and his team (Tseng et al. 1968; Tseng 1977) in Taiwan, Rosenberg (1974) in Chile, and Datta (1976) in India. Further studies have confirmed that chronic exposure increase the risk of dermal ailments, diabetes, vascular disease, and internal cancers (Smith et al. 1992; Navas-Acien et al. 2005, 2008; Chen et al. 2007; Wang et al. 2007; Jovanovic et al. 2013). However, parts of Indo-Bangladesh region are very badly affected by arsenic. An estimated approximately more than 30 million people in Bangladesh and West Bengal (India) are exposed to arsenic-contaminated ground water. Cases of liver fibrosis, in India, due to chronic exposure to arsenic-contaminated water, were reported in early 1978 from Chandigarh (Datta et al. 1979), however, a large number of such and more severe cases were reported from Kolkata, West Bengal in 1984 (Garai et al. 1984; Chakraborti et al. 2002). Since then, chronic arsenic toxicity has become a huge concern for most of the areas neighboring the lower middle, and upper Ganga and Brahmaputra plain (Chakraborti et al. 2002; Guha Mazumder 2008; Chetia et al. 2011; Lan et al. 2011; Iyer et al. 2016). Undeniably, the difficulty of billions of people exposed to iAs in Bangladesh and West Bengal has been described as a "public health emergency" (Smith et al. 2000) and as "the largest poisoning of a population in history" (Bhattacharjee 2007; Drobna et al. 2009). A recent pilot study on the pan India prevalence of arsenic in blood of 205,530 persons (with 111,737 males and 93,793 females) shows 1.37% cases detected with frequency of high (with blood arsenic levels of ≥ 5 μg L^{-1}) arsenic in the blood (with 1.47% males and 1.25% females) (Iyer et al. 2016).

Acute poisoning of arsenic, though rare, can cause cholera-like symptoms and death. However, chronic exposure in mild dose is converted into a less toxic form by the liver, and is ultimately mostly excreted through the urine. Very high exposure of arsenic gets accumulated in the body (Caroli 1996; Saha et al. 1999). Arsenic can affect protoplasmic enzymes, mitosis, respiration and post-translational histone modifications and related enzymes (Gordon and Quastel 1948; Chervona et al. 2012). Apart from poisoning, arsenic has long been in medicinal use for treatment of leukemia, asthma, syphilis, trypanosomiasis, lichen planus, verruca plenum, tropical eosinophilia, and psoriasis (Goodman and Gilman 1942; Nash 1960; Most 1972; Zachariac et al. 1974; Saha et al. 1999). Arsenic is also being used in production of insecticides, rodenticides, weedicides, etc.

3 Types of Arsenic Poisoning to Human

3.1 Acute Poisoning

Within approximately 30 min of exposure (most likely through ingestion), a person may have reddish rashes in the body and intense thirst but difficulty in swallowing, dry mouth with breath having slight garlicky odor. Clinical symptoms include weakness, muscular cramps and abdominal pain, nausea vomiting and profuse diarrhea. Effect on mucosal vascular supply leads to capillary damage, sloughing of tissue fragments, vasodilation, and vasogenic shock. Skin turn into cold due to circulatory failure beside decreased urine output and kidney damage. Cardiac indicators include sub-endocardial hemorrhages, acute cardiomyopathy, and electrocardiographic changes (with prolonged QT intervals and nonspecific ST-segment changes). Development of psychosis associated with hallucinations, paranoid delusions, and delirium with seizures, coma, and death might follow (Saha et al. 1999; Hughes et al. 2011; Naujokas et al. 2013).

3.2 Chronic Poisoning

Chronic nature of poisoning is more treacherous as accurate diagnosis of symptoms is very difficult. Arsenic related dermatosis often wrongly mixed up with other related indications. Skin, liver, lungs, and blood systems are mostly affected. Nonspecific but typical cutaneous vicissitudes with early persistent erythematous flush that leads to melanosis, hyperkeratosis, and desquamation is evident. Palms and soles are also affected with diffuse desquamation. Continuing skin problems lead to development of multicentric basal cell and squamous cell carcinomas (Saha et al. 1999). Anemia (normochromic and normocytic partial hemolysis) (Saha et al. 1999; IARC 2004; Guha Mazumder 2008) and leukopenia are nearly common symptoms. Again, symptoms like anemia, leukopenia are to be judiciously assessed as they are having common in both arsenic toxicity and malnutrition. Due to arsenicosis, a rare precancerous skin lesion (Bowen's disease), is evident; which is also reported in human papilloma virus HPV infection. However, both arsenic and HPV cause epithelial cancer, it is speculated that arsenic contamination cause activation of an oncogenic virus like HPV in human beings (Saha et al. 1999; Hughes et al. 2011; Naujokas et al. 2013).

4 Clinical Symptoms of Arsenic Poisoning

Initial phase of human arsenic poisoning is having a number of unspecific clinical symptoms. And hence, awareness of the problem is the basis of early appropriate diagnosis. There are typically two stages of arsenic poisoning—asymptomatic or preclinical and symptomatic or clinical.

4.1 Asymptomatic (Preclinical)

Asymptomatic or preclinical stage may demonstrate a transient or labile or blood phase with arsenic metabolites dimethylarsinic acid (DMAA) and trimethylarsinic acid (TMAA) are found in urine and a persistent or occult or tissue phase with tissues like hair, nails, and skin scales or other body tissues showing high arsenic concentrations with no obvious clinical symptoms.

4.2 Symptomatic (Clinical)

Higher level of arsenic in hair, nail, and skin scales are evident in symptomatic stage of the patient with arsenic toxicity. The features of arsenical toxicity progress slowly and may take 6 months to 10 years for onset. However, development of clinical features in the patients depends upon various factors like health conditions, nutritional status, and amount of daily intake of contaminated water. Dermatological signs are most important (Guha Mazumder 2000; Saha et al. 1999) in the course, which include:

1. Melano-keratosis: It is the main symptoms of arsenical dermatosis (ASD) having dry, coarse, spotted nodules in palms and/or soles with dark pigmentation (melanosis) and diffuse and/or spotted keratosis.
2. General melanosis: Initiation of skin darkening in the palm, that slowly spreads to the entire body.
3. Spotted melanosis (rain drop pigmentation): visible on back, chest, or limbs. It is a common symptom.

4. Diffuse and spotted and keratosis: Palms and soles are mostly affected showing cracks and fissures (hyperkeratosis).
5. Pigmentation and depigmentation spots (leucomelanosis): In the advance stage of arsenic toxicities, sometimes patients develop pigmented and depigmented spots in legs or trunk.
6. Other symptoms: mucus membrane melanosis (pigmentation in tongue, gums, lips, etc.), non-pitting edema (edema in feet but not pit on pressure), conjunctival congestion (reddish eye).
7. Fingernails develop whitish lines (Mees' lines) those are alike of traumatic injuries.

4.3 Stages of Other Health Issues and Malignancy

Internal complications like bronchitis, cough, breathlessness, expectoration, and restrictive asthma appear as non-dermatological indications along with dermatological symptoms. Gradually organs like muscles, eyes, vessels, liver (hepatomegaly), spleen (splenomegaly) are affected. Seizures (acute encephalopathy) are also common.

Development of cancers through arsenical poisoning may appear well after 10 years of onset of first symptoms. Skin cancers are common and are mostly monocentric but sometimes multicentric cases are also found. Malignancy could affect neighboring glands, lung, uterus, bladder, genitourinary tract, or other sites within 6 months and usually within a year patients may die.

5 Arsenic Toxicity to Humans

Experiments suggest the considerable less susceptibility of laboratory animals (like monkeys, dogs, and rats) when they are exposed to arsenite or arsenate at doses of 0.72–2.8 mg kg^{-1} day^{-1}. So quantitative dose-dependent data for animals should not be reflected as a consistent source for application to humans (Milton et al. 2001; EFSA European Food Safety Authority 2009; Naujokas et al. 2013). iAs related to development of symptoms and pathologies in human are very complex and possibly multifactorial. Due to stress and interference with different cellular components, production of free radicals in cells and thereby induction of a strong oxidative stress induces DNA damage, lipid peroxidation and decreased glutathione levels, regulation of transcription factors, and carcinogenesis (IARC 2004; Basu et al. 2005; Méndez-Gómez et al. 2008; Faita et al. 2013). Chronic exposure to iAs is totally hazardous as it leads to increased risk of varied diseases of almost all the major organs/systems, including skin, lung, kidney, endocrine, respiratory, cardiovascular, neurological, and gastrointestinal abnormalities (Table 1).

Table 1 Summary of different diseases related to arsenic exposure

Effecting organ/system	Effects	References
Lung, trachea etc. (respiratory system)	• Mucous membrane irritation of the upper respiratory tract; sore throat, voice hoarseness, chronic cough laryngitis, bronchitis, rhinitis; hemorrhages at mucosal and submucosal and alveolar region (due to arsenic trioxide inhalation); increased mortality from pulmonary tuberculosis, lung cancer.	Cebrian et al. 1983; De et al. 2004; Guha Mazumder 2007; Marshall et al. 2007; Heck et al. 2009; Smith et al. 2009, 2011; Naujokas et al. 2013
Cardiovascular	• Injury to the blood vessels or the heart. • Myocardial infarction and arterial thickening, increasing the risk of cardiovascular disease and death in humans.	Guha Mazumder et al. 1988; Nickson et al. 2007; Gong and O'Bryant 2012; Wiwanitkit 2015
Abdomen (gastrointestinal)	• Burning lips, painful swallowing. • Thirst, nausea, abdominal colic, vomiting and diarrhea; damage to the epithelial cells, and bleeding in esophageal varices, ascites.	Chakraborty and Saha 1987; Guha Mazumder et al. 1998; Wiwanitkit 2015
Hematological	• Anemia and leukopenia (malnutrition aggravates the toxicity); direct hemolytic (involves depletion of intracellular GSH) or cytotoxic effect on the blood cells; suppression of erythropoiesis; hemolytic debris causes renal damage (due to clogging of nephrons).	Chiou et al. 1997; Guha Mazumder 2008; Wiwanitkit 2015
Liver related	• Liver disease with an distended tender liver, jaundice; mitochondrial damage, affects porphyrin metabolism. • Cirrhosis of the liver, liver cancer.	Liu and Waalkes 2008; Yuan et al. 2010; Chen et al. 2010; Gibb et al. 2011
Kidney (renal) Effects	• Kidney damage includes capillaries, tubules, and glomeruli; oliguria common; arsine-induced hemolysis possibly cause tubular necrosis with partial or total renal failure; kidney cancer; bladder and other urinary cancers.	Yuan et al. 2010; Chen et al. 2010; Gibb et al. 2011; Naujokas et al. 2013; Peters et al. 2015
Dermal (skin) Effects	• Skin disorders; hyperkeratosis; warts or corns on the palms and soles; areas of hyperpigmentation interspersed with small areas of hypopigmentation on the neck, face, and back; skin cancer.	Guha Mazumder et al. 1998, 2005, 2008; Argos et al. 2011
Neurological (nervous system)	• The peripheral and central nervous system may be damaged; myopathy. • Encephalopathy; headache, lethargy, restless sleep, loss of libido, increased urinary urgency, mental confusion, hallucination, impaired intellectual function in children and adults; loss of reflexes, seizures and coma, impaired motor function.	Guha Mazumder 2008; Hamadani et al. 2011; Gong et al. 2011; Parvez et al. 2011

(continued)

Table 1 (continued)

Effecting organ/system	Effects	References
Developmental	• Increased infant mortality; reduced birth weight; congenital heart defects and arsenic exposure (drinking water): no overall association found; an association with coarctation of the aorta; higher risk of congenital malformations in babies born to women exposed to arsenic dusts during pregnancy; higher risk of miscarriages; early-life exposure associated with increased cancer risk as adults.	Rahman et al. 2010a; Intarasunanont et al. 2012; Kile et al. 2012; Dong and Su 2009; Yuan et al. 2010; Chen et al. 2010; Hamadani et al. 2011; Su et al. 2011
Reproductive effects	• Reproductive dysfunctions, testis weight reduction, effect on accessory sex organs, alterations of spermatogenesis, reduced level of testosterone and gonadotrophins.	Shen et al. 2013; Kim and Kim 2015
Immune system	• Alteration of cytokine expression and immune-related gene expression, inflammation; infant morbidity increased due to susceptibility of infectious diseases.	Spivey 2011; Ahmed et al. 2011; Kile et al. 2012
Endocrine system	• Diabetes; impaired glucose tolerance in pregnant women; disrupted thyroid hormone, retinoic acid, and glucocorticoid receptor pathways; inhibition of number of important enzymes; inhibition of insulin activation; phenylarsine oxide (PAO) blocks glucose transport; inhibition of NAD-linked oxidation of pyruvate or α-ketoglutarate.	Rahman et al. 2010b; Del Razo et al. 2011; Islam et al. 2012; Jovanovic et al. 2013
Genotoxicity, mutagenic and Carcinogenic Effect	• Chromosomal aberrations; clastogenesis causes inhibition of DNA repair; pre-cancerous dermal keratosis, epidermoid carcinoma; incidence of leukemia, cancer of respiratory system, etc.; squamous cell carcinomas, multiple basal cell carcinoma, epithelioid angiosarcoma	Rahman et al. 2010b; Guha Mazumder et al. 1998, Guha Mazumder 2008; Wiwanitkit 2015; Del Razo et al. 2011; Islam et al. 2012; Jovanovic et al. 2013

6 Arsenic Metabolism in Humans

Since 1980, the International Agency for Research on Cancer (IARC) has listed arsenic as a human carcinogen and several researchers have emphasized the possible human health hazard of arsenic when present in drinking water (Kapaj et al. 2006; IARC (International Agency for Research on Cancer monographs on the evaluation of carcinogenic risks to humans) 2012), and there is a growing interest to understand the distribution, metabolism, and potential modes of actions of iAs. Studies of iAs metabolism in microorganisms were started from the early nineteenth century (Cullen 2008). Gosio gas (trimethylarsine), a volatile product of microorganisms, and methylated metabolites were characterized. Frederick Challenger (Challenger 1945, 1951) presented a chemically probable plan for the methylation of inorganic arsenic. However, the typical postulated scheme is as follows:

$$iAs^{V} \rightarrow iAs^{III} \rightarrow MAs^{V} \rightarrow MAs^{III} \rightarrow DMAs^{V} \rightarrow DMAs^{III} \rightarrow TMAs^{V} \rightarrow TMAs^{III}$$

Methylation of arsenic is common in biota. Many of the organisms can biotransform iAs by (1) reducing As(V) to As(III) and (2) subsequent methylation by oxidation of trivalent As (Hughes et al. 2011). Similarly, the prime metabolic pathway of ingested As in humans is methylation (Gebel 2002; Styblo et al. 2002; Vahter 2002; Steinmaus et al. 2005). It is first methylated to monomethylarsonic acid (MMA5); MMA is reduced to monomethylarsonous acid (MMA3). MMA3 is subsequently methylated to dimethylarsinic acid (DMA5), which is reduced to dimethylarsinous acid (DMA3) (Steinmaus et al. 2005). Correspondingly, e.g., in a recent study on marine herbivorous fish (*Siganus fuscescens*), Zhang et al. (2016) showed biotransformation of both inorganic As(III) and As(V) to the less toxic arsenobetaine (AsB is an organoarsenic compound and main source of arsenic found in fish. It is the arsenic analog of trimethylglycine, commonly known as betaine) through typical reduction and subsequent methylation. Methylation of As in many species is an enzymatically catalyzed process, where methyltransferase (AS3MT) catalyzes both reduction and methylation reactions for iAs(III), and where AdoMet (S-Adenosyl-l-methionine) is the methyl donor (Lin et al. 2002). In addition, the action of AS3MT is correlated with reduction of thioredoxin (Tx: a small dithiol containing protein that provides reducing equivalents to AS3MT) by thioredoxin reductase (TR) (Waters et al. 2004). The activity of arsenic methylating enzyme may be linked to a multistep reaction that manages the accessibility of the prime methyl group donor in cells (Drobna et al. 2009) (Fig. 1).

There are marked variations among arsenic-exposed individuals on the average iAs, MMA, and DMA percentage urinary values, which is typically as follows: iAs—10–30%, MMA—10–20%, and DMA—60–70%. Along with age, primary methylation index (PMI), and the secondary methylation index (SMI) also vary with MMA% and PMI increased and DMA% and SMI decreased creating a potential risk for arsenicosis (Huang et al. 2008a, b; Zhang et al. 2014; Shen et al. 2016).

However, an alternative model has been proposed for As methylation where the element is bound to a cellular thiol (either a cellular protein or the reactive moiety in glutathione GSH), which is the substrate for methylation (Hayakawa et al. 2005; Naranmandura et al. 2006). Even though this substitute model is reliable with numerous experimental details, it is also suggested that the presence of thiol species is not an absolute prerequisite for biological methylation of As (Thomas et al. 2007; Drobna et al. 2009).

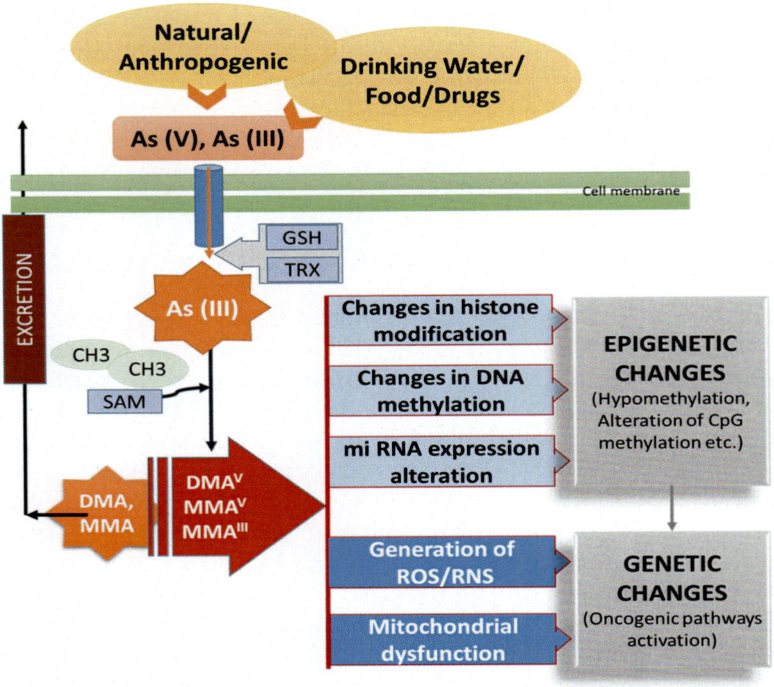

Fig. 1 Scheme of arsenic-induced carcinogenesis due to its biotransformation process, having effects at genetic and epigenetic levels (Modified after Hubaux et al. 2013)

7 Arsenic Methylation and Toxicity—The Interlinkages

It is interesting to note that arsenic toxicity is strongly correlated with its metabolism (i.e., methylation status and valence states of its metabolites). There are several factors like dietary history, body mass index (BMI), individual's ethnicity, age, sex, lifestyle, and inherited genetic characteristics that may affect the arsenic methylation (Loffredo et al. 2003; Gamble et al. 2005; Lindberg et al. 2008; Chen et al. 2012; Chen et al. 2013; Melak et al. 2014; Shen et al. 2016). A recent meta-analysis study on the available data of methylation status in human indicates that men have less methylation capacity than women. The factors like exposure dosage of arsenic, drinking and smoking habits, age, and BMI are negatively associated with arsenic methylation capacity (Shen et al. 2016); with no considerable variation between Asians and Americans and significant differences among Chinese, Mexicans, and Chileans in the distribution of urinary arsenic species (Hopenhayn-Rich et al. 1996; Loffredo et al. 2003; Shen et al. 2016). In children, there is a dilemma on methylation capacity as some authors suggested better (Chowdhury et al. 2003) while others showed lower capacity may be due to factors like organs, and their capability on metabolism and related enzymes (Shen et al. 2016).

iAs^{III} is more acutely cytotoxic than pentavalent methylated arsenicals ($MMAs^{V}$ and $DMAs^{V}$). Thus biotransformation of arsenicals is a process of detoxification within the body that led to excretion (Yamauchi and Fowler 1994). However, several reports suggest $DMAs^{V}$ is a teratogen, a nephrotoxic, a tumor promoter, and complete carcinogen in the rat (Rogers et al. 1981; Murai et al. 1993; Yamamoto et al. 1995; Wei et al. 1999). It is a matter of dispute whether, at all, methylation helps in detoxification in human body or exerts unique toxic effects forming more reactive species (Drobna et al. 2009).

Methylation of As in many species is an enzymatically catalyzed process, where, methyltransferase (AS3MT), catalyzes both reduction and methylation reactions for iAs(III), where, AdoMet is the methyl donor. In addition, the action of AS3MT is correlated with reduction of thioredoxin (Tx: a small dithiol containing protein that provides reducing equivalents to AS3MT) by thioredoxin reductase (TR) (Waters et al. 2004). The activity of arsenic methylating enzyme may be linked to a multistep reaction that manages the accessibility of the prime methyl group donor in cells (Drobna et al. 2009).

8 Biological Basis of Arsenic Methylation

Based on the studies that iAs is extensively methylated in humans and many other species, understanding the biological basis of As methylation has become a prime research area (Shen et al. 2016). It is demonstrated that methylation of As in many species is an enzymatically catalyzed process that is consistent with Challenger's original scheme for this process. A single protein, arsenic (+3 oxidation state) methyltransferase (AS3MT), catalyzes both reduction and methylation reactions for As. Arsenic is typical for methyltransferases; AdoMet is the methyl donor for these reactions. Hence, the activity of this As methylating enzyme can be linked to a multistep process in cells that controls the availability of this prime methyl group donor. This linkage is strengthened because MAs^{III}, an intermediate in the AS3MT-catalyzed pathway for As methylation, is a potent inhibitor of the activity of thioredoxin reductase (Lin et al. 1999, 2001). Integrating these different metabolic processes into a network of interconnected reactions will provide a more comprehensive understanding of the link between the metabolism of As in the cell and the effect of arsenicals on the molecular processes that underlie its actions as a toxin or carcinogen. The metadata analysis (databases include PubMed, Springer, Cochrane Library, Embase, and China National Knowledge Infrastructure) on the factors affecting arsenic methylation in arsenic-exposed humans suggests that arsenic exposure leads to increase in iAs, MMA, DMA, and total arsenic; however, arsenic methylation is found to be more efficient in women than in men (Shen et al. 2016).

There is a considerable variation in the patterns of arsenic metabolism among individual which is correlated with susceptibility to arsenic related toxicity and diseases (Lu et al. 2013). Amid the common factors those closely influence an individual's capability to metabolize arsenic are epigenetics, genetic polymorphisms. A considerable role of gut microflora of individual is being reported in metabolizing arsenic and triggering systemic responses in diverse organs (Lu et al. 2013, 2014). In a mice model metagenomics study showed perturbations of the gut microbial communities affect the spectrum of metabolizing (including reduction, methylation, and thiolation) of arsenic species and subsequent toxicological effects (Lu et al. 2013; Wang et al. 2015).

9 Arsenic and Cancer

Arsenic is a known human carcinogen. Common types of tumors associated with arsenic exposure are found in skin, bladder, liver, and lung (Mead 2005; IARC 2004; Hubaux et al. 2013; Kryeziu et al. 2016). iAs ingestion, absorption (absorbed by the gastrointestinal (GI) tract as soon as ingested) and biotransformation within the cellular level are the key factor linked with the carcinogenic property of arsenic in human (Ebert et al. 2011; Pan et al. 2016). Upon ingestion, iAs (or As(V) predominant form) goes into cells through membrane transporters, namely, inorganic phosphate transporters (PiT) and aquaporins (Hubaux et al. 2013). Within cell, using a glutathione-dependent reaction driven by polynucleotide phosphorylase and mitochondrial ATP synthase, arsenate (AsV) is reduced to more toxic arsenite (As III), which is then methylated as conjugated arsenicals (MMA, DMA) as a typical detoxification mechanism. Liver is an important organ to carry out the mechanism. The methylated arsenicals, however, are highly reactive and have been shown to induce damage in different organs once they are translocated from hepatocytes into bile as glutathione conjugates and subsequently travel through blood (Wang et al. 2004; Nemeti et al. 2010). The methylated conjugates are involved with glutathione depletion and generation of reactive oxygen species (ROS) (Cullen and Reimer 1989; Styblo et al. 2002; Hubaux et al. 2013). As(III) has high covalent reactivity towards thiol groups, that augment its toxicity even towards the interaction with proteins to induce their inactivation/degradation (Dilda and Hogg 2007). iAs is not regarded as mutagenic at nontoxic doses and does not directly damage DNA, however, as described earlier, methylated arsenicals are potent clastogens and mutagens (Rossman and Klein 2011). There are evidences that, along with the effect of other carcinogens such as UV light, *N*-methyl-*N*-nitrosourea, diepoxybutane, X-rays, methyl methane sulfonate, and tobacco, low doses of arsenic can also potentiate mutagenic effects (Lee et al. 1985; Wiencke and Yager 1992; Jha et al. 1992; Flora 2011; Hubaux et al. 2013).

Though As(V) is converted to As(III), interconversion of As(III) and As(V) is also evident that produces reactive species of oxygen (ROS) and nitrogen (RNS) and disrupts mitochondrial electron transport chain (Rossman 2003) (Fig. 2). Characteristically, monomethylarsonous acid (MMAIII)-mediated inhibition of mitochondrial complexes II and IV generates mitochondrial ROS, resulting in electron accumulation, subsequent electron leakage from complexes I and III (Naranmandura et al. 2011) and formation of superoxide anion radicals ($O2^{\bullet-}$), hydroxyl radicals ($OH^{\bullet}$), and hydrogen peroxide (H_2O_2) (Turrens 1997). These free-radical species has been related with varied DNA related problems like creation of adducts, DNA cross-linking, DNA double stranded breaks, DNA mutations and DNA deletions, and chromosomal aberrations (Halliwell 2007; Martinez et al. 2011; Hubaux et al. 2013).

Generation of arsenic induced reactive nitrogen species (RNS) like peroxynitrite has been reported to cause DNA deamination, alkylation, and oxidative DNA damage (Wink et al. 1991; Radi et al. 1991). The modes of actions are not entirely explicit; however, RNS occurs in a tissue-specific manner (Gurr et al. 2003). DNA repair processes are also hampered by the arsenic affecting both nucleotide excision repair (NER) and base-excision repair (BER) mechanisms (Fig. 2). Arsenic reduces the expression of NER associated genes and decreasing expression and protein levels (like complementation group C of xeroderma pigmentosum (XPC) (Andrew et al. 2006; Nollen et al. 2009; Hubaux et al. 2013). Further, methylated arsenicals (III) damage the activity of a promoter of NER (human PARP1), that becomes active in reaction to DNA damage (Walter et al. 2007). Metabolites of Arsenic, in addition, decrease gene expression and protein levels of BER-related mechanism, such as 8-oxoguanine DNA glycosylase 1 (hOGG1), DNA ligase IIIα (LIGIIIα), and X-ray cross complementing protein 1 (XRCC1) (Ebert et al. 2011). For example, in murine lung tissue, BER related genes, like apurinic/apyrimidinic, endonuclease/redox effector-1 (APE1), 8-oxoguanine DNA glycosylase (OGG1), ligase I,

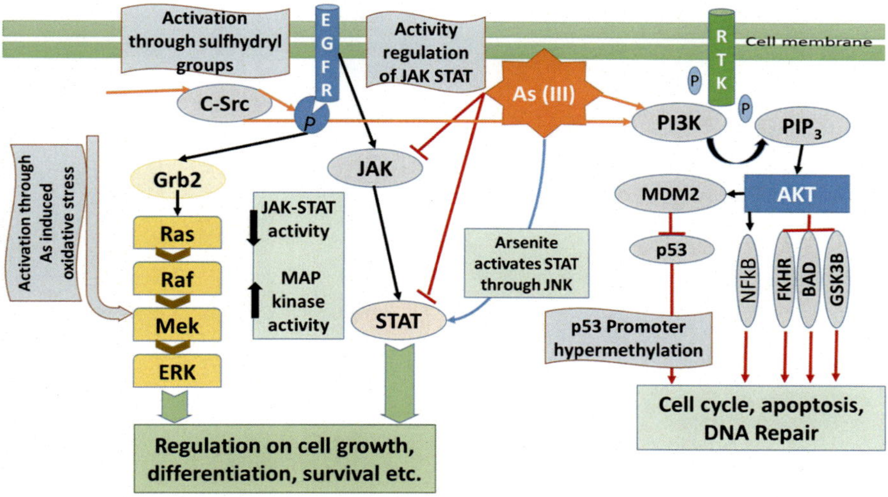

Fig. 2 Arsenic-mediated activation of EGFR and RTK signalling pathway. As^{III} can inhibit JAK, while it can activate STAT3 through JNK, contributing to disruption of PI3K/AKT signalling pathway, cell growth regulation, survival of cell, etc. There, the activated PI3K can phosphorylate to phosphatidylinositol 3,4,5-triphosphate (PIP3), in turn activating the kinase AKT, which is capable of phosphorylating a number of target proteins in the cytoplasm and nucleus. Further, methylation patterns at the promoter region of the p53 gene have been shown to be modified by arsenic, resulting in silencing of this tumor suppressor (Modified after Hubaux et al. 2013)

DNA, ATP-dependent (LIG1), and poly (ADPribose) polymerase 1 (PARP1) were reported to be elevated in nature (Osmond et al. 2010) (Table 2).

9.1 Arsenic Induces Chromosomal and Genomic Instability

Chromosomal and genomic instability are other issues related to arsenic toxicity. Considerable increment of micronuclei formations, chromosomal aneuploidy are likely due to an effect on sulfhydryl groups of tubulin and microtubule-associated proteins, that led to disruption of cell spindle assembly (Wen et al. 2008; Zhao et al. 2012). Couple of studies shown that, cell cycle progression in arsenic exposed cells is not being halted regardless of heavy DNA damage and genomic instability. This

Table 2 List of altered functions on exposure to arsenic (After Chen et al. 2011)

Functions altered	Category of protein	Disease or cell lines	Arsenic binding proteins
Tumor suppressor; probable transcription factor	Phosphoprotein; scaffold protein	APL	PML
Proteolysis	Ubiquitin E3 ligase	CML, K562 cells	c-CBL
			SIAH1
Redox regulation	Oxidoreductase	MCF-7 cells	Trx R
		Arsenic detoxification	GSR
		Fibrosarcoma cells	PDI
	Peroxidase	Ovary cells	TPX-2 II
Binding heavy metals	Metallothioneins (MT)	Arsenic detoxification	MTs
Activation of metallothionein transcription	Transcription factor		MTF1
Transcription regulation	Phosphoprotein	hepa1c1c7 cells	Keap1
Structural subunit of microtubules	Cell skeleton proteins	K562 cells	Tubulins
Cytoskeleton		K562 and MCF-7 cells	β-actin
Protein serine/threonine phosphatase; regulation of cell cycle	Phosphatase; oncoprotein	Malignancies	PPM1D
Cellular signaling	Dual specificity protein phosphatase	Carcinogenesis	JNK phosphatase
Regulation of NFκB pathway	Protein kinase	Inflammation and carcinogenesis	IκB
	Pyruvate kinase	Ovary cells	Galectin-1
Glycolysis	Phosphoprotein	MCF-7 cells	PKM2
Oxygen transport	Globin family	Red blood cells	Hemoglobin

abnormal phenomenon of cell cycle progression occurs due to inhibition of p53-dependent increase in p21 expression following DNA damage, which is otherwise typical in normal cell (Vogt and Rossman 2001; Tang et al. 2006; Huang et al. 2008a, b; Komissarova and Rossman 2010). Likewise, disruption of PARP1 activity causes genomic instability by permitting the survival of cells with considerable DNA lesions (Qin et al. 2012). Segmental DNA amplifications at 19q13.31 and 19q13.33 and segmental DNA losses at chromosomal locus 1q21 are found in lung tissue cells chronically exposed to arsenic. Interestingly, researchers found that a number of genes in 19q13.33 [like as Spleen focus forming virus (SFFV), proviral integration oncogene B (SPIB), and Nuclear receptor subfamily 1, group H, member 2 (NR1H2)] are oncogenic in mouse models (Venkatesan et al. 2007; Hubaux et al. 2013).

9.2 Arsenic-Induced Epigenetic Alterations and Aberrant DNA Methylation

S-Adenosyl methionine (SAM) is the main methyl donor require in arsenic detoxification. Chronic arsenic exposure leads to high demand of the SAM that sometimes overwhelms the cellular machinery of using methyl groups for different methylation activities leading to overall hypomethylation ("global hypomethylation" phenomenon) (Hubaux et al. 2013). Arsenic related epigenetic effects mainly occur due to deprivation of methyl (−CH3) groups of the cell following altered accessibility of SAM. Further, SAM is the main contributor of methyl group for DNA-methyltransferases (DNMT), depletion of methyl groups can lead to hypomethylation of the essential enzymes and changes in chromatin, alteration of CpG methylation status of promoters for specific genes (Deleted in Bladder Cancer 1 (DBC1), Death-Associated Protein Kinase 1 (DAPK), and P53 (Mazumder 2005; Tseng et al. 1996; Engel et al. 1994; Jensen et al. 2009). Again, due to hypomethylation, reduction in LINE-1 methylation and total 5-methyldeoxycytidine content in lymphoblastoid cells (Intarasunanont et al. 2012). Thus, reports show that epigenetic modifications can promote deregulation of gene expression and malignant transformation in different cell types (Marsit et al. 2006; Reichard and Puga 2010; Ziech et al. 2011). Low level of arsenic exposure resulted in DNA hypomethylation in rat models (Cui et al. 2006). However, hypermethylation in promoter regions of P53 and tumor suppressor CDKN2A and subsequent transcriptional silencing of the gene is reported when individuals chronically exposed to high yet nonlethal levels of arsenic (Chanda et al. 2006). Arsenic exposure can modify histone methylation patterns of H3K4, H3K9, and H3K27 in both nonmalignant and malignant lung cell lines, by primarily reducing the expression of genes linked with changes of DNA methylation and histone acetylation (Ren et al. 2011). Alteration of the chromatin landscape of arsenic induced cancer cells due to loss of the repressive histone modifications H3 triMe-K27 and H3 diMe-K9 and an increase in the levels of activating

Ac-H3 and diMe-K4 at the WNT5A locus is also evident, that ensuing in the ectopic expression of WNT family genes (Jensen et al. 2009; Hubaux et al. 2013).

9.3 *Arsenic and Micro-RNA Expression*

It is evident from a study using human bronchial epithelial cells (HBEC) that epithelial-to-mesenchymal transition (EMT) takes place in P53-knock down cells when chronically exposed to arsenic, which happens due to changes in micro-RNA (miR) expressions. Different reports suggest the expression reduction in miR-200 family, miR-9, miR-181b, miR-124, and miR-125b). Decrease of miR-181b and miR-9 lead to cell migration, tube formation and angiogenesis due to expression of Nrp1, which is otherwise under control in normal cell (Wang et al. 2011; Cui et al. 2012; Hubaux et al. 2013).

10 Development of Cancer: Receptors and Signaling Pathways

Epidermal growth factor receptor (EGFR) locus is an important cell receptor. Structural alterations (through mutation and/or amplification) of EGFR lead to its constitutive and ligand-independent active state by destabilizing the auto inhibitory loop (Yarden and Sliwkowski 2001; Andrew et al. 2009). Further, increased c-Src activity due to arsenic stimulation can physically activate EGFR (by Tyr845, Tyr1101 phosphorylation) that leads to ligand-independent EGFR phosphorylation and constitutive activation (Simeonova and Luster 2002). Overexpression of EGFR is reported in arsenic-exposed hepatocellular carcinoma cells (Sung et al. 2012). Arsenic related ROS activity can also trigger Ras, Raf, Mek, and ERK (components of the EGFR pathway in lung epithelial cells). Arsenite inactivates JAK and consequently STAT3 which play an important role in arsenic-associated pathogenesis (Cheng et al. 2004). On the contrary, it is reported that through c-Jun NH2 kinase (JNK), As(III) activates STAT3 contributing to Akt activation (Liu et al. 2012). As(III) exposed leukemia cell lines causes activation of Rac1 GTPases, that results in increased cell survival and proliferation (Verma et al. 2002; Hubaux et al. 2013).

10.1 *PI3K/AKT Signaling Pathway*

Activation (through binding extracellular growth factor) of receptor tyrosine kinases (RTKs) is important for subsequent activation and signaling through the PI3K/AKT pathway and varied target proteins for metabolism, cell growth, survival, and proliferation (Papadimitrakopoulou 2012). PI3K/AKT phosphorylation cascade may be stimulated by acute exposure to arsenite, which leads to cellular transformation (anchorage-independent growth and proliferation amplification (Dong 2002; Gao et al. 2004; Stueckle et al. 2012). Arsenic-induced AKT signaling can also be associated with induction of miR-190 that represses expression of the PH domain leucine-rich repeat protein phosphatase (PHLPP)—a negative regulator of AKT; AKT stimulation also leads to cell migration and discharge of vascular endothelial growth factor (VEGF) (Beezhold et al. 2011; Wang et al. 2012; Hubaux et al. 2013).

10.2 Nrf2-KEAP1 Signaling Pathway and Arsenic

Oxidative stress response in cell mainly dealt by the transcription factor nuclear factor erythroid-derived factor 2-related factor 2 (NRF2) that have a leucine-zipper DNA binding domain. This NRF2 is capable of binding electrophile response elements (EREs) and antioxidant response elements (AREs). In normal cell, sequestration of NRF2 using targeted proteolytic degradation is carried out by KEAP1. But in stressed cell, NRF2 detaches from KEAP1 and generate stress response. Further, NF-κB activation takes place as KEAP1 E3-ubiquitin ligase complex is affected by genetic disruption leading to tumorigenesis (Thu et al. 2011). Arsenite can also disrupt the NRF2-KEAP1-CUL3 complex, where NRF2 pathway provides protection against As^{III} and MMAIII induced toxic effects (Zhang et al. 2006; Hubaux et al. 2013).

11 Therapeutic Applications of Arsenic

As a drug, arsenic has a very old history. Realgar and orpiment pastes were first formulated by Hippocrates (Greek physician of the Age of Pericles, 460–370 BC) to treat ulcers. Chinese Nei Jing Treaty (263 BC) record shows use of arsenic pills to cure episodic fever (Huang and Wang 1993; Li 1982; Chen et al. 2011). China has a good history of use of arsenic to treat various ailments. Treatment of malaria like diseases was developed by Si-Miao Sun (581–682 AD) from realgar, orpiment, and ATO, and Shi-Zhen Li mentioned ATO as a solution for a range of diseases in his pharmacopedia (1518–1593 AD) (Li 1982; Chen et al. 2011). After the Avicennes (980–1037 AD) and Paracelsus (1493–1541 AD) who practiced arsenic therapy in Europe, Lefebure De Saint Ildefont (1774) introduced a paste of arsenic and claimed as an "established remedy to radically cure all cancers" (Zhu et al. 2002; Chen et al. 2011). In 1845 onwards, 1% potassium arsenite, $KAsO_2$ solution (Fowler solution) become popular medicine for treating a number of diseases, like psoriasis, eczematous eruptions, dermatitis herpetiformis, anemia, rheumatism, asthma, cholera, and syphilis. Fowler solution became the first (1865) chemotherapeutic agent used in the treatment of leukemia (Zhou et al. 2007) and Forkner and Scott (1931) of Boston

City Hospital used the same to treat chronic myeloid leukemia (CML) (Zhou et al. 2007; Chen et al. 2011). Arsenic treatment along with irradiation became the treatment of choice, until 1953, when busulfan (alkyl sulfonates—alkylating antineoplastic agents) came as an alternative (Chen et al. 2011).

11.1 Application on Acute Promyelocytic Leukemia (APML, APL)

Acute promyelocytic leukemia (APL) is the M3 subtype of acute myeloid leukemia (AML M3). APL is characterized by the precise chromosomal translocation [t(15;17)(q22;q21)], abnormal promyelocytes accumulation in blood and bone marrow, fibrinogenopenia, and disseminated intravascular coagulation (Wang and Chen 2008). First explain by Hillestad (1957), APL was considered the most fatal type of acute leukemia. APL leukemogenesis starts with formation PMLRARα fusion protein by the amalgamation of promyelocytic leukemia (*PML*) gene on 15q to the retinoic acid receptor α (*RAR*α) gene on chromosome 17 (Hillestad 1957; Wang and Chen 2008). Although, due to the development of drugs like anthracyclines and all-*trans* retinoic acid (ATRA), APL has become a highly curable disease, administration of ATO further extends the endurance of patients, especially those with relapsed or refractory disease (Hu et al. 2009; Bernard et al. 1973; Huang et al. 1988; Ades et al. 2010; Creutzig et al. 2010). The ATRA and ATO combination therapy enhances to clear PML-RARα transcript by triggering catabolism of this fusion protein (de The H and Chen de The and Chen 2010). Studies have shown that arsenic (ATO) triggers the targeted degradation of PML-RARα through its PML moiety (Chen et al. 1997; Chen et al. 2011).

11.2 Arsenic and Chronic Myelogenous Leukemia (CML) Treatment

Started from pluripotential hematopoietic stem cells, CML is a myeloproliferative neoplasms (MPNs) that generates a chimeric fusion protein BCR-ABL with constitutively activated tyrosine kinase activity due to translocation of [t(9;22)(q34;q11)] (Philadelphia chromosome) (Ren 2005). Recent drugs like imatinib mesylate (IM; or Gleevec, Glivec, or STI571), a BCR-ABL inhibitor, has shown notable clinical effectiveness, however, traditionally, ATO therapy (Fowler solution) was the initial chemotherapeutic intervention for CML. ATO can inhibit translation of *BCR/ABL* mRNA, ensuing in reduction of BCR/ABL levels and subsequent apoptosis of human leukemia cells (Nimmanapalli et al. 2003). Arsenic may target proteasomal degradation by ubiquitination of key lysine residues of BCR-ABL (Zhang et al. 2009).

11.3 Arsenic Therapy in Other Malignancies

Several studies suggest that, arsenic can be a therapeutic agent for various malignancies like multiple myeloma, myelodysplasia syndrome, lymphoid malignancies (include non-Hodgkin lymphoma) and T-cell leukemia/lymphoma (ATL). Synergizing with IFNα, ATO induce cell-cycle arrest and apoptosis in ATL cells through down-regulation of the HTLV-1 oncoprotein Tax and inactivation of NF-κB (Nasr et al. 2003; Chen et al. 2011). Recent transplantation studies have shown that ATO may promote catabolism of specific classes of oncoproteins (e.g., Tax degradation). Application of ATO in combination with chemotherapy has shown hopeful activity on a variety of solid tumors (lung cancer, osteosarcoma, Ewing sarcoma 180, hepatocellular carcinoma, and colorectal cancer). Although arsenic is a toxic substance, there is no major negative reports (like bone marrow depression or chemotherapy-associated secondary malignancy) for its application at therapeutic concentrations (Mathews et al. 2006). However, in patients with extreme susceptibility to arsenic toxicity, abnormal electrolyte levels, exposure to cardiotoxic agents (e.g., anthracyclines), etc. administration of ATO may cause problems (Mathews et al. 2006; Chen et al. 2011). ATO and/or ATRA may cause hyperleukocytosis due to increased chemokine production (Zhou et al. 2014; Liu et al. 2015).

11.4 Genetic Susceptibility

Epidemiological studies on genetic susceptibility have shown high interindividual variability to arsenic toxicity due to underlying genetic factors (Faita et al. 2013). Specific genetic polymorphisms can be found for the genes encoding enzymes involved in mechanisms of As metabolism and detoxification, including glutathione S-transferases (GST), arsenic(III) methyltransferase ($AS^{III}MT$), and methylenetetrahydrofolate reductase (MTHFR) enzymes (Kundu et al. 2011; Faita et al. 2013). Moreover, DNA repair pathways (e.g., genes like hOGG1, XRCC1, APE1, XRCC3 genes) to reduce the oxidative damage induced by iAs is also altered due to specific single nucleotide polymorphisms (SNPs) (Fujihara et al. 2011). Numerous studies have recognized the effect of genetic polymorphisms on vulnerability to arsenic over the modulation of As metabolism, detoxification and DNA repair. However, specific therapeutic strategies for the arsenic-related diseases can be developed by understanding of the mechanisms of action and dose–response relationships (as biomarker) (Fujihara et al. 2011; Faita et al. 2013).

12 Conclusion

For the past two decades or so, the research on arsenic has significantly expanded. The arsenic toxicity in different parts of the world has created a dreaded situation for millions of people. Though it is a natural component, anthropogenic interference has wreaked havoc through different exposures (through ingestion of contaminated food or drinking water, inhalation etc.) for most people. Arsenic biotransformation (of As^{V}, As^{III} and its methylated conjugates) plays a vital role in arsenic carcinogenicity at both epigenetic and genetic levels. Production of reactive oxygen and nitrogen species led to induction and/or suppression of expression of various metabolically active proteins/enzymes those are related with cell cycle regulation, repair of DNA damage, etc. As^{V} detoxification enzymes and DMT's competition for methyl groups contribute to epigenetic abnormalities.

Some oncogenic pathways (notably the EGFR, PI3K/AKT pathways) are governed by chronic arsenic exposure. More research on molecular mechanisms may help to produce more favorable translatable results for tumorigenesis. Further, on the other hand, research and applications on antitumoral effects of arsenic (like treatment of APL) will support the understanding of the extremely complicated pathways of toxicity and malignant transformation in the human body. To develop specific therapeutic strategies against arsenic, integrated study of environmental monitoring, exposure data, health surveillance, individual risk characterization, and human biomarkers will be able to provide mechanistic insight into the pathogenesis of disease processes.

References

Ades L, Guerci A, Raffoux E, Sanz M, Chevallier P, Lapusan S, Recher C, Thomas X, Rayon C, Castaigne S, Tournilhac O, de Botton S, Ifrah N, Cahn JY, Solary E, Gardin C, Fegeux N, Bordessoule D, Ferrant A, Meyer-Monard S, Vey N, Dombret H, Degos L, Chevret S, Fenaux P, European APL Group (2010) Very long-term outcome of acute promyelocytic leukemia after treatment with all-trans retinoic acid and chemotherapy: the European APL Group experience. Blood 115:1690–1696

Ahmed S, Mahabbat-e Khoda S, Rekha RS, Gardner RM, Ameer SS, Moore S, Ekström EC, Vahter M, Raqib R (2011) Arsenic-associated oxidative stress, inflammation, and immune disruption in human placenta and cord blood. Environ Health Perspect 119:258–264

Andrew AS, Burgess JL, Meza MM, Demidenko E, Waugh MG, Hamilton JW, Karagas MR (2006) Arsenic exposure is associated with decreased DNA repair in vitro and in individuals exposed to drinking water arsenic. Environ Health Perspect 114:1193–1198

Andrew AS, Mason RA, Memoli V, Duell EJ (2009) Arsenic activates EGFR pathway signalling in the lung. Toxicol Sci 109:350–357

Argos M, Pierce BL, Chen Y, Parvez F, Islam T, Ahmed A, Hasan R, Hasan K, Sarwar G, Levy D, Slavkovich V, Graziano JH, Rathouz PJ, Ahsan H (2011) A prospective study of arsenic exposure from drinking water and incidence of skin lesions in Bangladesh. Amer J Epidemiol 174:185–194

ATSDR-Agency for Toxic Substances and Disease Registry (2007) Toxicological profile for Arsenic. Public Health Service (US DHHS, http://www.atsdr.cdc.gov/toxprofiles/tp2-c3.pdf)

Basu A, Som A, Ghoshal S, Mondal L, Chaubey RC, Bhilwade HN, Rahman MM, Giri AK (2005) Assessment of DNA damage in peripheral blood lymphocytes of individuals susceptible to arsenic induced toxicity in West Bengal, India. Toxicol Lett 159:100–112

Beezhold K, Liu J, Kan H, Meighan T, Castranova V, Shi X, Chen F (2011) miR-190-mediated downregulation of PHLPP contributes to arsenic-induced Akt activation and carcinogenesis. Toxicol Sci 123:411–420

Bernard J, Weil M, Boiron M, Jacquillat C, Flandrin G, Gemon MF (1973) Acute promyelocytic leukemia: results of treatment by daunorubicin. Blood 41:489–496

Bhattacharjee Y (2007) A sluggish response to humanity's biggest mass poisoning. Science 315:1659–1661

Buchanan WD (1962) Toxicity of arsenic compounds. Elsevier, Amsterdam, pp 1–13

Caroli S (1996) Element speciation in bioinorganic chemistry. Wiley, New York.

Cebrian ME, Albores A, Aguilar M, Blakely E (1983) Chronic arsenic poisoning in the north of Mexico. Hum Toxicol 2:121–133

Chakraborti D, Rahman MM, Paul K, Sengupta MK, Chowdhury UK, Lodh D, Chanda CR, Saha KC, Mukherjee SC (2002) Arsenic calamity in India and Bangladesh sub-continent-whom to blame? Talanta 58:3–22

Chakraborty AK, Saha KC (1987) Arsenical dermatosis from tube-well water in West Bengal. Ind J Med Res 85:326–334

Challenger F (1945) Biological methylation. Chem Rev 36:315–361

Challenger F (1951) Biological methylation. Adv Enzymol 12:432–491

Chanda S, Dasgupta UB, Guhamazumder D, Gupta M, Chaudhuri U, Lahiri S, Das S, Ghosh N, Chatterjee D (2006) DNA hypermethylation of promoter of gene p53 and p16 in arsenic-exposed people with and without malignancy. Toxicol Sci 89:431–437

Chen CJ, Wang SL, Chiou JM, Tseng CH, Chiou HY, Hsueh YM, Chen SY, Wu M, Lai MS (2007) Arsenic and diabetes and hypertension in human populations: a review. Toxicol Appl Pharmacol 222:298–304

Chen CL, Chiou HY, Hsu LI, Hsueh YM, Wu MM, Wang YH, Chen CJ (2010) Arsenic in drinking water and risk of urinary tract cancer: a follow-up study from northeastern Taiwan. Cancer Epidemiol Biomark Prev 19:101–110

Chen GQ, Shi XG, Tang W, Xiong SM, Zhu J, Cai X, Han ZG, Ni JH, Shi GY, Jia PM, Liu MM, He KL, Niu C, Ma J, Zhang P, Zhang TD, Paul P, Naoe T, Kitamura K, Miller W, Waxman S, Wang ZY, de The H, Chen SJ, Chen Z (1997) Use of arsenic trioxide (As_2O_3) in the treatment of acute promyelocytic leukemia (APL), I: As_2O_3 exerts dose dependent dual effects on APL cells. Blood 89:3345–3353

Chen SJ, Zhou GB, Zhang XW, Mao JH, de The H, Chen Z (2011) From an old remedy to a magic bullet: molecular mechanisms underlying the therapeutic effects of arsenic in fighting leukemia. Blood 117:6425–6437

Chen JW, Wang SL, Wang YH, Sun CW, Huang YL, Chen CJ, Li WF (2012) Arsenic methylation, GSTO1 polymorphisms, and metabolic syndrome in an arseniasis endemic area of southwestern Taiwan. Chemosphere 88:432–438

Chen Y, Wu F, Liu M, Parvez F, Slavkovich V, Eunus M, Ahmed A, Argos M, Islam T, Rakibuz-Zaman M, Hasan R, Sarwar G, Levy D, Graziano J, Ahsan H (2013) A prospective study of arsenic exposure, arsenic methylation capacity, and risk of cardiovascular disease in Bangladesh. Environ Health Perspect 121:832–838

Cheng HY, Li P, David M, Smithgall TE, Feng L, Lieberman MW (2004) Arsenic inhibition of the JAK-STAT pathway. Oncogene 23:3603–3612

Chervona Y, Arita A, Costa M (2012) Carcinogenic metals and the epigenome: understanding the effect of nickel, arsenic, and chromium. Metallomics 4:619–627

Chetia M, Chatterjee S, Banerjee S, Nath MJ, Singh L, Srivastava RB, Sarma HP (2011) Groundwater arsenic contamination in Brahmaputra river basin: a water quality assessment in Golaghat (Assam), India. Environ Monit Assess 173:371–385

Chiou HY, Huang WI, Su CL, Chang SF, Hsu YH, Chen CJ (1997) Dose response relationship between prevalence of cardiovascular disease and ingested inorganic arsenic. Stroke 28:1717–1723

Chowdhury UK, Rahman MM, Sengupta MK, Lodh D, Chanda CR, Roy S, Quamruzzaman Q, Tokunage H, Ando M, Chakraborti D (2003) Pattern of excretion of arsenic compounds [arsenite, arsenate, MMA(V), DMA(V)] in urine of children compared to adults from an arsenic exposed area in Bangladesh. J Environ Sci Health A Tox Hazard Subst Environ Eng 38:87–113

Creutzig U, Zimmermann M, Dworzak M, Urban C, Henze G, Kremens B, Lakomek M, Bourquin JP, Stary J, Reinhardt D (2010) Favourable outcome of patients with childhood acute promyelocytic leukemia after treatment with reduced cumulative anthracycline doses. Bra J Haematol 149:399–409
Cui X, Wakai T, Shirai Y, Hatakeyama K, Hirano S (2006) Chronic oral exposure to inorganic arsenate interferes with methylation status of p16INK4a and RASSF1A and induces lung cancer in a/J mice. Toxicol Sci 91:372–381
Cui Y, Han Z, Hu Y, Song G, Hao C, Xia H, Ma X (2012) MicroRNA-181b and microRNA-9 mediate arsenic-induced angiogenesis via NRP1. J Cell Physiol 227:772–783
Cullen WR, Reimer KJ (1989) Arsenic speciation in the environment. Chem Rev 89:713
Cullen WR (2008) The sociochemistry of an element. Is arsenic an aphrodisiac? RSC Publishing, Cambridge, pp 130–214
Datta DV (1976) Arsenic and non-cirrhotic portal hypertension. Lancet 1:433
Datta DV, Mitra SK, Chhuttani PN, Chakravarti RN (1979) Chronic oral arsenic intoxication as a possible etiological factor in idiopathic portal hypertension (non-cirrhotic portal fibrosis) in India. Gut 20:378–384
De BK, Majumdar D, Sen S, Guru S, Kundu S (2004) Pulmonary involvement in chronic arsenic poisoning from drinking contaminated ground-water. J Assoc Physician Ind 52:395–400
de The H, Chen Z (2010) Acute promyelocytic leukaemia: novel insights into the mechanisms of cure. Nat Rev Cancer 10:775–783
Del Razo L, García-Vargas G, Valenzuela OL, Castellanos EH, Sánchez-Peña LC, Currier JM, Drobná Z, Loomis D, Stýblo M (2011) Exposure to arsenic in drinking water is associated with increased prevalence of diabetes: a cross-sectional study in the Zimapan and Lagunera regions in Mexico. Environ Health 10:73–84
Dilda PJ, Hogg PJ (2007) Arsenical-based cancer drugs. Cancer Treat Rev 33:542–564
Dong J, Su SY (2009) The association between arsenic and children's intelligence: a meta-analysis. Biol Trace Elem Res 103:88–93
Dong Z (2002) The molecular mechanisms of arsenic-induced cell transformation and apoptosis. Environ Health Perspect 5:757–759
Douer D, Tallman MS (2005) Arsenic trioxide: new clinical experience with an old medication in hematologic malignancies. J Clin Oncol 23:2396–2410
Drobna Z, Styblo M, Thomas DJ (2009) An overview of arsenic metabolism and toxicity. Overview of arsenic metabolism and toxicity. Curr Protoc Toxicol 42:4311–4316
Ebert F, Weiss A, Bultemeyer M, Hamann I, Hartwig A, Schwerdtle T (2011) Arsenicals affect base excision repair by several mechanisms. Mutat Res 715:32–41
EFSA (European Food Safety Authority) (2009) Scientific opinion on arsenic in food. EFSA J 7:60–71
Engel RR, Hopenhayn-Rich C, Receveur O, Smith AH (1994) Vascular effects of chronic arsenic exposure: a review. Epidemiol Rev 16:184–209
Faita F, Cori L, Bianchi F, Andreassi MG (2013) Arsenic-induced genotoxicity and genetic susceptibility to arsenic-related pathologies. Int J Environ Res Public Health 10:1527–1546
Flora SJ (2011) Arsenic-induced oxidative stress and its reversibility. Free Radic Biol Med 51:257–281
Forkner CE, McNAIR Scott TF (1931) Arsenic as a therapeutic agent in chronic myelogenous leukemia: preliminary report. JAMA 97:3–5
Fujihara J, Soejima M, Yasuda T, Koda Y, Kunito T, Iwata H, Tanabe S, Takeshita H (2011) Polymorphic trial in oxidative damage of arsenic exposed Vietnamese. Toxicol Appl Pharmacol 256:174–178
Gamble MV, Liu X, Ahsan H, Pilsner R, Ilievski V, Slavkovich V, Parvez F, Levy D, Factor-Litvak P, Graziano JH (2005) Folate, homocysteine, and arsenic metabolism in arsenic-exposed individuals in Bangladesh. Environ Health Perspect 113:1683–1688
Gao N, Shen L, Zhang Z, Leonard SS, He H, Zhang XG, Shi X, Jiang BH (2004) Arsenite induces HIF-1alpha and VEGF through PI3K, Akt and reactive oxygen species in DU145 human prostate carcinoma cells. Mol Cell Biochem 255:33–45

Garai R, Chakraborty AK, Dey SB, Saha KC (1984) Chronic arsenic poisoning from tube well water. J Indian Med Assoc 82:34–35
Gebel TW (2002) Arsenic methylation is a process of detoxification through accelerated excretion. Int J Hyg Environ Health 205:505–508
Gibb H, Haver C, Gaylor D, Ramasamy S, Lee JS, Lobdell D, Wade T, Chen C, White P, Sams R (2011) Utility of recent studies to assess the National Research Council 2001 estimates of cancer risk from ingested arsenic. Environ Health Perspect 119:284–290
Gong G, Hargrave KA, Hobson V, Spallholz J, Boylan M, Lefforge D, O'Bryant SE (2011) Low-level groundwater arsenic exposure impacts cognition: a project FRONTIER study. J Environ Health 74:16–22
Gong G, O'Bryant SE (2012) Low-level arsenic exposure, AS3MT gene polymorphism and cardiovascular diseases in rural Texas counties. Environ Res 113:52–57
Goodman L, Gilman A (1942) The pharmacology basis of therapeutics. MacMillan company, New York
Gordon JJ, Quastel GH (1948) Effect of organic arsenicals on enzyme system. Biochem J 42:337–350
Guha Mazumder DN (2000) Diagnosis and treatment of chronic arsenic poisoning (Downloaded from: http://www.who.int/water_sanitation_health/dwq/arsenicun4.pdf)
Guha Mazumder DN (2007) Arsenic and non-malignant lung disease. J Environ Sci Health 42:1859–1868
Guha Mazumder DN (2008) Chronic arsenic toxicity and human health. Ind J Med Res 128:436–447
Guha Mazumder DN, Chakraborty AK, Ghose A, Gupta JD, Chakraborty DP, Dey SB, Chattopadhyay N (1988) Chronic arsenic toxicity from drinking tube-well water in rural West Bengal. Bull World Health Organ 66:499–506
Guha Mazumder DN, Das Gupta J, Santra A, Pal A, Ghose A, Sarkar S (1998) Chronic arsenic toxicity in West Bengal—The worst calamity in the world. J Ind Med Assoc 96:4–7
Guha Mazumder DN, Steinmaus C, Bhattacharya P, von Ehrenstein OS, Ghosh N, Gotway M, Sil A, Balmes JR, Haque R, Hira-Smith MM, Smith AH (2005) Bronchiectasis in persons with skin lesions resulting from arsenic in drinking water. Epidemiology 16:760–765
Gurr JR, Yih LH, Samikkannu T, Bau DT, Lin SY, Jan KY (2003) Nitric oxide production by arsenite. Mut Res 533:173–182
Halliwell B (2007) Oxidative stress and cancer: have we moved forward? Biochem J 401:1–11
Hamadani JD, Tofail F, Nermell B, Gardner R, Shiraji S, Bottai M, Arifeen SE, Huda SN, Vahter M (2011) Critical windows of exposure for arsenic-associated impairment of cognitive function in pre-school girls and boys: a population-based cohort study. Int J Epidemiol 40:1593–1604
Hayakawa T, Kobayashi Y, Cui X, Hirano S (2005) A new metabolic pathway of arsenite: arsenic-glutathione complexes are substrates for human arsenic methyltransferase Cyt19. Arch Toxicol 79:183–191
Heck JE, Andrew AS, Onega T, Rigas JR, Jackson BP, Karagas MR, Duell EJ (2009) Lung cancer in a U.S. population with low to moderate arsenic exposure. Environ Health Perspect 117:1718–1723
Hillestad LK (1957) Acute promyelocytic leukemia. Acta Med Scand 159:189–194
Hopenhayn-Rich C, Biggs ML, Smith AH, Kalman DA, Moore LE (1996) Methylation study of a population environmentally exposed to arsenic in drinking water. Environ Health Perspect 104:620–628
Hu J, Liu YF, Wu CF, Xu F, Shen ZX, Zhu YM, Li JM, Tang W, Zhao WL, Wu W, Sun HP, Chen QS, Chen B, Zhou GB, Zelent A, Waxman S, Wang ZY, Chen SJ, Chen Z (2009) Long-term efficacy and safety of all-trans retinoic acid/arsenic trioxide based therapy in newly diagnosed acute promyelocytic leukemia. Proc Natl Acad Sci U S A 106:3342–3347
Huang B, Wang Y (1993) Thousand formulas and thousand herbs of traditional Chinese medicine. Heilongjiang Education Press, Harbin

Huang ME, Ye YC, Chen SR, Chai JR, Lu JX, Zhoa L, Gu LJ, Wang ZY (1988) Use of all trans retinoic acid in the treatment of acute promyelocytic leukemia. Blood 72:567–572

Huang Y, Zhang J, McHenry KT, Kim MM, Zeng W, Lopez-Pajares V, Dibble CC, Mizgerd JP, Yuan ZM (2008b) Induction of cytoplasmic accumulation of p 53:a mechanism for low levels of arsenic exposure to predispose cells for malignant transformation. Cancer Res 68:9131–9136

Huang YK, Huang YL, Hsueh YM, Yang MH, Wu MM, Chen SY, Hsu LI, Chen CJ (2008a) Arsenic exposure, urinary arsenic speciation, and the incidence of urothelial carcinoma: a twelve-year follow-up study. Cancer Causes Control 19:829–839

Hubaux R, Becker-Santos DD, Enfield KSS, Rowbotham D, Lam S, Lam WL, Martinez VD (2013) Molecular features in arsenic-induced lung tumors. Mol Cancer 12:20

Hughes MF, Beck BD, Chen Y, Lewis AS, Thomas DJ (2011) Arsenic exposure and toxicology: a historical perspective. Toxicol Sci 123:305–332

IARC (2004) Some drinking-water disinfectants and contaminants, including arsenic. Monographs on chloramine, chloral and chloral hydrate, dichloroacetic acid, trichloroacetic acid and 3-chloro-4-(dichloromethyl)-5-hydroxy-2(5H)-furanone. IARC Monogr Eval Carcino Risks Hum 84:269–477

IARC (International Agency for Research on Cancer monographs on the evaluation of carcinogenic risks to humans) (2012) Arsenic, metals, fibers, and dusts (Vol 100C) A review of human carcinogens; http://monographs.iarc.fr/ENG/Monographs/vol100C/mono100C.pdf. Accessed April 2016

Intarasunanont P, Navasumrit P, Woraprasit S, Chaisatra K, Suk WA, Mahidol C, Ruchirawat M (2012) Effects of arsenic exposure on DNA methylation in cord blood samples from newborn babies and in a human lymphoblast cell line. Environ Health 11:31

Islam R, Khan I, Hassan SN, McEvoy M, D'Este C, Attia J, Peel R, Sultana M, Akter S, Milton AH (2012) Association between type 2 diabetes and chronic arsenic exposure in drinking water: a cross sectional study in Bangladesh. Environ Health 11:38

Iyer S, Sengupta C, Velumani A (2016) Blood arsenic: pan-India prevalence. Clin Chim Acta 455:99–101

Jensen TJ, Wozniak RJ, Eblin KE, Wnek SM, Gandolfi AJ, Futscher BW (2009) Epigenetic mediated transcriptional activation of WNT5A participates in arsenical-associated malignant transformation. Toxicol Appl Pharmacol 235:39–46

Jha AN, Noditi M, Nilsson R, Natarajan AT (1992) Genotoxic effects of sodium arsenite on human cells. Mutat Res 284:215–221

Jolliffe DM (1993) A history of the use of arsenicals in man. J Roy Soc Med 86:287–289

Jovanovic D, Rasic-Milutinovic Z, Paunovic K, Jakovljevic B, Plavsic S, Milosevic J (2013) Low levels of arsenic in drinking water and type 2 diabetes in middle Banat region, Serbia. Int J Hyg Environ Health 216:50–55

Kapaj S, Peterson H, Liber K, Bhattacharya P (2006) Human health effects from chronic arsenic poisoning—a review. J Environ Sci Health A Tox Hazard Subst Environ Eng 41:2399–2428

Kile ML, Baccarelli A, Hoffman E, Tarantini L, Quamruzzaman Q, Rahman M, Mahiuddin G, Mostofa G, Hsueh YM, Wright RO, Christiani DC (2012) Prenatal arsenic exposure and DNA methylation in maternal and umbilical cord blood leukocytes. Environ Health Perspect 120:1061–1066

Kim YJ, Kim JM (2015) Arsenic toxicity in male reproduction and development. Dev Rep 19:167–180

Koedrith P, Kim H, Weon JI, Seo YR (2013) Toxicogenomic approaches for understanding molecular mechanisms of heavy metal mutagenicity and carcinogenicity. Int J Hyg Environ Health 216:587–598

Komissarova EV, Rossman TG (2010) Arsenite induced poly(ADP-ribosyl)ation of tumor suppressor P53 in human skin keratinocytes as a possible mechanism for carcinogenesis associated with arsenic exposure. Toxicol Appl Pharmacol 243:399–404

Kryeziu K, Pirker C, Englinger B, van Schoonhoven S, Spitzwieser M, Mohr T, Körner W, Weinmüllner R, Tav K, Grillari J, Cichna-Markl M, Berger W, Heffeter P (2016) Chronic arsenic trioxide exposure leads to enhanced aggressiveness via met oncogene addiction in cancer cells. Oncotarget 7:27379–27393

Kundu M, Ghosh P, Mitra S, Das JK, Sau TJ, Banerjee S, States JC, Giri AK (2011) Precancerous and non-cancer disease endpoints of chronic arsenic exposure: the level of chromosomal damage and XRCC3 T241 M polymorphism. Mutat Res 706:7–12

Lan CC, Yu HS, Ko YC (2011) Chronic arsenic exposure and its adverse health effects in Taiwan: a paradigm for management of a global environmental problem. Kaohsiung J Med Sci 27:411–416

Lee TC, Huang RY, Jan KY (1985) Sodium arsenite enhances the cytotoxicity, clastogenicity, and 6-thioguanine-resistant mutagenicity of ultraviolet light in Chinese hamster ovary cells. Mutat Res 148:83–89

Li SZ (1982) The compendium of Materia medica (originally published in the Ming dynasty of China, 1578). People's Medical Publishing House, Beijing

Lin S, Cullen WR, Thomas DJ (1999) Methylarsenicals and arsinothiols are potent inhibitors of mouse liver thioredoxin reductase. Chem Res Toxicol 12:924–930

Lin S, Del Razo LM, Styblo M, Wang C, Cullen WR, Thomas DJ (2001) Arsenicals inhibit thioredoxin reductase in cultured rat hepatocytes. Chem Res Toxicol 14:305–311

Lin S, Shi Q, Nix FB, Styblo M, Beck MA, Herbin-Davis KM, Hall LL, Simeonsson JB, Thomas DJ (2002) A novel S-adenosyl-L-methionine:arsenic(III) methyltransferase from rat liver cytosol. J Biol Chem 277:10795–10803.

Lindberg AL, Rahman M, Persson LA, Vahter M (2008) The risk of arsenic induced skin lesions in Bangladeshi men and women is affected by arsenic metabolism and the age at first exposure. Toxicol Appl Pharmacol 230:9–16

Liu J, Chen B, Lu Y, Guan Y, Chen F (2012) JNK-dependent Stat3 phosphorylation contributes to Akt activation in response to arsenic exposure. Toxicol Sci 29:363–371

Liu J, Waalkes MP (2008) Liver is a target of arsenic carcinogenesis. Toxicol Sci 105:24–32

Liu ZM, Tseng HY, Cheng YL, Yeh BW, Wu WJ, Huang HS (2015) TG-interacting factor transcriptionally induced by AKT/FOXO3A is a negative regulator that antagonizes arsenic trioxide-induced cancer cell apoptosis. Toxicol Appl Pharmacol 285:41–50

Loffredo CA, Aposhian HV, Cebrian ME, Yamauchi H, Silbergelda EK (2003) Variability in human metabolism of arsenic. Environ Res 92:85–91

Lu K, Abo RP, Schlieper KA, Graffam ME, Levine S, Wishnok JS, Swenberg JA, Tannenbaum SR, Fox JG (2014) Arsenic exposure perturbs the gut microbiome and its metabolic profile in mice: an integrated metagenomics and metabolomics analysis. Environ Health Perspect 122:284–291

Lu K, Cable PH, Abo RP, Ru H, Graffam ME, Schlieper KA, Parry NM, Levine S, Bodnar WM, Wishnok JS, Styblo M, Swenberg JA, Fox JG, Tannenbaum SR (2013) Gut microbiome perturbations induced by bacterial infection affect arsenic biotransformation. Chem Res Toxicol 26:1893–1903

Lubin JH, Pottern LM, Stone BJ, Fraumeni JF Jr (2000) Respiratory cancer in a cohort of copper smelter workers: results from more than 50 years of follow-up. Amer J Epidemiol 151:554–565

Mann KK, Wallner B, Lossos IS, Miller WH Jr (2009) Darinaparsin: a novel organic arsenical with promising anticancer activity. Expert Opin Investig Drugs 18:1727–1734

Marshall G, Ferreccio C, Yuan Y, Bates MN, Steinmaus C, Selvin S, Liaw J, Smith AH (2007) Fifty-year study of lung and bladder cancer mortality in Chile related to arsenic in drinking water. J Natl Cancer Inst 99:920–928

Marsit CJ, Eddy K, Kelsey KT (2006) MicroRNA responses to cellular stress. Cancer Res 66:10843–10848

Martinez VD, Vucic EA, Becker-Santos DD, Gil L, Lam WL (2011) Arsenic exposure and the induction of human cancers. J Toxicol 2011:431287

Mathews V, George B, Lakshmi KM, Viswabandya A, Bajel A, Balasubramanian P, Shaji RV, Srivastava VM, Srivastava A, Chandy M (2006) Single agent arsenic trioxide in the treatment of newly diagnosed acute promyelocytic leukemia: durable remissions with minimal toxicity. Blood 107:2627–2632

Mazumder DN (2005) Effect of chronic intake of arsenic-contaminated water on liver. Toxicol Appl Pharmacol 206:169–175

Mead MN (2005) Arsenic: in search of an antidote to a global poison. Environ Health Perspect 113:378–386

Melak D, Ferreccio C, Kalman D, Parra R, Acevedo J, Pérez L, Cortés S, Smith AH, Yuan Y, Liaw J, Steinmaus C (2014) Arsenic methylation and lung and bladder cancer in a case control study in northern Chile. Toxicol Appl Pharmacol 274:225–231

Méndez-Gómez J, García-Vargas GG, López-Carrillo L, Calderón-Aranda ES, Gómez A, Vera E, Valverde M, Cebrián ME, Rojas E (2008) Genotoxic effects of environmental exposure to arsenic and lead on children in region Lagunera, Mexico. Ann N Y Acad Sci 1140:358–367

Milton AH, Hasan Z, Rahman A, Rahman M (2001) Chronic arsenic poisoning and respiratory effects in Bangladesh. J Occup Health 43:136–140

Most H (1972) Drug therapy—treatment of common parasitic infection of man encountered in the United States. N Engl J Med 287:698

Muenyi CS, Ljungman M, States JC (2015) Arsenic disruption of DNA damage responses-potential role in carcinogenesis and chemotherapy. Biomolecules 5:2184–2193

Murai T, Iwata H, Otoshi T, Endo G, Horiguchi S, Fukushima S (1993) Renal lesions induced in F344/DuCrj rats by 4 weeks oral administration of dimethylarsenic acid. Toxicol Lett 66:53–61

Naranmandura H, Suzuki N, Suzuki KT (2006) Trivalent arsenicals are bound to proteins during reductive methylation. Chem Res Toxicol 19:1010–1018

Naranmandura H, Xu S, Sawata T, Hao WH, Liu H, Bu N, Ogra Y, Lou YJ, Suzuki N (2011) Mitochondria are the main target organelle for trivalent monomethylarsonous acid (MMA(III))-induced cytotoxicity. Chem Res Toxicol 24:1094–1103

Nash TA (1960) Review of the African trypanosomiasis problem. Trop Dis Bull 57:973–1013

Nasr R, Rosenwald A, El-Sabban ME, Arnulf B, Zalloua P, Lepelletier Y, Bex F, Hermine O, Staudt L, de Thé H, Bazarbachi A (2003) Arsenic/interferon specifically reverses 2 distinct gene networks critical for the survival of HTLV-1-infected leukemic cells. Blood 101:4576–4582

Naujokas MF, Anderson B, Ahsan H, Aposhian HV, Graziano JH, Thompson C, Suk WA (2013) The broad scope of health effects from chronic arsenic exposure: update on a worldwide public health problem. Environ Health Perspect 121:295–302

Navas-Acien A, Sharrett AR, Silbergeld EK, Schwartz BS, Nachman KE, Burke TA, Guallar E (2005) Arsenic exposure and cardiovascular disease: a systematic review of the epidemiologic evidence. Amer J Epidemiol 162:1037–1049

Navas-Acien A, Silbergeld EK, Pastor-Barriuso R, Guallar E (2008) Arsenic exposure and prevalence of type 2 diabetes in US adults. JAMA 300:814–822

Nemeti B, Regonesi ME, Tortora P, Gregus Z (2010) Polynucleotide phosphorylase and mitochondrial ATP synthase mediate reduction of arsenate to the more toxic arsenite by forming arsenylated analogues of ADP and ATP. Toxicol Sci 117:270–281

Ng JC, Wang J, Shraim A (2003) A global health problem caused by arsenic from natural sources. Chemosphere 52:1353–1359

Nickson R, Sengupta C, Mitra P, Dave SN, Banerjee AK, Bhattacharya A, Basu S, Kakoti N, Moorthy NS, Wasuja M, Kumar M, Mishra DS, Ghosh A, Vaish DP, Srivastava AK, Tripathi RM, Singh SN, Prasad R, Bhattacharya S, Deverill P (2007) Current knowledge on the distribution of arsenic in groundwater in five states of India. J Environ Sci Health 42:1707–1718

Nimmanapalli R, Bali P, O'Bryan E, Fuino L, Guo F, Wu J, Houghton P, Bhalla K (2003) Arsenic trioxide inhibits translation of mRNA of bcr-abl, resulting in attenuation of bcr-abl levels and apoptosis of human leukemia cells. Cancer Res 63:7950–7958

Nollen M, Ebert F, Moser J, Mullenders LH, Hartwig A, Schwerdtle T (2009) Impact of arsenic on nucleotide excision repair: XPC function, protein level, and gene expression. Mol Nutr Food Res 53:572–582

Nriagu JO, Azcue JM (1990) Food contamination with arsenic in the environment. Adv Environ Sci Technol 23:121–143

Osmond MJ, Kunz BA, Snow ET (2010) Age and exposure to arsenic alter base excision repair transcript levels in mice. Mutagenesis 25:517–522

Pan C, Zhu D, Zhuo J, Li L, Wang D, Zhang CY, Liu Y, Zen K (2016) Role of signal regulatory protein α in arsenic trioxide-induced promyelocytic leukemia cell apoptosis. Sci Rep 6:23710

Papadimitrakopoulou V (2012) Development of PI3K/AKT/mTOR pathway inhibitors and their application in personalized therapy for non-small-cell lung cancer. J Thorac Oncol 7:1315–1326

Parvez F, Wasserman GA, Factor-Litvak P, Liu X, Slavkovich V, Siddique AB, Sultana R, Sultana R, Islam T, Levy D, Mey JL, van Geen A, Khan K, Kline J, Ahsan H, Graziano JH (2011) Arsenic exposure and motor function among children in Bangladesh. Environ Health Perspect 119:1665–1670

Peters BA, Hall MN, Liu X, Slavkovich V, Ilievski V, Alam S, Siddique AB, Islam T, Graziano JH, Gamble MV (2015) Renal function is associated with indicators of arsenic methylation capacity in Bangladeshi adults. Environ Res 143:123–130

Qin XJ, Liu W, Li YN, Sun X, Hai CX, Hudson LG, Liu KJ (2012) Poly(ADP-ribose)polymerase-1 inhibition by arsenite promotes the survival of cells with unrepaired DNA lesions induced by UV exposure. Toxicol Sci 127:120–129

Radi R, Beckman JS, Bush KM, Freeman BA (1991) Peroxynitrite oxidation of sulfhydryls. The cytotoxic potential of superoxide and nitric oxide. J Biol Chem 266:4244–4250

Rahman A, Persson LA, Nermell B, El Arifeen S, Ekstrom EC, Smith AH, Vahter M (2010a) Arsenic exposure and risk of spontaneous abortion, stillbirth, and infant mortality. Epidemiology 21:797–804

Rahman A, Vahter M, Ekstrom EC, Persson LA (2010b) Arsenic exposure in pregnancy increases the risk of lower respiratory tract infection and diarrhea during infancy in Bangladesh. Environ Health Perspect 119:719–724

Reichard JF, Puga A (2010) Effects of arsenic exposure on DNA methylation and epigenetic gene regulation. Epigenomics 2:87–104

Ren R (2005) Mechanisms of BCR-ABL in the pathogenesis of chronic myelogenous leukaemia. Nat Rev Cancer 5:172–183

Ren X, McHale CM, Skibola CF, Smith AH, Smith MT, Zhang L (2011) An emerging role for epigenetic dysregulation in arsenic toxicity and carcinogenesis. Environ Health Perspect 119:11–19

Rogers EH, Chernoff N, Kavlock RJ (1981) The teratogenic potential of cacodylic acid in the rat and mouse. Drug Chem Toxicol 4:49–61

Rosenberg HG (1974) Systemic arterial disease and chronic arsenicism in infants. Arch Pathol 97:360–365

Rossman TG, Klein CB (2011) Genetic and epigenetic effects of environmental arsenicals. Metallomics 3:1135–1141

Rossman TG (2003) Mechanism of arsenic carcinogenesis: an integrated approach. Mutat Res 533:37–65

Saha JC, Dikshit AK, Bandyopadhyay M, Saha KC (1999) A review of arsenic poisoning and its effects on human health. Crit Rev Environ Sci Technol 29:281–313

Sanchez TR, Perzanowski M, Graziano JH (2016) Inorganic arsenic and respiratory health, from early life exposure to sex-specific effects: a systematic review. Environ Res 147:537–555.

Schwarz RS (2004) Paul Ehrlich's magic bullets. N Engl J Med 350:1079–1080

Shen H, Xu W, Zhang J, Chen M, Martin FL, Xia Y, Liu L, Dong S, Zhu YG (2013) Urinary metabolic biomarkers link oxidative stress indicators associated with general arsenic exposure to male infertility in a Han Chinese population. Environ Sci Technol 47:8843–8851

Shen H, Niu Q, Xu M, Rui D, Xu S, Feng G, Ding Y, Li S, Jing M (2016) Factors affecting arsenic methylation in arsenic-exposed humans: a systematic review and meta-analysis. Int J Environ Res Public Health 13:205

Simeonova PP, Luster MI (2002) Arsenic carcinogenicity: relevance of c-Src activation. Mol Cell Biochem 234–235:277–282

Smith AH, Ercumen A, Yuan Y, Steinmaus CM (2009) Increased lung cancer risks are similar whether arsenic is ingested or inhaled. J Expo Sci Environ Epidemiol 19:343–348

Smith AH, Hopenhayn-Rich C, Bates MN, Goeden HM, Hertz-Picciotto I, Duggan HM, Wood R, Kosnett MJ, Smith MT (1992) Cancer risks from arsenic in drinking water. Environ Health Perspect 97:259–267

Smith AH, Lingas EO, Rahman M (2000) Contamination of drinking-water by arsenic in Bangladesh: a public health emergency. Bull World Health Organ 78:1093–1103

Smith AH, Marshall G, Yuan Y, Liaw J, Ferreccio C, Steinmaus C (2011) Evidence from Chile that arsenic in drinking water may increase mortality from pulmonary tuberculosis. Amer J Epidemiol 173:414–420

Spivey A (2011) Arsenic and infectious disease: a potential factor in morbidity among Bangladeshi children. Environ Health Perspect 119:218

Steinmaus C, Carrigan K, Kalman D, Atallah R, Yuan Y, Smith AH (2005) Dietary intake and arsenic methylation in a U.S. population. Environ Health Perspect 113:1153–1159

Stueckle TA, Lu Y, Davis ME, Wang L, Jiang BH, Holaskova I, Schafer R, Barnett JB, Rojanasakul Y (2012) Chronic occupational exposure to arsenic induces carcinogenic gene signaling networks and neoplastic transformation in human lung epithelial cells. Toxicol Appl Pharmacol 261:204–216

Styblo M, Drobna Z, Jaspers I, Lin S, Thomas DJ (2002) The role of biomethylation in toxicity and carcinogenicity of arsenic: a research update. Environ Health Perspect 110:767

Su CC, Lu JL, Tsai KY, Lian Ie B (2011) Reduction in arsenic intake from water has different impacts on lung cancer and bladder cancer in an arseniasis endemic area in Taiwan. Cancer Causes Control 22:101–108

Sung TI, Wang YJ, Chen CY, Hung TL, Guo HR (2012) Increased serum level of epidermal growth factor receptor in liver cancer patients and its association with exposure to arsenic. Sci Total Environ 424:74–78

Tang F, Liu G, He Z, Ma WY, Bode AM, Dong Z (2006) Arsenite inhibits p53 phosphorylation, DNA binding activity, and p53 target gene p21 expression in mouse epidermal JB6 cells. Mol Carcinog 45:861–870

Thomas DJ, Li J, Waters SB, Xing W, Adair BM, Drobna Z, Devesa V, Styblo M (2007) Arsenic (+3 oxidation state) methyltransferase and the methylation of arsenicals. Exp Biol Med 232:3–13

Thu KL, Pikor LA, Chari R, Wilson IM, Macaulay CE, English JC, Tsao MS, Gazdar AF, Lam S, Lam WL, Lockwood WW (2011) Genetic disruption of KEAP1/CUL3 E3 ubiquitin ligase complex components is a key mechanism of NF-kappaB pathway activation in lung cancer. J Thor Oncol 6:1521–1529

Tseng CH, Chong CK, Chen CJ, Tai TY (1996) Dose–response relationship between peripheral vascular disease and ingested inorganic arsenic among residents in black foot disease endemic villages in Taiwan. Atherosclerosis 120:125–133

Tseng WP (1977) Effect of dose-response relationships of skin cancer and Blackfoot disease with arsenic. Environ Health Perspect 19:109–119

Tseng WP, Chu HM, How SW, Fong JM, Lin CS, Yeh S (1968) Prevalence of skin cancer in an endemic area of chronic arsenicism in Taiwan. J Natl Cancer Inst 40:453–463

Turrens JF (1997) Superoxide production by the mitochondrial respiratory chain. Biosci Rep 17:3–8

Vahter M (2002) Mechanisms of arsenic biotransformation. Toxicology 181:211–217

Venkatesan RN, Treuting PM, Fuller ED, Goldsby RE, Norwood TH, Gooley TA, Ladiges WC, Preston BD, Loeb LA (2007) Mutation at the polymerase active site of mouse DNA polymerase delta increases genomic instability and accelerates tumorigenesis. Mol Cell Biol 27:7669–7682

Verma A, Mohindru M, Deb DK, Sassano A, Kambhampati S, Ravandi F, Minucci S, Kalvakolanu DV, Platanias LC (2002) Activation of Rac1 and the p38 mitogen-activated protein kinase pathway in response to arsenic trioxide. J Biol Chem 277:44988–44995

Vogt BL, Rossman TG (2001) Effects of arsenite on p53, p21 and cyclin D expression in normal human fibroblasts—a possible mechanism for arsenite's comutagenicity. Mut Res 478:159–168

Walter I, Schwerdtle T, Thuy C, Parsons JL, Dianov GL, Hartwig A (2007) Impact of arsenite and its methylated metabolites on PARP-1 activity, PARP-1 gene expression and poly(ADP-ribosyl)ation in cultured human cells. DNA Repair 6:61–70

Waters SB, Devesa V, Del Razo LM, Styblo M, Thomas DJ (2004) Endogenous reductants support the catalytic function of recombinant rat cyt19, an arsenic methyltransferase. Chem Res Toxicol 17:404–409

Wang CH, Hsiao CK, Chen CL, Hsu LI, Chiou HY, Chen SY, Hsueh YM, Wu MM, Chen CJ (2007) A review of the epidemiologic literature on the role of environmental arsenic exposure and cardiovascular diseases. Toxicol Appl Pharmacol 222:315–326

Wang WL, Xu SY, Ren ZG, Tao L, Jiang JW, Zheng SS (2015) Application of metagenomics in the human gut microbiome. World J Gastroenterol 21:803–814

Wang Y, Fang J, Leonard SS, Rao KM (2004) Cadmium inhibits the electron transfer chain and induces reactive oxygen species. Free Radic Biol Med 36:1434–1443

Wang Z, Yang J, Fisher T, Xiao H, Jiang Y, Yang C (2012) Akt activation is responsible for enhanced migratory and invasive behavior of arsenic transformed human bronchial epithelial cells. Environ Health Perspect 120:92–97

Wang Z, Zhao Y, Smith E, Goodall GJ, Drew PA, Brabletz T, Yang C (2011) Reversal and prevention of arsenic-induced human bronchial epithelial cell malignant transformation by microRNA-200b. Toxicol Sci 121:110–122

Wang ZY, Chen Z (2008) Acute promyelocytic leukemia: from highly fatal to highly curable. Blood 111:2505–2515

Wei M, Wanibuchi H, Yamamoto S, Li W, Fukushima S (1999) Urinary bladder carcinogenicity of dimethylarsinic acid in male F344 rats. Carcinogenesis 20:1873–1876

Wen G, Calaf GM, Partridge MA, Echiburu-Chau C, Zhao Y, Huang S, Chai Y, Li B, Hu B, Hei TK (2008) Neoplastic transformation of human small airway epithelial cells induced by arsenic. Mol Med 14:2–10

Wiencke JK, Yager JW (1992) Specificity of arsenite in potentiating cytogenetic damage induced by the DNA crosslinking agent diepoxybutane. Environ Mol Mutagen 19:195–200

Wink DA, Kasprzak KS, Maragos CM, Elespuru RK, Misra M, Dunams TM, Cebula TA, Koch WH, Andrews AW, Allen JS (1991) DNA deaminating ability and genotoxicity of nitric oxide and its progenitors. Science 254:1001–1003

Wiwanitkit V (2015) Contamination of arsenic species in rice and the calculation for risk of cancer. J Cancer Res Ther 11:1044

Yamamoto S, Konishi Y, Matsuda T, Murai T, Shibata MA, Matsui-Yuasa I, Otani S, Kuroda K, Endo G, Fukushima S (1995) Cancer induction by an organic arsenic compound, dimethylarsinic acid (cacodylic acid), in F344/DuCrj rats after pretreatment with five carcinogens. Cancer Res 55:1271–1276

Yamauchi H, Fowler BA (1994) Toxicity and metabolism of inorganic and methylated arsenicals. In: Nriagu JO (ed) Arsenic in the environment, part II: human health and ecosystem effects. Wiley, New York, pp 35–43

Yarden Y, Sliwkowski MX (2001) Untangling the ErbB signalling network. Nat Rev Mol Cell Biol 2:127–137

Yuan Y, Marshall G, Ferreccio C, Steinmaus C, Liaw J, Bates M, Smith AH (2010) Kidney cancer mortality: fifty-year latency patterns related to arsenic exposure. Epidemiology 21:103–108

Zachariac H, Sogard H, Nyfors A (1974) Liver biopsy in psoriatic previously treated with potassium arsenite. A controlled study. Acta Derm Venereol 54:235–236
Zhang QY, Mao JH, Liu P, Huang QH, Lu J, Xie YY, Weng L, Zhang Y, Chen Q, Chen SJ, Chen Z (2009) A systems biology understanding of the synergistic effects of arsenic sulfide and imatinib in BCR/ABL-associated leukemia. Proc Natl Acad Sci U S A 106:3378–3383
Zhang W, Chen L, Zhou Y, Wu Y, Zhang L (2016) Biotransformation of inorganic arsenic in a marine herbivorous fish *Siganus fuscescens* after dietborne exposure. Chemosphere 147:297–304
Zhang Y, Bhatia D, Xia H, Castranova V, Shi X, Chen F (2006) Nucleolin links to arsenic-induced stabilization of GADD45alpha mRNA. Nucleic Acids Res 34:485–495
Zhang Q, Wang D, Zheng Q, Zheng Y, Wang H, Xu Y, Li X, Sun G (2014) Joint effects of urinary arsenic methylation capacity with potential modifiers on arsenicosis: a cross-sectional study from an endemic arsenism area in Huhhot Basin, northern China. Environ Res 132:281–289
Zhao Y, Toselli P, Li W (2012) Microtubules as a critical target for arsenic toxicity in lung cells in vitro and in vivo. Int J Environ Res Public Health 9:474–495
Zhou G, Zhang J, Wang Z, Chen S, Chen Z (2007) Treatment of acute promyelocytic leukaemia with all-trans retinoic acid and arsenic trioxide: a paradigm of synergistic molecular targeting therapy. Philos Trans Roy Soc B 362:959–971
Zhou L, Hou J, Chan CFG, Sze DMY (2014) Arsenic trioxide for non-acute promyelocytic leukemia hematological malignancies: a new frontier. J Blood Disord 1:1018
Zhu J, Chen Z, Lallemand-Breitenbach V, de The H (2002) How acute promyelocytic leukaemia revived arsenic. Nat Rev Cancer 2:705–713
Ziech D, Franco R, Pappa A, Panayiotidis MI (2011) Reactive oxygen species (ROS)–induced genetic and epigenetic alterations in human carcinogenesis. Mutat Res 711:167–173

Epigenetics in Arsenic Toxicity: Mechanistic Response, Alterations, and Regulations

Dibyendu Talukdar

Contents

1 Introduction 68
2 Epigenetic Regulation and Alterations in Animal and Human System During Mechanistic Response of Cellular As Toxicity 69
 2.1 DNA Methylation and As Toxicity 70
  2.1.1 Modulation of Global DNA Methylation Footprint Under As Exposure 71
  2.1.2 As Toxicity, Gene-Specific DNA Methylation and Tumorigenesis/Oncogenesis 73
  2.1.3 As-Induced Hypomethylation in Gene Promoters 74
  2.1.4 As-Induced Hypermethylation in Gene Promoters 75
 2.2 Histone Modifications Under As Toxicity 76
  2.2.1 Histone Acetylation Under As Exposure 76
  2.2.2 Histone Methylation During As Exposure 78
  2.2.3 Histone Phosphorylation During As-Induced Epigenetic Regulations 79
 2.3 As-Induced Epigenetic Alterations of miRNAs 79
 2.4 Cross Talk Among Epigenetic Regulations During As Exposure 82
 2.5 As Toxicity, Foetal Growth and Epigenetic Memory 82
3 Epigenetic Response to As Toxicity in Plants 83
 3.1 Induced DNA Methylation in Plants Under As Exposure 83
 3.2 Histone Modification During Stress Response in Plants 86
 3.3 The miRNAs During Metal(loid) Toxicity in Plant Cells 86
4 Challenges and Future Research 89
References 93

D. Talukdar (✉)
Department of Botany, R.P.M. College, Hooghly 712258, India
e-mail: dibyendutalukdar9@gmail.com

D.K. Gupta, S. Chatterjee (eds.), *Arsenic Contamination in the Environment*,
DOI 10.1007/978-3-319-54356-7_4

1 Introduction

Arsenic (As) is a ubiquitous toxic, co-mutagenic, and carcinogenic metalloid. More than 300 million people are now affected by As-contaminated drinking water, and the worst scenario is prevalent in the Indian state of West Bengal, parts of Bihar and Uttar Pradesh, and in the neighboring country Bangladesh (Sengupta et al. 2008; Flanagan et al. 2012; Dey et al. 2014). Inorganic species of As is more toxic than organic form of As and is abundant in natural environment. As poisoning or arsenicosis is a condition caused by the continuous ingestion, absorption, or inhalation of dangerous levels of As in drinking water, i.e., markedly above the WHO guideline level of 10 $\mu g\ L^{-1}$ (WHO 2006; Centeno et al. 2007). In humans, arsenicosis-related health problems are mainly arise through (a) drinking water, (b) contaminated plant food due to toxic water–soil–plant system, (c) animal biomagnifying in meat, poultry, fish, and even in seafood (Guha Mazumder 2008; Rana et al. 2008; Datta et al. 2010), and also through occupational hazards like industrial workers engaged in glass productions, wood industries, smelting, and production of arsenical pesticides (Datta et al. 2010).

Inorganic As has been considered as a Group 1 human carcinogen (IARC 2004). Prolonged exposure to As contaminated toxic drinking water and food, without taking any nutritional fortifications and/or any other preventive measure, may result in development of arsenicosis in human beings (Guha Mazumder 2008). Pain or dizziness coupled with confusion, headache, rash, terrible diarrhea, and swellings are some of the predominant symptoms of arsenicosis. Melanosis, transverse white striae (Mees' lines) in nails, keratosis, convulsions, fingernail pigmentation (leukonychia), body weight loss, upset stomach, cramped muscle, hair loss, vomiting, excess saliva in mouth, and blood in the urine may trigger complexity of symptoms at advanced stages, followed by multi-organ failure, coma, shock, and subsequently death (Guha Mazumder 2008). The pharmacological complications of prolonged exposure include cancer, diabetes, cardiovascular disease, hepatic disorder, chronic dermal symptoms, encephalopathy, bronchitis, pulmonary fibrosis, hepatosplenomegaly, peripheral vascular disease/"blackfoot disease," atherosclerosis, nervous system complications, peripheral neuropathy, digestive difficulties, neurotoxicity, and miscarriage (Flora et al. 2007; Das and Sengupta 2008; Rana et al. 2008; Datta et al. 2010). There is accumulating evidence suggesting that inorganic As has been associated with kidney, liver, skin, lung, and urinary cancers (WHO 2006; Guha Mazumder 2008; Straif et al. 2009). The primary mechanisms related to As-mediated carcinogenesis involve perturbations in cellular signalling pathway, induction of oxidative stress, numerical and structural chromosomal aberrations, and serious impediment in DNA damage response and repair mechanisms (Kitchin and Wallace 2008; Kitchin and Conolly 2010; Ray et al. 2014). It is however clear that As cannot induce point mutation but induces tumorigenesis through deletion mutation and subsequent instability in genomes. An interaction between genomic and epigenomic events is thus likely involved in regulation of As toxicity and carcinogenicity (Hei and Filipic 2004; Fournier et al. 2007).

Environmental toxicant like As has the ability to mediate onset and mechanistic response of disease states through modulation of key cellular signalling molecules and pathways via differential gene expressions and consequent alterations in protein structure and functions (Ray et al. 2014). In the last two decades, growing evidences are indicating that apart from influence from nuclear gene epigenetic mechanisms play major roles in response, regulations, and alterations of cellular As toxicity (Cheng et al. 2012). Despite non-redox active, As is known to perturb cellular oxidative balance by generating excess reactive oxygen species (ROS) and consequent alteration in redox homeostasis (Bailey and Fry 2014a) which has been recognized as one of the major events during epigenetic regulations of As toxicity in metal-exposed aerobic cells (Cerda and Weitzman 1997). There has been growing recognition of the importance of epigenomic mechanisms in maintaining cellular homeostasis and genomic imprinting. Epigenetic alteration which is certainly not a genotoxic event denotes processes that governs mitotically and/or meiotically heritable changes with chromosomal marks in genomic expression that does not involve changes in the nuclear DNA sequence (Feinberg and Tycko 2004; Cortessis et al. 2012; Ray et al. 2014), and therefore, it can serve as a potentially reversible DNA alteration and a priming mechanism for next generations to better tolerate biotic and abiotic stresses (Ren et al. 2011; Yong-Villalobos et al. 2015). Epigenetic mechanisms include but are not limited to DNA methylation, the covalent, posttranslational modifications of histone proteins, and small noncoding RNAs or miRNAs and regulate number of homeostatic and inducible gene expressions (Weake and Workman 2010). Each of these components plays specific roles in regulation of gene expression. Yet they interact with each other coordinately to mediate cellular response to environmental stressors (Bollati and Baccarelli 2010). Being an environmental toxicant and carcinogenic agent, As can induce epigenetic alterations and the epigenome can exert its own influence over the transcriptome and proteomes during onset of disease. Much attention has been paid over epigenetic regulations by environmental stimuli in deciphering the roles of epigenome in human diseases (Skinner 2011). In plants, epigenetic mechanisms have been envisaged as major targets of different environmental stressors (Labra et al. 2002; Cheng et al. 2012). Changes in DNA methylation pattern and alteration in chromatin structure are now being considered as potential biomarkers of epigenotoxic effects of heavy metals and metalloid toxicity in plants (Ge et al. 2012; Erturk et al. 2014).

2 Epigenetic Regulation and Alterations in Animal and Human System During Mechanistic Response of Cellular As Toxicity

It is becoming increasingly evidenced that events associated with epigenetic mechanisms can play distinct roles in cellular As toxicity in animal and human systems (Arita and Costa 2009; Bailey et al. 2013). Exposure to inorganic As can lead to plethora of epigenetic regulations and alterations in global and gene-specific DNA methylation, histone modifications, and miRNA expressions. Streams of evidences indicate that these alterations can trigger cascading changes in downstream cellular and metabolic events which can induce carcinogenesis and non-cancer end points (Ren et al. 2011; Bailey and Fry 2014a, b). Studies with animal model suggest that onset of As-induced toxicity and/or disease development was not associated with changes in DNA sequence but persistent changes in expression and epigenetic landscape of vital genes responsible for chronic disease onset (Bailey and Fry 2014a).

2.1 DNA Methylation and As Toxicity

DNA methylation is the most common epigenetic mechanism studied extensively during As exposure. Methylation occurs through the transfer of a methyl group to 5th position of cytosine pyrimidine ring, forming 5-methyl cytosine and *S*-adenosyl homocysteine. The DNA methyltransferase (DNMT) family preferentially drives methylation of cytosine in CpG dinucleotides through the transfer of a single methyl group from S-adenosyl methionine (SAM) to cytosine. So far, three different families and their several sub-families of DNMTs have been identified with diverse functions, as is enlisted in Table 1. The CPG is a palindromic sequence which during cell division is used by DNMT 1 as a template strand for copying DNA methylation from parental strand to transfer it to unmethylated daughter strand. Nearly 30% of CpGs are observed in GC-rich regions, the so-called "CpG islands" or "CpG island shores," which normally is unmethylated (euchromatin region) and located in promoter regions of tissue-specific genes, cellular housekeeping genes, some oncogenes, and tumor-suppressor genes (Hermann et al. 2004). Rest of the CpGs are normally methylated in transcriptionally silenced heterochromatic regions. However, during As exposure, 18% of the CpG islands get methylated and a linear relationship is found between magnitude of CpG methylation and total urinary As concentration in Bangladeshi population (Koestler et al. 2013). Apart from oncogene suppression, DNA methylation in normal cells has been involved in governing gene expression related to cell cycle progression, cell proliferation and differentiation, normal developmental process with fidelity in parental imprinting, inactivation of X chromosomes, and maintenance of chromosomal structural as well as functional integrity through inactivation of repetitive elements and transposable elements (Hermann et al. 2004; Stefanska et al. 2011; Szyf 2011). Silencing of transcriptional processes by CpG methylation can occur through (1) direct methylation of recognition sequence of transcription factor which can block their binding with DNA and suppress transcription process, (2) blockage of access of transcription factors to regulatory regions of promoters by methylated DNA-binding domain proteins (MBDs). MBDs can bind to methylated DNA and mask recognition element, and (3) changes in chromatin structure via MBD-mediated recruitment of chromatin modifiers such as histone methyltransferases and histone deacetylases,

Table 1 DNA methyltransferases (DNMT) and their functions in epigenetic regulations

DNMT1 (vertebrates)	Most abundant; DNA methylation (maintenance), requires hemi-methylated DNA (DNA with only one stand methylated), at CpG sties; directly involved in DNA damage repair in a DNA methylation-independent manner; demethylase functions.
MET1 (plants)	METI methyltransferase (MET I, IIA, IIB, III in *Arabidopsis*, MET 1a, 2b in *Brassica rapa*, MET 1–1, 1–2 in *Oryza sativa*; MET1 in *Zea mays*, *Nicotiana tabacum*, MET in *Lycopersicum esculentum, Pisum sativum*) shows a preference for cytosines (C5) in CpG and CpNpG trinucleotides sequences.
DNMT2	Participates in the recognition of DNA damage, DNA recombination, and mutation repair; execute methylation of anticodon loop of transfer RNA^{Asp} at cytosine 38; subfamilies like DNMT2L in *Arabidopsis*, maize, rice.
DNMT3A	DNA methylation de novo, crucial for the establishment of gene imprinting and silencing of retrotransposons; can mediate methylation-independent gene repression; does not require hemi-methylated DNA to bind; demethylase functions; DNMT3L in plants.
DNMT3B	DNA methylation de novo, crucial for the establishment of gene imprinting and silencing of retrotransposons; specifically required for stabilization of pericentromeric satellite repeats; does not require hemi-methylated DNA to bind; demethylase functions; preferential methylation through selective binding of DNMT3B to the bodies of transcribed genes in mouse stem cells requires SETD2-mediated methylation of lysine 36 on histone H3 and a functional PWWP domain.
DNMT3L	Does not possess any inherent enzymatic activity, regulate the activity and substrate recognition by DNMT3A and DNMT3B.
Chromomethylases	Unique to plants, preferentially methylate cytosines in CpNpG sequences.

resulting in inactive chromatin site and transcriptional repression (Newell-Price et al. 2000; Stefanska et al. 2011; Ray et al. 2014).

2.1.1 Modulation of Global DNA Methylation Footprint Under As Exposure

Any aberrant change in DNA methylation process by external stimuli either in the form of hypermethylation or hypomethylation can lead to series of epigenetic modifications at genome level. The changes in DNA methylation footprint can serve as hallmark or epigenomic biomarker of several diseases. Diverse modulation of methylation process has been observed following exposure to environmental toxicants and DNA methylation is the most studied phenomenon during As-exposure (Reichard and Puga 2010; Ren et al. 2011).

Mechanistic study confirmed that As-induced changes in DNA-methylation pattern is responsible for significant decline of human urinary defensin, beta 1 (DFB1) protein level in US and Chile population which possibly occurs through As-mediated targeted gene silencing (Sun et al. 2006; Hegedus et al. 2008; Ren et al. 2011).

Another interesting feature is the relationship between inorganic As biotransformation and the DNA methylome. In mammalian system, both arsenite [As(III)] and arsenate [As(V)] can undergo biotransformation to trivalent and pentavalent methylated forms, respectively, and the primary enzyme identified in reduction of As(V) to As(III) and methylation process is arsenic methyltransferase (AS3MT), which utilizes SAM as a methyl donor (Thomas et al. 2007; Drobna et al. 2009; Bailey and fry 2014a). Using the "Illumina Infinium Human Methylation 450K Bead Chip," Engstrom et al. (2013) studied methylation status of a few CpG regions located at the 5′ *AS3MT* sites in the samples collected from Bangladeshi and Argentinian population affected with As toxicity. DNA methylation and biotransformation of inorganic As both employ SAM as a methyl donor, suggesting a close link between As-induced carcinogenesis and DNA methylation (Ren et al. 2011). SAM serves as unique methyl group donor in more than 40 metabolic reactions. Thus, biomethylation of As for prolonged exposure may lead to SAM deficiency and altered DNA methylome through global hypomethylation (Coppin et al. 2008; Ren et al. 2011). As(III) is more toxic to As(V) due to its high affinity to sulfhydryl group present in numerous cellular biomolecules like cysteine and glutathione, and can induce DNA hypomethylation by depleting SAM reserve in As-exposed cells (Coppin et al. 2008; Nohara et al. 2011; Ren et al. 2011). Furthermore, methionine, an essential amino acid, is exclusively required for SAM synthesis. In several resource-poor countries like India and Bangladesh, dietary intake of sulfur-containing amino acids like cysteine and methionine are extremely poor. This may exacerbate effects of arsenic toxicity on DNA methylation pattern and antioxidant defense. Besides SAM deficiency, As can directly interact with DNMT and impede their activities, although exact mechanism is still unclear.

Conflicting relationships are available regarding As-exposure and DNA methylation pattern. Several in vitro and in vivo studies indicate that As exposure results in dose-dependent decrease of mRNA levels and DNMT1, DNMT3A, and DNMT3B activities (Ahlborn et al. 2008; Reichard et al. 2007; Fu et al. 2010), although instances of As-induced global hypermethylation are also not rare (Davis et al. 2000; Ray et al. 2014). In vitro studies with cell lines revealed global genomic DNA hypomethylation due to As(III) exposure (Sciandrello et al. 2004; Benbrahim-Tallaa et al. 2005; Reichard et al. 2007; Coppin et al. 2008). Bagnyukova et al. (2007) reported that As(III) exposure in fish for 1, 4, or 7 days led to sustained reduction of DNA methylation in comparison to non-exposed fish. Similarly, mice and rats exposed to As(III) for several weeks resulted in global hepatic DNA hypomethylation (Chen et al. 2004; Xie et al. 2007; Uthus and Davis 2005). In contrast to these findings, genome-wide DNA hypermethylation was reported in peripheral blood leucocytes from Indian and Bangladeshi population. Majumdar et al. (2010) found inconsistent DNA methylation pattern in 64 individual exposed to As-contaminated water. While people exposed to 250–500 $\mu g\ L^{-1}$ As exhibited DNA hypermethylation, individual exposed to >500 $\mu g\ L^{-1}$ As had DNA methylation comparable to individuals belonging to lowest exposed group (Majumdar et al. 2010). Pilsner et al. (2007, 2009), however, observed positive relationship between As exposure and DNA methylation with increasing As exposure leads to elevated level of DNA methylation. Addition of folate at adequate amount may fine tune this increase in methylation in peripheral blood leukocyte which can adapt As exposure and reduce risk of skin lesion (Pilsner et al. 2009). Folate-mediated regulated levels of DNA hypermethylation has immense significance as DNA hypomethylation in many cases are related to As-induced skin lesions and onset of carcinogenesis (Wilson et al. 2007; Pilsner et al. 2009). DNA hypermethylation has been found as a common mechanism of silencing of tumor-suppressor genes in several types of cancer. Whereas hypermethylation in glutathione S-transferase pi 1(*GSTP1*) gene can trigger prostate cancer, glioma results when O 6-methylguanine-DNA methyltransferase gene gets hypermethylated (Esteller et al. 2001).

2.1.2 As Toxicity, Gene-Specific DNA Methylation and Tumorigenesis/Oncogenesis

Onset of cellular As toxicity has been found associated with differential methylation of *AS3MT* gene, responsible for encoding the enzyme arsenic methyltransferase for inorganic As biotransformation. Increased methylation of *AS3MT* and differential methylation of several other genes located on chromosome band 10q24 along with *AS3MT* was observed in Argentinian Andes Women population heavily exposed with As(III) (Hossain et al. 2012). This study observed positive correlation between *p16* methylation and the slow-metabolizing *AS3MT* haplotype in the blood of chronically exposed Andean women. The group also observed that *p16* and nonpolyposis type 2 (*E. coli*) (*MLH1*) methylation were positively associated with urinary inorganic As concentration. However, the urinary concentrations of methylated monomethyl arsenic (MMA) and dimethyl arsenite (DMA) were negatively associated with *p16* methylation. The results indicate that inorganic As exerts more toxic effects on p16 DNA methylation than methylated arsenicals (Hossain et al. 2012).

Growing evidences suggest that As toxicity modifies DNA methylation pattern of genes involved in tumor initiation (Bailey and Fry 2014a). Besides genome-wide DNA methylation, As exposure can also trigger gene specific DNA methylation. Notable among this is tumor suppressor cyclin-dependent kinase inhibitor 2A (*CDKN2A/p16)* promoter which gets methylated in As-exposed population of West Bengal, India and Guizhou, China (Chanda et al. 2006; Zhang et al. 2007). Methylation of Ras association (RalGDS/AF-6) domain family member 1 (*RASSF1A)*, tumor suppressors protease serine 3 (*Homo sapiens*; *PRSS3*) promoter and death-associated protein kinase 1 (*DAPK)* was correlated with As exposure in Taiwanese population having urothelial carcinoma (Marsit et al. 2006a; Chen et al. 2007). Besides *CDKN2A*, *RASSF1A* and *DAPK*, several other tumor suppressor genes such as *tumor protein (TP) 53*, *reversion-inducing-cysteine-rich protein with kazal motifs* (*RECK*), and *von Hippel–Lindau tumor suppressor* (*VHL*) get methylated due to As exposure (Davis et al. 2000; Bailey and Fry 2014a). In Bangladeshi population, positive correlation was found between methylation of Long Interspersed Nuclear Element 1 (LINE-1) and total urinary As level and some CpG sites within the *p16* promoter (Bailey and Fry 2014a). Several oncogenes such as estrogen receptor alpha (*ER*-α), members of the RAS family of small G-proteins such as *HRAS* and *KRAS*, and cyclin D1 (*CCND1*) are methylated during As exposure (Chen et al. 2004; Waalkes et al. 2004; Benbrahim-Tallaa et al. 2005). No correlation between As toxicity and methylation of *CDKN2A/p16* DNA was, however, found in As-exposed population of New Hampshire (Marsit et al. 2006a, b).

2.1.3 As-Induced Hypomethylation in Gene Promoters

As-induced changes in DNA methylation status have also been reported in gene promoter regions. As-induced hypomethylation and hypermethylation of promoters have been observed in human skin cancer (Chanda et al. 2006) and bladder cancer (Marsit et al. 2006a; Chen et al. 2007). Hypomethylation and hypermethylation of genes could mediate oncogenesis by upregulating oncogene expression or down-regulating tumor suppressor genes, respectively (Ren et al. 2011). In vitro studies showed As-induced hypomethylation of promoter region and altered mRNA expressions of oncogene *Hras1* and *c-myc* (Takahashi et al. 2002). Likewise, mice treated with sodium As(III) manifested hypomethylation of the promoter region of oncogenic *Hras1* and an elevated *Hras1* mRNA level (Okoji et al. 2002). Promoter of the estrogen receptor 1 (*Esr 1*) gene also gets methylated during As exposure, resulting in over expression of *Esr 1* gene and induction of As-induced hepatocarcinogenesis (Chen et al. 2004; Waalkes et al. 2004). Fu et al. (2010) observed that exposure of malignant cell lines to arsenic trioxide reduced hypermethylation of cyclin dependent kinase 2A/2B (*CDKN2A/CDKN2B*; *p16/p15*) promoters and elevated mRNA expressions. Similar findings were reported in *RASSF1A*, *p16, GSTP1* lines after treated with arsenic trioxide (Cui et al. 2006b). Gribble et al. (2013) studied relationship between As species and methylome of *AS3MT* promoter in native Americans. Around 30 hypomethylated CpGs were identified out of total 48 CpGs assayed in *AS3MT* promoter in the moderate total urinary As group compared to the low total urinary As group. Expression of *AS3MT* was induced in presence of sodium As(III) exposure which reduced DNA methylation of the *AS3MT* promoter. Additionally, *AS3MT* knockdown inhibited As-induced global DNA hypomethylation in peripheral blood mononuclear cells and MMA and DMA levels have no relationship with *AS3MT* promoter methylation events (Gribble et al. 2013). It is however not fully clear that how this DNA methylation in *AS3MT* promoter regulates its expression and consequently, mediates As-induced disease development in human population. In As(III) and MMA(III)-transformed human urothelial cell lines, a correlation was found between focal DNA methylation and gene expressions. The focal DNA methylation pattern being stable has been focused as one of the epigenomic biomarkers of As toxicity and risk assessment process (Ray et al. 2014).

2.1.4 As-Induced Hypermethylation in Gene Promoters

Hypermethylation of genes can mediate carcinogenesis by downregulating tumor suppressor genes (Ren et al. 2011). Promoters of several tumor suppressor genes like *p15*, *p16*, *p53*, and death-associated protein kinase (*DAPK*) get hypermethylated by As exposure as dose-dependent way (Chanda et al. 2006; Zhang et al. 2007; Fu et al. 2010; Ren et al. 2011). DAPK positively mediates γ-interferon-induced programmed cell death. Hypermethylation of *DAPK* promoter was found associated with 13 of 17 tumors in patients living in As-contaminated areas and suffering from urothelial carcinoma in comparison to 8 of 21 tumors in patients living in As-free areas (Chen et al. 2007). As exposure induced DAPK hypermethylation in immortalized human uroepithelial cells too (Chai et al. 2007). Earlier, Marsit et al. (2006b) studied As exposure and DNA methylation pattern in human bladder cancer and observed that As exposure induced hypermethylation in promoter of *RASSF1A* and *PRSS3* genes which subsequently triggers invasive cancer. Similar observation was reported in A/J mice where As-exposure provoked hypermethylation of *RASSF1A* promoter which repressed its expression and subsequently caused As-induced lung carcinogenesis (Cui et al. 2006a). As-induced increase in DNA methylation was also observed in promoter regions in *p16* and *p53* which subsequently induced malignant transformation in several human cell lines (Chanda et al. 2006; Zhang et al. 2007; Jensen et al. 2009a; Ren et al. 2011). Bailey and Fry (2014a) reported identification of 183 genes whose promoters were differentially methylated in Mexican women having inorganic As-induced skin lesions, diabetes mellitus, cardiovascular disease, and cancer. DNA hypermethylation in human lung adenocarcinoma A549 cells, promoter region of tumour suppressor gene *p53* and in *p16* (CDKN2A) gene suggested epigenetic silencing of key tumor suppressor may be an important mechanism of As-promoted cancer initiation (Chanda et al. 2006). This was evident from strong positive correlation between DNA hypermethylation of *p53* and *p16* and As levels and increased cellular sensitivity to As cytotoxicity after mutation in *p53*. This suggests epi-mutagenic involvement of As in tumorigenesis (Lu et al. 2014). An inverse relationship between DNA hypermethylation of promoters and mRNA expression was noticed in G0/G1 switch 2 (*G0S2*), keratin 7 (*KRT7*), zinc finger, and SCAN domain containing 12 (*ZSCAN12*), thioesterase superfamily member 4 (*THEM4*), tumor necrosis factor related protein 6 (*C1QTNF6*),C1Q, and epiregulin (*EREG*) during As exposure (Jensen et al. 2008, 2009b). In contrast, positive association was found between varying levels of As exposure and *tumor protein (TP) 53* promoter methylation pattern in cord blood of a Thailand population. Interestingly, many of the identified genes associated with As-induced changes in DNA methylome show common mechanistic response and thus "occupancy" of transcription factor may influence accessibility of DNMT enzymes to particular regions of the genome under As exposure (Bailey and Fry 2014a).

2.2 *Histone Modifications Under As Toxicity*

Eukaryote genomes are usually large in size. Thus, DNA is required to be packaged into chromatin to accommodate genome size. Chromatin may be either in highly condensed, inert and inaccessible form (heterochromatin) or in "relaxed" and accessible form (euchromatin). Modulation of chromatin structure is fundamental for proper functioning of any cellular events such as DNA replication and repair, and transcription. In the later case, chromatin remodelling is essential to access the large transcriptional components to binding elements within the gene promoter. The basic structural unit of chromatin is nucleosome which is composed of an octamer of pairs of four core histones H2A, H2B, H3, H4 around which 147 base pairs of DNA are wrapped in ~1.7 super left-handed superhelical turns (kinner et al. 2008). The N-terminal tail, protruding from nucleosome, in the core histone contains positively charged amino acids like lysine, arginine and serine which interact with the negatively charged phosphate groups of the DNA, resulting in compaction of the chromatin (Kinner et al. 2008; Bannister and Kouzarides 2011). Histone tails, particularly those of histones H3 and H4, are subject to numerous posttranslational and covalent modifications which include methylation, arginine methylation and deamination, lysine acetylation, serine and threonine phosphorylation, ubiquitylation, sumoylation, biotinylation, proline isomerization, and ADP ribosylation (Wang et al. 2007; Kinner et al. 2008; Bannister and Kouzarides 2011; Ray et al. 2014). The "histone code" composed of specific combinations of histone variants and core histones N-tail posttranslational modifications offers huge combinational possibilities of chromatin remodelling and regulation of gene expressions (Chinnusamy et al. 2008; Kinner et al. 2008). While histone acetylation, phosphorylation, and ubiquitination activate transcription, biotinylation and sumoylation repress gene expression (Chinnusamy et al. 2008; Kinner et al. 2008; Ray et al. 2014). There is accumulating evidence indicating that histone posttranslational modifications are pivotal in establishing global epigenetic status of a cell during As toxicity (Ramirez et al. 2008; Chervona et al. 2012).

2.2.1 Histone Acetylation Under As Exposure

Acetylation is one the best understood processes of the "histone code" and is a dynamic reversible process. Histone acetyltransferases catalyze the histone acetylation by transferring one acetyl group from acetyl-coenzyme A to ε-amino groups in the positively charged lysine residues. This leads to neutralization of the positive charge and subsequent loosening of the histone-DNA compacts, decondensation of chromatin and accession of the transcriptional machinery to DNA. The whole process can be reversed by histone deacetylase and is often associated with transcriptional downregulation/repression (Kinner et al. 2008; Bannister and Kouzarides 2011).

There is growing evidence indicating a strong relationship between altered histone acetylation and As-induced toxicity. In the early 1980s, Arrigo (1983) reported reduced level of histone acetylation in *Drosophila* under As exposure. As exposure can also induce global and focal histone acetylation in cultured human cell lines. Ramirez et al. (2008) observed that sodium arsenite increased mobility of the nucleosome-associated high mobility group N proteins and inhibited histone deacetylase activity which triggered hyperacetylation of histone protein in human cell line, accelerating chromatin opening and consequent deregulation of gene expression during As exposure. The human MYST1 is a histone 4 lysine 16 (H4K16) acetyltransferase, the acetylation of which is required for As(III) resistance through modulation of chromatin sites (Jo et al. 2009). This was evident in human bladder epithelial cells in which knocking down of *MYST1* gene reduced acetylation of H4K16 and triggered sensitivity to As(III) and to its more toxic species monomethylarsonous acid [MMA(III)]. Both the As(III) and MMA(III) exposure decrease global H4K16 acetylation and increase As toxicity in human bladder cells (Jo et al. 2009). As trioxide can directly bind to histone acetyltransferase hMOF (human male absent on first) protein via C2HC (Cys2HisCys—a conserved region of hMOF protein) zinc finger domain and reduces H4 acetylation at lysine 16 residue in human cell by inhibiting histone acetyltransferase activity (Liu et al. 2015b). Available evidence suggests that As can exert its toxic effects through direct interaction with thiol group of cysteine residues (Ren et al. 2011; Chang et al. 2012; Liu et al. 2015b). As trioxide can directly bind to three cysteine residues of hMOF proteins and interferes with cellular As resistance, as revealed by MALDI-TOF mass spectrometry and UV absorbance detection study (Liu et al. 2015b). The hMOF forms two distinct cellular complexes—MSL and NSL, and both of which has the capacity to acetylate H4K16. As trioxide exposure, however, has the capacity to inhibit the process. This was associated with increased deacetyltransferase HDAC4 expression in arsenic trioxide-exposed HeLa or HEK293T cells. Overexpression of histone acetyltransferase hMOF effectively reversed reduced H4K16 acetylation and reduces As-induced necrosis (Liu et al. 2015b). Knocking down of hMOF in human cervical carcinoma HeLa cell under As trioxide significantly increased As-induced damage and necrosis. Both MOF and H4K16 acetylation are also implicated in DNA damage response and double strand break repair (Sharma et al. 2010). Thus, compromising MOF activity and H4 acetylation due to As exposure may impede repair process and accelerate As-induced cellular damage. Füllgrabe et al. (2013) suggested that regulation of histone acetylation by hMOF can cause autophagy of tumor cell under As trioxide exposure which warranted further study to reveal more mechanistic details regarding regulation of histone acetylation and As-induced cellular toxicity.

Gene specific histone acetylation by As has also been reported. Phosphoacetylation of H3 by As has been found to be associated with induction of apoptosis-related cysteine peptidase (*CASP10*), caspase 10, and the human proto-oncogenes *C-Jun* and *C-Fos* (Li et al. 2002; Li et al. 2003). Elevated acetylation of histone is probably mediated via As-induced inhibition of histone deacetylase gene (Ramirez et al. 2008). Jensen et al. (2008) studied epigenetic remodelling in As-transformed urothelial cells. The investigation revealed differential H3 acetylation of several promoters and positive correlation between H3 acetylation and mRNA expression of the hypoacetylated *DBC1, FAM83A, ZSCAN12,* and *C1QTNF6* genes, as well as the hyperacetylated neurofilament, light polypeptide (*NEFL*) gene (Jensen et al. 2008). Using genome-wide analysis Cronican et al. (2013) suggests that As exposure during embryonic development caused global hypo-acetylation at H3K9.

2.2.2 Histone Methylation During As Exposure

Histone methylation is a dynamic and reversible process and unlike acetylation, which targets only lysine residues at the tail region, methylation occurs on both lysine and arginine residues of histone (Martin and Zhang 2005; Wysocka et al. 2006). Furthermore, while acetylation involves addition of single acetyl group, histone methylation can occur in the monomethyl, symmetrical as well as asymmetrical dimethyl states and also in the trimethyl group states (Ren et al. 2011). Whereas dimethylation of H3K9 and H3K36 downregulates transcription, trimethylation of H3K4 upregulates transcription (Chinnusamy et al. 2008). However, data on As-induced histone methylation and gene expression, as well as its relationship with As-induced carcinogenesis is extremely limited. In *Drosophila* cells, Arrigo (1983) first reported complete demethylation of H3 and H4 under As(III) exposure. However, the process is more complex in mammalian cells. In human lung A549 cells, As(III) exposure triggered differential effects on H3 lysine methylation pattern; while H3 lysine 9 dimethylation (H3K9me2) and H3 lysine 4 trimethylation (H3K4me3) underwent hypermethylation, H3 lysine 27 trimethylation (H3K27me3) got hypomethylated (Zhou et al. 2008). Methylation of lysine residue is usually catalyzed by two different classes of proteins, the SET-domain-containing protein family and the non-SET-domain proteins DOT1/DOT1L (Martin and Zhang 2005). With further progress of the study, Zhou et al. (2009) found increased methylation of H3K4me3 and H3K9me2 due to increased levels of histone methyltransferase G9a protein under As(III) treatment. Significantly, H3K4me3 remained elevated perhaps through inherited cell division even 7 days after removal of As(III) and repressed transcriptional process as well as the silencing of tumor suppressers in the cancer cell lines (McGarvey et al. 2006; Esteve et al. 2007; Zhou et al. 2009). In human keratinocytes, As exposure leads methylation of H4R3 (H4 arginine 3) and H3R17 (H3 arginine 17) which induces expression of antioxidant gene ferritin (Huang et al. 2013). Arginine is generally methylated via arginine *N*-methyltransferase. The study revealed that transient knockdown of the methyltransferases PRMT1 and PRMT4 decreased histone methylation and ferritin expression which was accompanied by the decrease in the transcription factor Nrf2 binding to the ferritin promoter. This suggests that As exposure can trigger cascading effects through histone methylation, and recruitment of transcription factors and promoter binding are some of the prime events towards As-mediated epigenetic regulation of gene expressions (Huang et al. 2013).

2.2.3 Histone Phosphorylation During As-Induced Epigenetic Regulations

Phosphorylation is one of the posttranslational modifications of all four core histone proteins, H2A, H2B, H3, and H4, and the linker histone H1. Several kinases such as cyclin-dependent protein kinases and casein kinase are believed to be involved in histone phosphorylation which plays pivotal roles in cell cycle progression and regulation of gene expressions (Ren et al. 2011). Epigenetic regulations through histone phosphorylation have been implicated in As-induced carcinogenesis. In mammals, nearly 25% of total H2A pool is composed of H2AX. Phosphorylated H2AX was found to be upregulated during As trioxide induced apoptosis and perhaps, one of the prime mechanisms by which As trioxide can act as an antineoplastic agent (Zykova et al. 2006). Earlier investigations have revealed that As(III)-exposure can induce H3 phosphorylation (H3 serine 10) via the activation of JNK (c-jun N-terminal kinase) and p38/Mpk2 kinase by inhibition of the corresponding protein phosphatases and upregulate the expressions of the oncogenes c-*fos* and c-*jun* (Li et al. 2003) and a protoapoptotic factor, caspase 10 (Li et al. 2002). In mammalian cells, zeste homologue 2 (EZH2), a methyltransferase, overexpressed during cancer development and thus, its activity is regulated at different levels, one of which is its phosphorylation on serine 21 (S21). In human bronchial epithelial cell, As exposure can stimulate S21 phosphorylation via JNK-STAT3-Akt signaling cascade (Chen et al. 2013). There is growing evidence that this phosphorylation of EZH2 is partially dependent on As-induced ROS generation, a mechanism which is inhibited by antioxidant (like N-acetyl-L-cysteine) but is promoted by hydrogen peroxide (itself a ROS) (Chen et al. 2013).

A coordinated response of different histone modifications has been observed during As exposure. As(III) and MMA(III)-induced malignant transformation in uroepithelial cells involved upregulation of *WNT5A* gene expression. This process comprises induction of permissive modifications of histone proteins in one hand and decline of repressive modifications in the *WNT5A* promoter region in other hand (Jensen et al. 2009b). Similarly, transcriptional induction comprises acetylation of H3K9 and H3K14 and dimethylation of H3K4 while transcriptional repression correlates dimethylation of H3K9 and trimethylation of H3K27 during As exposure (Ren et al. 2011). In primary human keratinocytes, Herbert et al. (2014) reported that As(III) at 0.5 μM accelerated the acetylation of histone H4 at lysine 16 (H4K16). An extended exposure (>5 weeks) disrupted the pattern of CpG methylation at histone deacetylase SIRT1 which altered the expression of *SIRT1* gene.

2.3 *As-Induced Epigenetic Alterations of miRNAs*

The miRNAs (miR) are noncoding RNAs (transcribed from DNA but not translated into proteins), nearly 20–23 nucleotide long, and located in specifically associated promoter sequences. The miRNAs binds complementarily to 3′ or 5′ region of untranslated mRNA and/or binds directly to promoter sequences to induce transcriptional and posttranscriptional alterations of gene expressions (Zardo et al. 2012; Ray et al. 2014). In the cell nucleus, miRNAs are transcribed by RNA polymerase II and the long primary miRNA transcripts thus yielded are cleaved to precursor miRNA (pre-miRNA), a 70-nucleotide long hair-pin like structure. The pre-miRNAs are then transported to the cytoplasm via exportin-5 and subsequently processed by the endonuclease Dicer to 22-nucleotide long duplexes of mature miRNA which is loaded into RNA-induced Silencing Complex (RSIC) for elimination of one of its strands. The RSIC guided by the remaining strand of miRNA then regulates translation by facilitating mRNA-miRNA complementary binding, a process which is induced or repressed by number of environmental toxicants, carcinogens, and epigenetic regulators (Ray et al. 2014). ARGONAUTE (AGO) proteins are an integral part of RSIC which associates with miRNA and inhibits mRNA targets by mRNA cleavage and/or translational repression and plays important roles in RNA-targeted epigenetic regulations under diverse abiotic stress conditions (Li et al. 2012).

Involvement of As in regulation of miRNA expression and subsequent gene regulation has been shown in a number of eukaryote genomes (Ray et al. 2014). Using genome-wide expression analysis, Marsit et al. (2006b) reported that expressions of miR-22, miR-34a, miR-210, miR-221, and miR-222 in human immortalized lymphoblasts are altered under 2 μM $NaAsO_2$ exposure for 6 days. While expression was repressed for miR-210, elevated expression was found in rest of the cases (Marsit et al. 2006b). In human bladder (T24) and hepatocellular carcinoma (Hep-G2) cells, miR-19a and miR-29a respectively mediate As-induced apoptosis (Cao et al. 2011; Meng et al. 2011). A targeted approach revealed that miR-19a expression is decreased while miR-222 is induced in T24 cells exposed to 4 μM As trioxide (Cao et al. 2011). In the same concentration, miR-24, miR-29a, miR-30a, and miR-210 overexpressed in Hep-G2 human cell (Meng et al. 2011). Upregulation of several miRNAs such as miR-16, miR-17, miR-20a, miR-20b, miR-26b, miR-96, miR-98, miR-107, miR-126, miR-195, and miR-454 was also observed in newborn cord blood cells of a Mexican population exposed to varying doses of As (Rager et al. 2014). Upregulation of a specific set of miRNAs such as miR-9, miR-125b, and miR-128 has been suspected during As-induced ROS generation and subsequent toxicity (Ren et al. 2011). In acute promyelocytic leukaemia (APL), application of As trioxide induces the relocalization and degradation of the promyelocytic leukaemia (PML) protein (a nuclear body protein), as well as the degradation of PML–retinoic acid receptor-α (PML-RARα). The APL-associated PML-RAR oncogene downregulates the transcription of a group of miRNA in the patients treated with all-trans retinoic acid (Saumet et al. 2009), indicating As-mediated alteration and regulation of miRNAs in APL patients (Ren et al. 2011). As-induced malignant transformation of $TP53^{-}/^{-}$ cells was found correlated with decreased expressions of miR-200b in human bronchial epithelial cells (Wang et al. 2011). In contrast, As was found to induce miR-190 (derived from an intron region of the *talin 2* gene on chromosome 15 in human and a regulator of cell cycle and cell proliferation) in a concentration-dependent manner which targets mRNA of the PH domain leucine-rich repeat protein phosphatase (*PHLPP*) gene. PHLPP is known to target phosphorylated serine 473 specifically for dephosphorylation and inactivation of Akt kinase, leading to an increased apoptotic potential, and thereby causing tumor suppression (Beezhold et al. 2011). As-induced upregulation of miR-190 represses the PHLPP protein expression which enhances the malignant characteristics in bronchial cells (Beezhold et al. 2011). As at 10 and 20 μM induced the miR-190 expression through activating its parent talin 2 genes by about 1.5–2.0 fold. The downregulation of *PHLPP* gene is occurred through direct binding of the miR-190 with the 3′-UTR (untranslated region) of the PHLPP mRNA (223 and 217 nt downstream of the stop code for PHLPP), resulting in a decreased PHLPP protein level. Consequently, activation of Akt via phosphorylation of its serine 473 as well as subsequent expression of vascular endothelial growth factor, an Akt-regulated protein, is enhanced, resulting in cell proliferation and carcinogenic transformation. Besides, As-induced miR-190 upregulation itself is able to enhance proliferation and malignant transformation of the cells. The study suggest that induction of miR-190 is one of the prime events in As(III) induced carcinogenesis (Beezhold et al. 2011). Additional line of evidences suggest that As can also induce expression of miR-21, a well-known oncomiR, through JNK-dependent STAT3 activation (Chen et al. 2013). In another study, effects of As(III) on epigenetic regulations of histone deacetylase SIRT1 and its targeting microRNA, miR-34a was studied in primary human keratinocytes (Herbert et al. 2014). As(III) at 0.5 μM alters epigenetic regulation of SIRT1 expression via chromatin remodelling at the *miR-34a* gene promoter in the initial 24 h of exposure. Extended exposure (>5 weeks) modulated CpG methylation pattern both in *miR-34a* promoter and *SIRT1* gene and altered their expression in a cumulative way (Herbert et al. 2014). In As-exposed chick embryos, miR-9 and miR-181b were implicated as promoting abnormal angiogenesis (Cui et al. 2012). Microarray expression profiles of chick embryos injected with 100 nM sodium arsenite revealed that expressions of miR-9, miR-181b, miR-124, miR-10b, and miR-125b were downregulated. Several miRNAs, including miR-9 and miR-181b, might target several key genes, such as neuropilin-1 (Nrp1) involved in As-induced developmental toxicity. The target of NRP1 by miR-9 and miR-181b is subsequently involved in the inhibition of the As-induced EA.hy926 cell migration and tube formation, thus inducing angiogenesis by altering the expression of miRNAs and their cognate mRNA targets (Cui et al. 2012).

Research findings indicate that As-induced excess ROS and concomitant oxidative burst can mediate changes in cell behaviour, and activation of proliferative signalling is one of such behaviours. In human embryo lung fibroblast cells, chronic exposure of As(III) induces excess ROS generation which in turn activates ROS-sensitive pathways involving extracellular signal-regulated kinase mitogen-activated protein kinase (ERK/MAPK) and transcription factor nuclear factor kappa-light-chain-enhancer of activated B cells (NF-κB). NF-κB p65 regulated miR-21 expression by binding directly to the promoter of miR-21. ROS-mediated activation of ERK/NF-κB pathway in turn upregulated the expression of miR-21 which promoted the malignancy as was evident from increase in anchorage-independent growth of As(III)-transformed cells (Ling et al. 2012). Li and Chen (2016) proposed that As^{3+} can induce carcinogenesis via epigenetic modulations which is induced through MAP and Akt kinase cascade in either ROS-dependent or direct way.

2.4 Cross Talk Among Epigenetic Regulations During As Exposure

Jensen et al. (2008) reported DNA hypermethylation in a number of the hypoacetylated promoters identified in the study, suggesting that As coordinately targets genes through dysregulation of different epigenetic mechanisms contributing to malignant transformation. The decreased expression of miR-200b which facilitated As-induced malignant transformation of TP-53 cells in human bronchial epithelial cells was found correlated with increased DNA methylation of miR-200 promoter, suggesting that As exposure can transcriptionally suppress miR-200 expression through DNA methylation (Wang et al. 2011).

2.5 As Toxicity, Foetal Growth and Epigenetic Memory

As triggers detrimental effects on the foetal growth because both As(III) and MMA readily cross the placenta (Ahmed et al. 2011; Bailey et al. 2013) possibly by transplacental glucose transporter Glut1 which has sulfhydryl functional group with strong affinity to arsenicals (Liu et al. 2006). As can accumulate in the placenta, producing oxidative stress in placental tissues and interfering with nutrient transport to the foetus, thereby affecting foetal growth (Liu et al. 2006).

DNA methylation is the most critical epigenetic mechanism of foetal programming. This has been discussed during healthy foetal development and healthy life thereafter, the so-called Barker effect (Langley-Evans 2006) and epigenomic roles in embryonic development, tissue differentiation, and development of arsenicosis by controlling methylation related gene expression (Nye et al. 2014). As induces hypomethylation in foetus and new-born liver which coupled with poor nutritional status may initiate cancer development later in life. Interestingly, DNA methylation genes are upregulated significantly in children, adolescents and in pregnant women than adult men, suggesting effects of age and sex hormones on arsenicosis (Vahter 2007). Estrogen can trigger enhanced production of endogenous choline in pregnant women, by inducing phosphatidylethanolamine N-methyltransferase (PEMT) gene. PEMT regulates the de novo synthesis of phosphatidylcholine, allowing premenopausal women to synthesize more choline for remethylation of homocysteine to methionine and further to SAM, critical for methylation during fatal development (Vahter 2007). In this way, epigenomic alterations induced by As can change gene regulations and can serve as an interface between environmental toxicants like As and phenotypic manifestations by translating the As-generated environmental signals to phenotypic responses through altered gene expressions. This can be transmitted as a heritable epigenetic profile in transgenerational ways through germ line and is marked as epigenetic imprinting (Mirbahai and Chipman 2014). Interestingly, a potential sex-specific relationship was observed between global cord blood DNA methylation and maternal U-tAs levels in newborns of Bangladeshi population (Pilsner et al. 2012). This relationship was positive for males and negative for females, suggesting a sex-specific differential relationship may influence differential susceptibility to As toxicity in males and females. As exposure influencing histone marks such as H3K27me3, H3K4me3, H3K18ac, and H3K27ac in a sex-specific way also represents an interesting and significant outcome. The sex differences occurred in both methylation and acetylation marks of histones. But, the direction of the association with As by gender differed by the type of mark. For example, H3K18ac and H3K27ac increased in males and decreased in females, whereas H3K27me3 and H3K4me3 increased in females and decreased in males (Chervona et al. 2012). During the reprogramming, As-induced-epigenetic alterations in imprinted genes at the critical germ cell stage can influence both sex determination and can potentially be inherited leading to transgenerational modifications, despite the fact that several epigenetic marks are removed and reset in higher eukaryotes (Iwasaki and Paszkowski 2014; Mirbahai and Chipman 2014). This is indeed observed in As-exposed various cohorts. In plants, DNA methyltransferase MET1 regulates DNA methylation in maternally imprinted genes. During female gametogenesis, MET1 is suppressed in the central cell line and DNA is hypomethylated in maternally expressed genes. In contrast, DNA demethylase DEMETER (DME) possessing DNA glycosylase activity removes methylated cytosines from maternally expressed genes such as MEA, FIS2, and FWA in central cells, resulting in transcriptional activation of their maternal alleles (Jullien et al. 2008; Iwasaki and Paszkowski 2014). Owing to its property to interact with both methylation and demethylation process, environmental toxicant like As exposure may alter the maternally expressed genes (Iwasaki and Paszkowski 2014; Mirbahai and Chipman 2014). This indicates that even low to moderate levels of As exposure can impact the DNA methylation landscape and molecular memory. Some of these effects may be sex-dependent and suggest a strong interaction between As metabolism and DNA methylation.

3 Epigenetic Response to As Toxicity in Plants

Being sessile, plants are more prone to the deleterious effects of environmental stress in comparison to animal and human beings. Regulation and alterations of epigenetic mechanisms have been revealed in different model and crop plants under diverse environmental stresses (Steward et al. 2002; Yong-Villalobos et al. 2015). Scanty information, however, is available regarding modulations of plant epigenome under metal(loid) toxicity (Cheng et al. 2012).

3.1 Induced DNA Methylation in Plants Under As Exposure

Like human and animal system, changes in DNA methylation pattern is the most extensively studied epigenetic events in plants under metal(loid) stress. In plants, status of DNA methylation depends on the types of heavy metals and the plant species. As was shown known to cause oxidative damage through generation of excess ROS which induces DNA hypomethylation in plant cells (Cerda and Weitzman 1997). The mechanisms by which ROS can induce DNA hypomethylation in plants are diverse: (1) ROS can induce DNA damage by single-strand breaks through activation of endonuclease and thus, DNA becomes poor acceptor of methyl groups, (2) As-induced ROS in the vicinity of DNA produce pre-mutagenic adduct 8-oxo-2′-deoxyguanosine whose presence in the CpG sequences strongly induce hypomethylation of adjacent C residues, and (3) ROS-induced oxidative stress can overproduce nicotinamide and its metabolites like trigonelline (in plants) and over activity of poly (ADP-ribose) polymerase activity which has adverse effects on DNA methylation (Aina et al. 2004).

Generally, modulations in DNA methylation is involved in three different sequence contexts (CG, CHG, and CHH, where H = A, C, or T), and the coordination among different pathways is necessary for the establishment, modification, and maintenance of DNA methylation patterns in a dynamic mode of epigenetic regulations and alterations (Chinnusamy and Zhu 2009; Mirouze and Paszkowski 2011; Yong-Villalobos et al. 2015). Methylated cytosines account for more than 30% of the nucleotides in plants. In *Arabidopsis*, METHYL-CpG BINDING PROTEIN 7 (AtMBD7) and METHYL-CpG BINDING PROTEIN 5–7 (AtMBD5–AtMBD7) binds to arginine methyltransferase (PRMT11) and DDM1 protein, respectively, and control gene expression through chromatin modifications during stress response (Yaish 2013). Dynamic changes in methylation level of some genes in response to environmental stresses like metalloid toxicity are some of the events plants undertake to endure stress conditions by preventing unfavourable genetic rearrangement at a specific locus (Chinnusamy and Zhu 2009; Boyko et al. 2010). Changes in DNA methylation pattern have been observed in model and crop plants under different abiotic stressors including heavy metal toxicity. Li et al. (2015) found increased 18S rDNA cytosine methylation to the tune of 11.22–18.88% and variation in genomic DNA methylation pattern under nickel exposure. Toxic metal cadmium can enhance the DNA demethylation in red seaweed *Gracilaria dura* (Kumar et al. 2012), while cobalt compounds could change both methylation and demethylation status in *Vicia faba* seeds (Rancelis et al. 2012). Using slot-blot technique, Aina et al. (2004) found that global DNA methylation level was much higher in roots of metal-tolerant *Cannabis sativa* than that in metal-sensitive *Trifolium repens*. However, due to reduction in 5′-methyl cytosine level, DNA underwent hypomethylated in both plant species with increasing concentration of metals (Aina et al. 2004). Methylation-sensitive amplification polymorphism (MSAP) analysis revealed that heavy metal-induced changes in DNA methylation involved hypomethylation events in 5′-CCGG-3′ containing sequences which enhanced polymorphism with increasing concentration of metals in both plant species (Aina et al. 2004). Global DNA methylation pattern was found changed in *A. thaliana* seedlings with nearly 1.5–1.8-fold high level of 5-methyl cytosine in phosphate-starved roots and shoots (Yong-Villalobos et al. 2015). Maize seedlings undergoing cold stress exhibited a periodic demethylation and methylation in nucleosome. Direct methylation mapping revealed that cold stress could induce genome wide demethylation in root tissues. The 150 bp long hypomethylated regions are actually alternated with 50 bp long hypermethylated regions which corresponds to nucleosome core and linker region, respectively (Steward et al. 2002). Using coupled restriction enzyme digestion-random amplification (CRED-RA) technique, Erturk et al. (2014) detected increasing global DNA methylation in maize seedlings and maize seeds with increasing concentration of chromium nitrate and zinc. This was accompanied by enhanced percentage of mitotic cell aberration and decreased level of growth promoting phytohormone levels (Erturk et al. 2015).

The biological significance of changes in DNA methylation pattern in plants is conceivable. Changes in methylation pattern can cause alterations in chromatin structure and hence, gene expression. Thus, wide demethylation can serve as "common switch" of many plant genes which are simultaneously governed under environmental cues (Steward et al. 2002). Comparative analysis of the methylome and gene expression in poplar under different abiotic stressors revealed differences in cytosine methylation pattern (Song et al. 2016). Annotation revealed that out of 1376 stress-specific differentially methylated region (SDMRs), 81.61% encodes proteins while rest encodes miRNA and long noncoding RNA (lncRNA) genes (Song et al. 2016). The study pointed out that recovery from abiotic stress conditions corresponds to the decline of cytosine methylation by about 15.3–35.0% (Song et al. 2016). Secondly, heritable changes in plant phenotype may occur due to demethylation and therefore, changes in DNA methylation pattern may result in epigenetic inheritance, altering the gene expression without changing the nucleotide sequence (Steward et al. 2002; Tan 2010; Erturk et al. 2014)

Methylation-sensitive amplified polymorphism (MSAP) unravelled that osmotic stress in maize leaves induced methylation in retrotransposon Gag–Pol protein genes (Tan 2010). The investigation revealed gene-specific changes in methylation pattern of maize heavy metal-induced transporter, heat shock protein HSP82, dehydration-responsive element-binding (DREB) factor, Lipoxygenase, Poly [ADP-ribose] polymerase 2, casein kinase (CK2), and first intron of maize protein phosphatase 2C (zmPP2C) under salinity and/or drought stress. While expression of intron methylation of root zmPP2C was significantly repressed, demethylation of leaf zmGST weakly upregulated its expression under NaCl stress (Tan 2010). MSAP technology was also used to reveal variation in DNA methylation pattern of 18S rDNA methylation in *Arabidopsis* under nickel stress (Li et al. 2015). In *A. thaliana*, DNA methyltransferase genes METHYLTRANSFERASE 1 (MET1), DOMAINS REARRANGED METHYLASE 1 (DRM1), CHROMOMETHYLASE 3 (CMT3), DOMAINS REARRANGED METHYLASE 2 (DRM2) and the DNA demethylases REPRESSOR OF SILENCING 1(ROS1), DEMETER LIKE 2 (DML2), and DEMETER LIKE 3 (DML3) are supposed to be involved in regulation of DNA methylome during abiotic stress response (Boyko et al. 2010; Yong-Villalobos et al. 2015). Significantly, the study unravelled (i) differential pattern of hypermethylation and hypomethylation of DNA bases under temporal and spatial regulation of differentially expressed genes and gene related transposable genetic elements, and (ii) both cytosine-guanine (CG) and non-CG methylation are necessary for correct response of *Arabidopsis* plants to abiotic stress (Boyko et al. 2010; Yong-Villalobos et al. 2015).

3.2 Histone Modification During Stress Response in Plants

In plants, modifications in histone octamer can modulate stress responses by coordinating "open" or "closed" chromatin conformations (Dhar et al. 2014). Studies on *Arabidopsis* revealed that the major histone modifiers are histone acetyltransferases, histone deacetylases, histone methyltransferases, and histone demethylases which besides *Arabidopsis*, have been isolated in several plants, including tomato, rice, barley, grapevine, *Brassica*, and *Brachypodium* (Kim et al. 2015). In some cases, chromatin changes are steady and autonomous as a result of heritable epialleles that induce phenotypic alteration (Dhar et al. 2014). In plants, DNA methylation and histone acetylation comprise dual epigenetic pathways during developmental processes. For example, AtMBD9 modulates *Arabidopsis* flowering and hormonal level involving both regulations under stress (Yaish et al. 2011; Yaish 2013). In *Arabidopsis*, decrease in DNA methylation in the *ddm1* mutant (DECREASE IN DNA METHYLATION1) is associated with hypermethylation in H3K4 and loss of methylation in H3K9 which make the plants stress sensitive (Yaish 2013). However, *A. thaliana* seedlings grown under salt stress exhibit that DNA methylation is associated with methylation gain in H3K9 and depletion of acetylation in H3K9 (Bilichak et al. 2012). In transgenic *Arabidopsis*, overexpression of the histone deacetylase, AtHD2C resulted in abscisic acid (ABA) insensitivity and showed tolerance to salt and drought stresses (Dhar et al. 2014). Changes in both histone acetylation and methylation pattern have been reported in *A. thaliana* seedlings under cold, drought, salt, and biotic exposures, but very little information is available during metal (loid) exposure (Dhar et al. 2014). As can modulate the *Arabidopsis* polycomb complex PRC2, harboring MEA, FIE, FIS2, and MSI1, through histone (H3K27) which subsequently can suppress the expression of paternally or maternally expressed genes and along with additional epigenomic imprinting factors can sustain and inherit the As-induced changes in subsequent generations (Iwasaki and Paszkowski 2014). Another route of histone modification by As is through generation of excess ROS and consequent induction of oxidative damage. As-induced DNA damage and modifications in dynamics of histone modifications is now known. Posttranslational modifications of histone proteins are involved in detection and repair of DNA damage in several plants. Studies with alfalfa, soybean and *Arabidopsis* revealed acetylation of H3 and H4 N-terminal peptides (via lysine) and increase in H3 acetylation as well as hypoacetylation of H4 (likely via k16 and k12) with increase in DNA damage (Drury et al. 2012).

3.3 The miRNAs During Metal(loid) Toxicity in Plant Cells

Emerging data indicate potential roles of miRNAs in metal stress responses in plants (Sunkar et al. 2006; Zhou et al. 2010; Li et al. 2012). In *Arabidopsis*, several mutants of the miRNA biogenesis pathway, such as hyponastic leaves1, serrate, dicer-like1, and cap-binding protein (cbp20 and cbp80) have lower miRNA levels but higher pre-miRNA in relation to their wild type relative (Laubinger et al. 2008). The miR-395 is one of the first miRNAs discovered in *Arabidopsis* during metal stress response. Three roles of miRNAs namely, (a) in metal complexation and subsequent sequestration, (b) in regulation of oxidative balance, and (c) in signal transduction process involving kinase cascade, hormonal pathway, and transcription factors have generally been proposed in *Arabidopsis*, rice, *Medicago truncatula*, *Brassica napus*, *Phaseolus vulgaris*, and *Nicotiana tabacum* (reviewed Gielen et al. 2012). Heavy metal-induced overexpression of miR-398 has been reported in copper-starved *A. thaliana* seedlings and their subsequent downregulation of target genes is necessary for nutrient reallocation (Jagadeeswaran et al. 2009). Similarly, miR-168 and its several homologues and miR-156, miR-166, miR-171, miR-390, miR-396, miR-397, miR-398, and miR-408 have emerged as prime players in heavy metal stress responsive networks of *Arabidopsis*, rice, *Brassica*, poplar, tobacco, and maize (Ding et al. 2009; Zhou et al. 2010; Sharma et al. 2015). However, compared to other heavy metals, information regarding involvement of miRNAS in As-exposed plants is rather scanty. Using an in silico approach, Tuli et al. (2010) predicted putative As(III)- and As(V)-responsive miRNAs, such as miR-160f, miR-168a/b, miR-169q, miR-319a, miR-416, and miR-1427 from rice. Yu et al. (2012) identified significant alterations in 36 miRNAs responsive to As(III) in rice plants in which 14 miRNAs are found involved in transcriptional regulation of gene expressions in metabolism. Thirty miRNAs belonging to 23 families were observed in the shoots, while 25 miRNAs (22 families) were observed in the roots. Among the root miRNAs, 12 miRNAs were downregulated of which miR-156j was repressed by16–50-fold in the roots. In contrast, miR-162a, miR-168a, miR-390, miR-393, miR-394, miR-535, and miR-2106 were upregulated in roots under As(III) exposure. In shoots, miR-156j, three miR166 subfamily members (miR-166h, miR-166l, and miR-166n) and miR-319b were differentially downregulated, whereas three miRNAs (miR-812j, miR-1428e-3p, and miR-1876) and miR-394 and miR-1876 were specifically upregulated under different As(III) treatment regimes (Yu et al. 2012). Several potential miRNA targets have been identified in jasmonic acid pathway, lipid as well as oligopeptide transporters, and lipid metabolism under As(III) stress in rice cv. Nipponbare (Yu et al. 2012). In another study, 67 As (III)-responsive miRNAs belonging to 26 miRNA families have been identified in rice out of which 54 miRNAs were downregulated while only 13 were upregulated (Liu and Zhang 2012). All miR-159, miR-164, miR-167, and miR-169 members were reportedly downregulated in this study (Liu and Zhang 2012). Compared to control, forty-four miRNAs were found to be differentially regulated in rice cv. IR-64 exposed to 150 μM As(III) (Pandey et al. 2015). Out of this 44 miRNAs, 18 mRNAs exhibited contrasting response between As and selenium (Se) treatments. While miR-395 and miR-1433 family was overexpressed in As and As + Se treatment, its expression was repressed in Se-treatment alone. Targets of miR-395 involve mainly up-stream thiol cascade in reductive sulfate assimilation pathway, namely ATP sulfurylase (APS) and sulfate transporter (SULTR2;1). Studies indicate that induction of miR-395 is significantly correlated with increased thiol assimilation, antioxidant activity, higher thiol metabolites like GSH and phytochelatins, higher chlorophyll content, and better performances of plant species under As-induced oxidative stress (Pandey et al. 2015). Similarly, miR-399d was downregulated during As and As + Se exposure but was induced in Se treatment alone (Pandey et al. 2015). Venn diagram constructed in this study identified only two miRNAs common to control, As-, Se-, and As + Se treatments whereas 15 miRNAs are common to As vs. control and As + Se vs. control treatment in upregulated category. In downregulated category, 25 miRNAs are common to all the treatment regimes. Stem-loop RT-PCR analysis in accordance with array revealed that miR-159, miR-171, miR-396, miR-398, miR-399, and miR-415 were downregulated in As(III) treated seedlings but their levels were normalized in As + Se treated plants (Pandey et al. 2015). Notably, miR-159, miR-171, miR-396, miR-399, miR-812, and miR-815 have emerged as common As-responsive miRNAs in rice (Liu and Zhang 2012; Pandey et al. 2015). Transgenic rice plants overexpressing miR528 (Ubi::MIR528) in both roots and leaves exhibited strong alterations in antioxidant potentials, amino acid profiling as well as in As uptake, translocations, and detoxification mechanisms due to strong correlation between overexpressions of Ubi::MIR528 and downregulation of many target genes. This led to excess ROS generation and consequent oxidative stress in Ubi::MIR528 rice plants (Liu et al. 2015a). Sharma et al. (2015) detected several differentially regulated miRNAs belonging to 92 miRNA families in two contrasting (high As-accumulating vs. low As-accumulating) natural accessions group of rice, responsive to 25 μM As(III) and 50 μM As(V) treatments. Out of these, 114 differentially regulated miRNAs belonging to 30 miRNA families are from high As-accumulating group while 166 belonging to 62 families are from low As-accumulating germplasms. The miRNA array revealed upregulation of expressions of miR-396, miR-399, miR-408, miR-528, miR-1861, miR-2102, and miR-2907 families in response to As(III) and As(v) stress in both accessions. In contrast, As exposure led to downregulation of members of the miR-164, miR-171, miR-395, miR-529, miR-820, miR-1432, and miR-1846 families (Sharma et al. 2015). Validation of expression and bioinformatic analysis attributed differential expressions of rice accessions to As(III) and As(V) to As-speciation and miRNAs specific to rice accessions. Further study on proximal promoter sequence of the As-responsive miRNAs revealed occurrence of metal-responsive cis-acting motifs which are accompanied by other elicitor and hormonal related motifs. The results suggest that As-species specific differential stress response in rice accessions is actually miRNA-dependent (Sharma et al. 2015). Microarray data revealed that COPPER/ZINC SUPEROXIDE DISMUTASE1/2 (CSD1/2) and Cytochrome b5-like heme/steroid binding domain containing protein are some of the prime targets of stress responsive miRNAs. Targets of miR-398, CSD1 and CSD2 are involved in ROS scavenging mechanisms, preventing deleterious effects of metalloid-induced oxidants. *Brassica juncea* (Indian mustard) roots exposed to As for 1 h and 4 h revealed altered expression profiling of 69 miRNAs belonging to 18 plant miRNA families (Srivastava et al. 2012). Time- and organ-dependent changes were observed for As-responsive miRNAs. Notable among which were six miRNAs (156, 159, 162, 167, 838, and 854) controlling root and shoot development, sulfur metabolism, hormone biosynthesis and metabolism, as well as in miRNA processing itself (Srivastava et al. 2012). While miR-156 and miR-159 overexpressed from 6–72 h treatment period in both shoot and root, miR-162 repressed in both organs until 24 h, and then showed upregulation at 72 h. The miR-167 and miR-838 upregulated in shoots but downregulated in roots across the time scale. Expression of miR-854 was initially upregulated in roots but downregulated at 72 h, while it was downregulated throughout the time scale in shoot (Srivastava et al. 2012). Several transcription factors and genes of metabolic pathways (Table 2) are targets of miRNAs and their mRNA expression changed inversely with miRNA expressions under As exposure (Srivastava et al. 2012). *Arabidopsis* seedlings overexpressing a miR-398-resistant form of CSD2 show increased resistance to oxidative stresses (Sunkar et al. 2006). An interesting expression pattern of miRNAs has been observed in jute (*Corchorus olitorius* var. O-9897) under 250 μM As(V) treatment. Expression of miR-159 and miR-167 was downregulated at initial period of As treatment but upregulated after 24 h of treatment. Inverse expression pattern was observed for mRNA expressions controlling ATP-binding cassette transporter (ABC) and myeloblastosis transcription factor (MYB), the targets of miR-159, and auxin response factor (ARF6 and ARF8), the target of miR-167. While repressed expression of ABC led to decreased phytochelatin-mediated metal detoxification, downregulation of ARFs resulted in decline in lateral root elongation as well as low biomass and metal adaptation capacity of jute seedlings (Hauqe et al. 2014). A list of miRNA-target genes and cis-acting elements responsive to As-stress in plants has been presented in Table 2.

4 Challenges and Future Research

Accepting the caveat that the roles of epigenetic mechanisms in As toxicity are yet to be fully understood, it is conceivable to believe from abovementioned information that tremendous progress has been made in recent years to underpin the regulations and alterations of epigenomic mechanisms in development of As toxicity in both plants and animal biology. The epigenome can serve as a link between individuals or plants' genomic makeup and their response to environmental toxicants like As. The changes in epigenomic mechanisms responsive to As exposure have huge predictive values in temporal and spatial toxicokinetic analysis. The potential of heritability of epigenetic modifications through mitosis and germ line provides unique possibilities in the future risk assessment process. This epigenetic memory has paved the way to determine interindividual differences in susceptibility of As-induced toxicity by studying epigenetic reprogramming and its disruption from ancestral background. An effective disease forecast for later life health is now likely. However, the epigenetic data generated in both plant and animal system needs to be correlated with target gene and protein expression data, the holistic approach for which is currently lacking. Furthermore, the outcome of various animal experimental protocols and human clinical data on As effect on epigenetic mechanisms are complicated and huge stray data carrying conflicting results are further

Table 2 Predicted targets and regulation of their functions by As-responsive microRNAs (miR) in different plants

Plant species	miR	Target genes	Functions
Oryza sativa (Os)/*Arabidopsis thaliana* (At)	AtmiR-156[a]	SPL2,3,4, 6,9. 10, Galactosyltransferase	Stress responsive, leaf, floral meristem development, fruit initiation, sporogenesis, GA-signaling, sugar metabolism.
	Os/At miR-159a,c,e,f[a]	MYB 33, 65, 104, 120, DUO1 (DUO POLLEN 1), DNA binding/transcription factor, ACS8, myb-like DNA-binding domain containing protein, UDP-glucuronosyl and UDP-glucosyl transferase family protein, TCP family transcription factor containing protein, retrotransposon protein, putative, Ty3-gypsy subclass	Cell cycle control, Anthocyanin biosynthesis; phenylpropanoid metabolism, cell shape, petal morphogenesis, shoot (trichome) and root hair patterning, seed coat differentiation, circadian rhythm, responsive to biotic and abiotic stress; controlling 1-amino-cyclopropane-1-carboxylic acid synthase activity in ethylene biosynthesis, hormone signalling, signal transduction, chromatin configuration, sugar metabolism, transposon translation.
	OsmiR-159a[a]	AtOPT1	Oligopeptide transporter 1.
	AtmiR-162[a]	DCL1(Dicer-like 1), CYP96A1 (Cytochrome p450, family 96, subfamily A, polypeptide 1)	ATP-dependent helicase/ribonuclease III, Oxygen binding in cytochrome P450, control development and stress tolerance.
	Os/At miR-164[a]	NAC/NAM (TF) domain protein like CUC1/2 and NAC1	Lateral root development; expansion of the boundary domains in meristems; controlling taxadien-5-alpha-olO-acetyl transferase activity.
	AtmiR-165a,b[a]	REV/PHV/ATHB-8,15	Leaf architecture; REVOLUTA; DNA binding/lipid binding, PHAVOLUTA/TF.
	Os miR-166	HD-ZIPIII (Homodomain -leucine zipper III)	Abiotic stress response, deetiolation, auxin signalling, embryogenesis.
	Os/At miR-167[a]	ARF 6,8	Regulation of auxins biogenesis, auxin signalling, root development, other multiple developments, reproduction.
	OsmiR-169[a]	HAP2A,2B, NFY, MIR169D, CCAAT-binding factor	Photoperiodic flowering, circadian rhythms, light signalling, high-affinity metal transport, nuclear transcription.
	OsmiR-171	GRAS domain transcription factors (SCARECROW-like)	Glycosyl hydrolase; pyridoxin biosynthesis protein ER1, phase transition and floral meristem determination.

Plant species	miR	Target genes	Functions
	Os/At miR-172[a]	AP2-LIKE	Controls transitions from juvenile to vegetative to reproductive phase, zinc-finger, 1,3-beta glucan synthase activity, ribosome structure.
	Os/At miR-319[a]	TCP4,10, MYB33,65, 104, ALDH22a1	Controlling leaf development by targeting TF, aldehyde dehydrogenase, 3-chloroallyl aldehyde dehydrogenase activity.
	OsmiR-390[a]	TAS3 (ta-siRNA), Q9T9F9	Multi-sensor signal transduction, NADH-ubiquinone oxidoreductase chain 3.
	OsmiR-393	TIRI/AFB (Transport Inhibitor-Response 1/Auxin F-Box)	Auxin perception and structure–activity relationship.
	Os/AtmiR-395a	Cytochrome b5-like heme/steroid binding domain containing protein	Controls lateral roots formation, response to environmental cues, sulfate uptake, metabolism and nutrition stress.
	Os/AtmiR-396/e-3p	WRKY/Growth-regulating factor 1	Cell division and differentiation during leaf development; plant abiotic stress response, plant immunity.
	Os miR-397, 408, 528	LAC	Plant development, sugar transport.
	Os/At miR-398b	Cu/Zn SOD	Antioxidant defense via superoxide scavenging.
	Os/At miR-399a	Ubiquitin-conjugating enzyme protein	Phosphate-starved stress; regulating Pi homeostasis by targeting phosphate2 (PHO2), an ubiquitin-conjugating E2 enzyme.
	OsmiR-415	40S ribosomal protein S10	Positively regulate cell proliferation, profiling under oxidative stress.
	OsmiR-426[a]	Ty3-gypsy subclass, monocopper oxidase-like protein SKS1 precursor	Retroposon proteins, metal metabolism.
	Os/AtmiR-444	MADS-boxTF	Reproductive development; floral meristem identity; nutrient homeostasis.
	OsmiR-535[a]	cyclinB2, expressed proteins	

(continued)

Table 2 (continued)

Plant species	miR	Target genes	Functions
	Os/AtmiR-812, 818	Kinase cascade	Regulate Ca-dependent protein kinase activity, serine/threonine kinase activity.
	OsmiR-819	Ligase	Regulation of ubiquitin/protein ligase activity.
	AtmiR-824[a]	Agamous-like 16 (AGL16), MADS-box protein	Zn ion binding; pre-mRNA cleavage; stomatal complex development.
	AtmiR-838[a]	DCL1, armadillo/β-catenin	Expressing armadillo/β-catenin, ketoacyl-CoA synthase, fatty acid elongase, ethylene signalling, helicase, DNA strand ligation, DNA methylase, protein dimerization, oligopeptide transport, sulfur transport.
	AtmiR-854[a]	At UBP1b, CLV2, C3HC4, AtPUP14, AtSERAT2;2, ACA5, PRA1.B1, HKT1	Oligouridylate binding, RNA binding, Zn-finger, purine transport, reductive sulfur assimilation by serine acetyl transferase activity, Alpha carbonic anhydrase, carbonate dehydratase, Prenylated RAB acceptor 1.B1, high affinity K+ transport, transmembrane Na transport.
	OsmiR-1318, 1432	EF-hand protein	Plant developmental process.
	OsmiR-1436[a]	P450 94A2, MCM	Kinase cascade, electron transport, minichromosome maintenance.
	OsmiR-1561[a]	SPL9, 11	Teosinte glume architecture, SBP domain-containing protein, leucine-rich repeat receptor protein kinase EXS precursor, AP-1 complex subunit sigma-2.
	OsmiR-1875	CDP-diacylglycerol-inositol 3-phosphatidyltransferase1, putative, expressed	Signal transducer, GPCR and tyrosine kinase activity.
	AtmiR-3979	bHLH TF	Basic helix-loop-helix.

Abbreviation used: SPL—Squamosa promoter binding protein-like, TF—Transcription factor, MYB—Myeloblastosis, NAC—(no apical meristem or NAM)/ATAF1–2 (*Arabidopsis* transcription activation factor)/CUC (cup-shaped cotyledon), NFY—Nuclear factor-Y, AP2——Apetala2, TCP—Teosinte branched1 cycloidea proliferating cell factor, TAS—Trans acting short, TIRI—Transport inhibitor-response 1, AFB—Auxin F-box, LAC—Laccase, SOD—Superoxide dismutase, HAP—Heme activator protein, ARF—Auxin response factor

[a]Expressed also in *Brassica juncea* during As exposure (Srivastava et al. 2012)

complicating our understanding on epigenetic response to toxic environmental cues. Owing to lack of alternate methods, introduction of 3R principles (refinement, reduction, and replacement of animal experimentation) proposed by Russell and Burch (1959) for toxicological analysis is also not possible. Similarly, our knowledge on the mechanistic response of "histone codes" and its interaction with DNA methylation pattern under As exposure are still nascent in animal model and poor in various plant species. No reports are still available whether As-induced epigenetic regulations and alterations are part of the causative pathways that trigger disease development. A precise and accurate regulation of stress responses is of major importance for both animal and plants to be able to complete their life cycle. Some epigenetic modifications may be adaptive or stochastic responses that do not cause disease development. Likewise, global and/or gene-specific changes in DNA methylation patterning and histone posttranslational modifications may not be reflective of functional changes that are manifested at the protein level. Whether these identified metal-regulated epigenomic components are specifically altered in their gene expression for adjustment and tolerance to the As stress or if these altered expressions are secondary consequences of a disturbed cellular homeostasis due to As stress remains to be explored in future studies. The identification of entire sets of metal-regulated DNA-methylation pattern, histone modifications and miRNAs and their interactions and targets in a tissue-specific manner is needed. Emerging epigenomic technologies such as chromatin immunoprecipitation (ChIP)-on-chip and ChIP sequencing (ChIP-seq), global methylation, and miRNA microarrays, as well as whole genome DNA sequencing platforms will greatly facilitate As-induced modifications of genome wide DNA methylation pattern, posttranslational histone modification, and miRNA expression in vitro and in vivo. Detection of the genes dysregulated by As-induced epigenetic mechanisms can elucidate the associated biological processes and disease states. Proteomics using both conventional "bottom-up" and newer cutting-edge "top-down" mass spectrometry approaches can also be helpful to detect labile As-induced posttranslational modifications. As-induced epigenetic alterations are different between plant, animal, and human systems, and even between various tissues and cell types at different environmental, diet, and age backgrounds. A comprehensive study is therefore need of the hour to identify and validate the levels and patterns of epigenetic manifestations and fully understand the epigenetic mechanisms of cellular As toxicity.

References

Ahlborn GJ, Nelson GM, Ward WO, Knapp G, Allen JW, Ouyang M, Roop BC, Chen Y, O'Brien T, Kitchin KT, Delker DA (2008) Dose response evaluation of gene expression profiles in the skin of K6/ODC mice exposed to sodium arsenite. Toxicol Appl Pharmacol 227:400–416

Ahmed S, Mahabbat-e Khoda S, Rekha RS, Gardner RM, Ameer SS, Moore S, Ekstrom EC, Vahter M, Raqib R (2011) Arsenic-associated oxidative stress, inflammation, and immune disruption in human placenta and cord blood. Environ Health Perspect 119:258–264

Aina R, Sgorbati S, Santagostino A, Labra M, Ghiani A, Citterio S (2004) Specific hypomethylation of DNA is induced by heavy metals in white clover and industrial hemp. Physiol Plant 121:472–480

Arita A, Costa M (2009) Epigenetics in metal carcinogenesis: nickel, arsenic, chromium and cadmium. Metallomics 1:222–228

Arrigo AP (1983) Acetylation and methylation patterns of core histones are modified after heat or arsenite treatment of *Drosophila* tissue culture cells. Nucl Acid Res 11:1389–1404

Bagnyukova TV, Luzhna LI, Pogribny IP, Lushchak VI (2007) Oxidative stress and antioxidant defenses in goldfish liver in response to short-term exposure to arsenite. Environ Mol Mutagen 48:658–665

Bailey K, Fry R (2014a) Arsenic-associated changes to the epigenome: what are the functional consequences? Curr Environ Health Rep 1:22–34

Bailey K, Fry RC (2014b) Long-term health consequences of prenatal arsenic exposure: links to the genome and the epigenome. Rev Environ Health 29:9–12

Bailey KA, Wu MC, Ward WO, Smeester L, Rager JE, Garcia-Vargas G, Del Razo LM, Drobna Z, Styblo M, Fry RC (2013) Arsenic and the epigenome: inter individual differences in arsenic metabolism related to distinct patterns of DNA methylation. J Biochem Mol Toxicol 27:106–115

Bannister AJ, Kouzarides T (2011) Regulation of chromatin by histone modifications. Cell Res 21:381–395

Beezhold K, Liu J, Kan H, Meighan T, Castranova V, Shi X, Chen F (2011) miR-190-mediated downregulation of PHLPP contributes to arsenic-induced Akt activation and carcinogenesis. Toxicol Sci 123:411–420

Benbrahim-Tallaa L, Waterland RA, Styblo M, Achanzar WE, Webber MM, Waalkes MP (2005) Molecular events associated with arsenic-induced malignant transformation of human prostatic epithelial cells: aberrant genomic DNA methylation and K-ras oncogene activation. Toxicol Appl Pharmacol 206:288–298

Bilichak A, Ilnystkyy Y, Hollunder J, Kovalchuk I (2012) The progeny of *Arabidopsis thaliana* plants exposed to salt exhibit changes in DNA methylation, histone modifications and gene expression. PLoS One 7:e30515

Bollati V, Baccarelli A (2010) Environmental epigenetics. Heredity (Edinb) 105:105–112

Boyko A, Boyko A, Blevins T, Yao Y, Golubov A, Bilichak A, Ilnytskyy Y, Hollander J, Meins F Jr, Kovalchuk I (2010) Transgenerational adaptation of *Arabidopsis* to stress requires DNA methylation and the function of dicer-like proteins. PLoS One 5:e9514

Cao Y, Yu SL, Wang Y, Guo GY, Ding Q, An RH (2011) MicroRNA-dependent regulation of PTEN after arsenic trioxide treatment in bladder cancer cell line T24. Tumour Biol 32:179–188

Centeno JA, Tseng CH, Vander Voet GB, Finkelman RB (2007) Global impacts of geogenic arsenic: a medical geology research case. Ambio 36:78–81

Cerda S, Weitzman SA (1997) Influence of oxygen radical injury on DNA methylation. Mutat Res 386:141–152

Chai CY, Huang YC, Hung WC, Kang WY, Chen WT (2007) Arsenic salts induced autophagic cell death and hypermethylation of DAPK promoter in SV-40 immortalized human uroepithelial cells. Toxicol Lett 173:48–56

Chanda S, Dasgupta UB, Guhamazumder D, Gupta M, Chaudhuri U, Lahiri S, Das S, Ghosh N, Chatterjee D (2006) DNA hypermethylation of promoter of gene p53 and p16 in arsenic-exposed people with and without malignancy. Toxicol Sci 89:431–437

Chang YY, Kuo TC, Hsu CH, Hou DR, Kao YH, Huang RN (2012) Characterization of the role of protein-cysteine residues in the binding with sodium arsenite. Arch Toxicol 86:911–922

Chen H, Li S, Liu J, Diwan BA, Barrett JC, Waalkes MP (2004) Chronic inorganic arsenic exposure induces hepatic global and individual gene hypomethylation: implications for arsenic hepatocarcinogenesis. Carcinogenesis 25:1779–1786

Chen WT, Hung WC, Kang WY, Huang YC, Chai CY (2007) Urothelial carcinomas arising in arsenic-contaminated areas are associated with hypermethylation of the gene promoter of the death-associated protein kinase. Histopathology 51:785–792

Chen B, Liu J, Chang Q, Beezhold K, Lu Y, Chen F (2013) JNK and STAT3 signaling pathways converge on Akt-mediated phosphorylation of EZH2 in bronchial epithelial cells induced by arsenic. Cell Cycle 12:112–121

Cheng TF, Choudhuri S, Muldoon-Jacobs K (2012) Epigenetic targets of some toxicologically relevant metals. J Appl Toxicol 32:643–653

Chervona Y, Hall MN, Arita A, Wu F, Sun H, Tseng HC, Ali E, Uddin MN, Liu X, Zoroddu MA, Gamble MV, Costa M (2012) Associations between arsenic exposure and global posttranslational histone modifications among adults in Bangladesh. Cancer Epidemiol Biomark Prev 21:2252–2260

Chinnusamy V, Zhu JK (2009) Epigenetic regulation of stress responses in plants. Curr Opin Plant Biol 12:133–139

Chinnusamy V, Gong Z, Zhu JK (2008) Abscisic acid-mediated epigenetic processes in plant development and stress responses. J Integr Plant Biol 50:1187–1195

Coppin JF, Qu W, Waalkes MP (2008) Interplay between cellular methyl metabolism and adaptive efflux during oncogenic transformation from chronic arsenic exposure in human cells. J Biol Chem 283:19342–19350

Cortessis VK, Thomas DC, Levine AJ, Breton CV, Mack TM, Siegmund KD, Haile RW, Laird PW (2012) Environmental epigenetics: prospects for studying epigenetic mediation of exposure-response relationships. Hum Genet 131:1565–1589

Cronican AA, Fitz NF, Carter A, Saleem M, Shiva S, Barchowsky A, Koldamva R, Schug J, Lefterov L (2013) Genome-wide alteration of histone H3K9 acetylation pattern in mouse offspring prenatally exposed to arsenic. PLoS One 8:e53478

Cui X, Wakai T, Shirai Y, Hatakeyama K, Hirano S (2006a) Chronic oral exposure to inorganic arsenate interferes with methylation status of p16INK4a and RASSF1A and induces lung cancer in A/J mice. Toxicol Sci 91:372–381

Cui X, Wakai T, Shirai Y, Yokoyama N, Hatakeyama K, Hirano S (2006b) Arsenic trioxide inhibits DNA methyltransferase and restores methylation-silenced genes in human liver cancer cells. Hum Pathol 37:298–311

Cui Y, Han Z, Hu Y, Song G, Hao C, Xia H, Ma X (2012) MicroRNA-181b and microRNA-9 mediate arsenic-induced angiogenesis via NRP1. J Cell Physiol 227:772–783

Das NK, Sengupta SR (2008) Arsenicosis: diagnosis and treatment. Ind J Dermatol Venereol Leprol 74:571–581

Datta BK, Mishra A, Singh A, Sar TK, Sarkar S, Bhatacharya A, Chakraborty AK, Mandal TK (2010) Chronic arsenicosis in cattle with special reference to its metabolism in arsenic endemic village of Nadia district West Bengal India. Sci Total Environ 409:284–288

Davis CD, Uthus EO, Finley JW (2000) Dietary selenium and arsenic affect DNA methylation *in vitro* in Caco-2 cells and *in vivo* in rat liver and colon. J Nutr 130:2903–2909

Dey TK, Banerjee P, Bakshi M, Kar A, Ghosh S (2014) Groundwater arsenic contamination in West Bengal: current scenario, effects and probable ways of mitigation. Int Lett Nat Sci 13:45–58

Dhar MK, Vishal P, Sharma R, Kaul S (2014) Epigenetic dynamics: role of epimarks and underlying machinery in plants exposed to abiotic stress. Int J Genom 2014:187146

Ding D, Zhang LF, Wang H, Liu ZJ, Zhang ZX, Zheng YL (2009) Differential expression of miRNAs in response to salt stress in maize roots. Annl Bot (Lond) 103:29–38

Drobna Z, Styblo M, Thomas DJ (2009) An overview of arsenic metabolism and toxicity. In: Hodgson E (ed) Current protocols in toxicology, supplement 42: techniques for analysis of chemical biotransformation. John Wiley and Sons, Chapel Hill, NC

Drury GE, Dowle AA, Ashford DA, Waterworth WM, Thomas J, West CE (2012) Dynamics of plant histone modifications in response to DNA damage. Biochem J 445:393–401

Engstrom KS, Hossain MB, Lauss M, Ahmed S, Raqib R, Vehter M, Broberg K (2013) Efficient arsenic metabolism–the AS3MT haplotype is associated with DNA methylation and expression of multiple genes around AS3MT. PLoS One 8:e53732

Erturk FA, Agar G, Arslan E, Nardemir G, Sahin Z (2014) Determination of genomic instability and DNA methylation effects of Cr on maize (*Zea mays* L.) using RAPD and CRED-RA analysis. Acta Physiol Planta 36:1529–1537

Erturk FA, Agar G, Arslan E, Nardemir G (2015) Analysis of genetic and epigenetic effects of maize seeds in response to heavy metal (Zn) stress. Environ Sci Pollut Res 22:10291–10297

Esteller M, Fraga MF, Guo M, Garcia-Foncillas J, Hedenfalk I, Godwin AK, Trojan J, Vaurs-Barrière C, Bignon YJ, Ramus S, Benitez J, Caldes T, Akiyama Y, Yuasa Y, Launonen V, Canal MJ, Rodriguez R, Capella G, Peinado MA, Borg A, Aaltonen LA, Ponder BA, Baylin SB, Herman JG (2001) DNA methylation patterns in hereditary human cancers mimic sporadic tumorigenesis. Hum Mol Genet 10:3001–3007

Esteve PO, Chin HG, Pradhan S (2007) Molecular mechanisms of transactivation and doxorubicin-mediated repression of survivin gene in cancer cells. J Biol Chem 282:2615–2625

Feinberg AP, Tycko B (2004) The history of cancer epigenetics. Nat Rev Cancer 4:143–153

Flanagan SV, Johnston RB, Zheng Y (2012) Arsenic in tube well water in Bangladesh: health and economic impacts and implications for arsenic mitigation. Bull World Health Organ 90:839–846

Flora SJS, Bhadauria S, Kannan GM, Singh N (2007) Arsenic induced oxidative stress and the role of antioxidant supplementation during chelation: a review. J Environ Biol 28:333–347

Fournier A, Florin A, Lefebvre C, Solly F, Leroux D, Callanan MB (2007) Genetics and epigenetics of 1q rearrangements in hematological malignancies. Cytogenet Genome Res 118:320–327

Fu HY, Shen JZ, Wu Y, Shen SF, Zhou HR, Fan LP (2010) Arsenic trioxide inhibits DNA methyltransferase and restores expression of methylation-silenced CDKN2B/CDKN2A genes in human hematologic malignant cells. Oncol Rep 24:335–343

Füllgrabe J, Lynch-Day MA, Heldring N, Li W, Struijk RB, Ma Q, Hermanson O, Rosenfeld MG, Klionsky DJ, Joseph B (2013) The histone H4 lysine 16 acetyltransferase hMOF regulates the outcome of autophagy. Nature 500:468–471

Ge CL, Yang XY, Liu XN, Sun H, Luo SS, Wang ZG (2012) Effect of heavy metal on levels of methylation in DNA of rice and wheat. J Plant Physiol Mol Biol 28:363–368

Gielen H, Remans T, Vangronsveld J, Cuypers A (2012) MicroRNAs in metal stress: specific roles or secondary responses? Int J Mol Sci 13:15826–15847

Gribble MO, Tang WY, Shang Y, Pollak J, Umans JG, Francesconi KA, Goessler W, Silbergeld EK, Guallar E, Cole SA, Fallin MD, Navas-Acien A (2013) Differential methylation of the arsenic (III) methyltransferase promoter according to arsenic exposure. Arch Toxicol 88:275–282

Guha Mazumder DN (2008) Chronic arsenic toxicity and human health. Ind J Med Res 128:436–447

Hauqe S, Ferdous AS, Hossain K, Islam Md T, Khan H (2014) Identification and expression profiling of microRNAs and their corresponding targets related to phytoremediation of heavy metals in jute (*Corchorus olitorius* var. O-9897). Proc 5th International Conference in Environment, Bangladesh ID E78, pp 116–119

Hegedus CM, Skibola CF, Warner M, Skibola DR, Alexander D, Lim S, Dangleben NL, Zhang L, Clark M, Pfeiffer RM, Steinmaus C, Smith AH, Smith MT, Moore LE (2008) Decreased urinary beta-defensin-1 expression as a biomarker of response to arsenic. Toxicol Sci 106:74–82

Hei TK, Filipic M (2004) Role of oxidative damage in the genotoxicity of arsenic. Free Radic Biol Med 37:574–581

Herbert KJ, Holloway A, Cook AL, Chin SP, Snow ET (2014) Arsenic exposure disrupts epigenetic regulation of SIRT1 in human keratinocytes. Toxicol Appl Pharmacol 281:136–145

Hermann A, Gowher H, Jeltsch A (2004) Biochemistry and biology of mammalian DNA methyltransferases. Cell Mol Life Sci 61:2571–2587

Hossain MB, Vahter M, Concha G, Broberg K (2012) Environmental arsenic exposure and DNA methylation of the tumor suppressor gene p16 and the DNA repair gene MLH1: effect of arsenic metabolism and genotype. Metallomics 4:1167–1175

Huang BW, Ray PD, Iwasaki K, Tsuji Y (2013) Transcriptional regulation of the human ferritin gene by coordinated regulation of Nrf2 and protein arginine methyltransferases PRMT1 and PRMT4. FASEB J 27:3763–3774

International Agency for Research on Cancer (IARC) (2004) Some drinking-water disinfectants and contaminants, including arsenic. Volume 84. Available online at: http://monographs.iarc.fr/ENG/ Monographs/vol84/mono84-1.pdf

Iwasaki M, Paszkowski J (2014) Epigenetic memory in plants. EMBO J 33:1987–1998

Jagadeeswaran G, Saini A, Sunkar R (2009) Biotic and abiotic stress downregulate miR398 expression in *Arabidopsis*. Planta 229:1009–1014

Jensen TJ, Novak P, Eblin KE, Gandolfi AJ, Futscher BW (2008) Epigenetic remodelling during arsenical-induced malignant transformation. Carcinogenesis 29:1500–1508

Jensen TJ, Novak P, Wnek SM, Gandolfi AJ, Futscher BW (2009a) Arsenicals produce stable progressive changes in DNA methylation patterns that are linked to malignant transformation of immortalized urothelial cells. Toxicol Appl Pharmacol 241:221–229

Jensen TJ, Wozniak RJ, Eblin KE, Wnek SM, Gandolfi AJ, Futscher BW (2009b) Epigenetic mediated transcriptional activation of WNT5A participates in arsenical-associated malignant transformation. Toxicol Appl Pharmacol 235:39–46

Jo WJ, Ren X, Chu F, Aleshin M, Wintz H, Burlingame A, Smith MT, Vulpe CD, Zhang L (2009) Acetylated H4K16 by MYST1 protects UROtsa cells from arsenic toxicity and is decreased following chronic arsenic exposure. Toxicol Appl Pharmacol 241:294–302

Jullien PE, Mosquna A, Ingouff M, Sakata T, Ohad N, Berger F (2008) Retinoblastoma and its binding partner MSI1 control imprinting in *Arabidopsis*. PLoS Biol 6:e194

Kim JM, Sasaki T, Ueda M, Sako K, Seki M (2015) Chromatin changes in response to drought, salinity, heat, and cold stresses in plants. Front Plant Sci 6:114

Kinner A, Wu W, Staudt C, Iliakis G (2008) Gamma-H2AX in recognition and signalling of DNA double-strand breaks in the context of chromatin. Nucl Acid Res 36:5678–5694

Kitchin KT, Conolly R (2010) Arsenic-induced carcinogenesis–oxidative stress as a possible mode of action and future research needs for more biologically based risk assessment. Chem Res Toxicol 23:327–335

Kitchin KT, Wallace K (2008) The role of protein binding of trivalent arsenicals in arsenic carcinogenesis and toxicity. J Inorg Biochem 102:532–539

Koestler DC, Avissar-Whiting M, Houseman EA, Karagas MR, Marsit CJ (2013) Differential DNA methylation in umbilical cord blood of infants exposed to low levels of arsenic *in utero*. Environ Health Perspect 121:971–977

Kumar M, Bijo AJ, Baghel RS, Reddy CRK, Jha B (2012) Selenium and spermine alleviate cadmium induced toxicity in the red seaweed *Gracilaria dura* by regulating antioxidants and DNA methylation. Plant Physiol Biochem 51:129–138

Labra M, Vannini C, Sala F, Bracale M (2002) Methylation changes in specific sequences in response to water deficit. Plant Biosyst 136:269–276

Langley-Evans SC (2006) Developmental programming of health and disease. Proc Nutr Soc 65:97–105

Laubinger S, Sachsenberg T, Zeller G, Busch W, Lohmann JU, Rätsch G, Weigel D (2008) Dual roles of the nuclear cap-binding complex and SERRATE in pre-mRNA splicing and microRNA processing in *Arabidopsis thaliana*. Proc Natl Acad Sci U S A 105:8795–8800

Li L, Chen F (2016) Oxidative stress, epigenetics, and cancer stem cells in arsenic carcinogenesis and prevention. Curr Pharmacol Rep 2:57–63

Li J, Chen P, Sinogeeva N, Gorospe M, Wersto RP, Chrest FJ, Barnes J, Liu Y (2002) Arsenic trioxide promotes histone H3 phosphoacetylation at the chromatin of CASPASE-10 in acute pro myelocytic leukemia cells. J Biol Chem 277:49504–49510

Li J, Gorospe M, Barnes J, Liu Y (2003) Tumor promoter arsenite stimulates histone H3 phosphoacetylation of proto-oncogenes c-fos and c-jun chromatin in human diploid fibroblasts. J Biol Chem 278:13183–13191

Li W, Cui X, Meng Z, Huang X, Xie Q, Wu H, Jin H, Zhang D, Liang W (2012) Transcriptional regulation of *Arabidopsis* MIR168a and ARGONAUTE1 homeostasis in abscisic acid and abiotic stress responses. Plant Physiol 158:1279–1292

Li Z, Chen X, Li S, Wang Z (2015) Effect of nickel chloride on *Arabidopsis* genomic DNA and methylation of 18S rDNA. Electron J Biotechnol 18:51–57

Ling M, Li Y, Xu Y, Pang Y, Shen L, Jiang R, Zhao Y, Yang X, Zhang J, Zhou J, Wang X, Liu Q (2012) Regulation of miRNA-21 by reactive oxygen species-activated ERK/NF-κB in arsenite-induced cell transformation. Free Radic Biol Med 52:1508–1518

Liu Q, Zhang H (2012) Molecular identification and analysis of arsenite stress-responsive miRNAs in rice. J Agric Food Chem 60:6524–6536

Liu Z, Sanchez MA, Jiang X, Boles E, Landfear SM, Rosen BP (2006) Mammalian glucose permease GLUT1 facilitates transport of arsenic trioxide and methylarsonous acid. Biochem Biophys Res Commun 351:424–430

Liu Q, Hu H, Zhu L, Li R, Feng Y, Zhang L, Yang Y, Liu X, Zhang H (2015a) Involvement of miR528 in the regulation of arsenite tolerance in rice (*Oryza sativa* L.) J Agric Food Chem 63:8849–8861

Liu D, Wu D, Zhao L, Yang Y, Ding J, Dong L, Hu L, Wang F, Zhao X, Cai Y, Jin J (2015b) Arsenic trioxide reduces global histone H4 acetylation at lysine 16 through direct binding to histone acetyltransferase hMOF in human cells. PLoS One 10:e0141014

Lu G, Xu H, Chang De Wu Z, Yao X, Zhang S, Li Z, Bai J, Cai Q, Zhang W (2014) Arsenic exposure is associated with DNA hypermethylation of the tumour suppressor gene p16. J Occup Med Toxicol 9:42

Majumdar S, Chanda S, Ganguli B, Mazumder DN, Lahiri S, Dasgupta UB (2010) Arsenic exposure induces genomic hypermethylation. Environ Toxicol 25:315–318

Marsit CJ, Karagas MR, Danaee H, Liu M, Andrew A, Schned A, Nelson HH, Kelsey KT (2006a) Carcinogen exposure and gene promoter hypermethylation in bladder cancer. Carcinogenesis 27:112–116

Marsit CJ, Karagas MR, Schned A, Kelsey KT (2006b) Carcinogen exposure and epigenetic silencing in bladder cancer. Annl N Y Acad Sci 1076:810–821

Martin C, Zhang Y (2005) The diverse functions of histone lysine methylation. Nat Rev Mol Cell Biol 6:838–849

McGarvey KM, Fahrner JA, Greene E, Martens J, Jenuwein T, Baylin SB (2006) Silenced tumour suppressor genes reactivated by DNA demethylation do not return to a fully euchromatic chromatin state. Cancer Res 66:3541–3549

Meng XZ, Zheng TS, Chen X, Wang JB, Zhang WH, Pan SH, Jiang HC, Liu LX (2011) microRNA expression alteration after arsenic trioxide treatment in HepG-2 cells. J Gastroenterol Hepatol 26:186–193

Mirbahai L, Chipman JK (2014) Epigenetic memory of environmental organisms: a reflection of lifetime stressor exposures. Mutat Res 764–765:10–17

Mirouze M, Paszkowski J (2011) Epigenetic contribution to stress adaptation in plants. Curr Opin Plant Biol 14:267–274

Newell-price J, Clark AJ, King P (2000) DNA methylation and silencing of gene expression. Trend Endocrinol Metab 11:142–148

Nohara K, Baba T, Murai H, Kobayashi Y, Suzuki T, Tateishi Y, Matsumoto M, Nishimura N, Sano T (2011) Global DNA methylation in the mouse liver is affected by methyl deficiency and arsenic in a sex-dependent manner. Arch Toxicol 85:653–661

Nye MD, Fry RC, Hoyo C, Murphy SK (2014) Investigating epigenetic effects of prenatal exposure to toxic metals in newborns: challenges and benefits. Med Epigenet 2:53–59

Okoji RS, Yu RC, Maronpot RR, Froines JR (2002) Sodium arsenite administration via drinking water increases genome-wide and Ha-ras DNA hypomethylation in methyl-deficient C57BL/6 J mice. Carcinogenesis 23:777–785

Pandey C, Raghuram B, Sinha AK, Gupta M (2015) miRNA plays a role in the antagonistic effect of selenium on arsenic stress in rice seedlings. Metallomics 7:857–866

Pilsner JR, Liu X, Ahsan H, Ilievski V, Slavkovich V, Levy D, Factor-Livak P, Graziano JH, Gamble MV (2007) Genomic methylation of peripheral blood leukocyte DNA: influences of arsenic and folate in Bangladeshi adults. Amer J Clin Nutr 86:1179–1186

Pilsner JR, Liu X, Ahsan H, Ilievski V, Slavkovich V, Levy D, Factor-Livak P, Graziano JH, Gamble MV (2009) Folate deficiency, hyperhomocysteinemia, low urinary creatinine, and hypomethylation of leukocyte DNA are risk factors for arsenic-induced skin lesions. Environ Health Perspect 117:254–260

Pilsner JR, Hall MN, Liu X, Ilievski V, Slavkovich V, Levy D, Factor-Litvak P, Yunus M, Rahman M, Graziano JH, Gamble MV (2012) Influence of prenatal arsenic exposure and newborn sex on global methylation of cord blood DNA. PLoS One 7:e37147

Rager JE, Bailey KA, Smeester L, Miller SK, Parker JS, Laine JE (2014) Prenatal arsenic exposure and the epigenome: altered microRNAs associated with innate and adaptive immune signalling in new born cord blood. Environ Mol Mutagen 55:196–208

Ramirez T, Brocher J, Stopper H, Hock R (2008) Sodium arsenite modulates histone acetylation, histone deacetylase activity and HMGN protein dynamics in human cells. Chromosoma 117:147–157

Rana T, Sarkar S, Mandal TK, Batabyal S (2008) Haematobiochemical profiles of affected cattle at arsenic prone zone in Haringhata block of Nadia District of West Bengal in India. Int J Haematol 4:1642–1657

Rancelis V, Cesniene T, Kleizaite V, Zvingila D, Balciuniene L (2012) Influence of cobalt uptake by *Vicia faba* seeds on chlorophyll morphosis induction, SOD polymorphism, and DNA methylation. Environ Toxicol 27:32–41

Ray PD, Yosim A, Fry RC (2014) Incorporating epigenetic data into the risk assessment process for the toxic metals arsenic, cadmium, chromium, lead, and mercury: strategies and challenges. Front Genet 5:201

Reichard JF, Puga A (2010) Effects of arsenic exposure on DNA methylation and epigenetic gene regulation. Epigenomics 2:87–104

Reichard JF, Schnekenburger M, Puga A (2007) Long term low-dose arsenic exposure induces loss of DNA methylation. Biochem Biophys Res Commun 352:188–192

Ren X, McHale CM, Skibola CF, Smith AH, Smith MT, Zhang L (2011) An emerging role for epigenetic dysregulation in arsenic toxicity and carcinogenesis. Environ Health Perspect 119:11–19

Russell WMS, Burch RL (1959) The principles of humane experimental technique. Methuen and Co Ltd., London

Saumet A, Vetter G, Bouttier M, Portales-Casamar E, Wasserman WW, Maurin T, Mari B, Barbry P, Vallar L, Friederich E, Arar K, Cassinat B, Chomienne C, Lecellier CH (2009) Transcriptional repression of microRNA genes by PML-RARA increases expression of key cancer proteins in acute promyelocytic leukemia. Blood 113:412–421

Sciandrello G, Caradonna F, Mauro M, Barbata G (2004) Arsenic-induced DNA hypomethylation affects chromosomal instability in mammalian cells. Carcinogenesis 25:413–417

Sengupta SR, Das NK, Datta PK (2008) Pathogenesis, clinical features and pathology of chronic arsenicosis. Ind J Dermatol Venereol Leprol 74:559–570

Sharma GG, So S, Gupta A, Kumar R, Cayrou C, Avvakumov N, Bhadra U, Pandita RK, Porteus MH, Chen DJ, Cote J, Pandita TK (2010) MOF and histone H4 acetylation at lysine 16 are critical for DNA damage response and double-strand break repair. Mol Cell Biol 30:3582–3595

Sharma D, Tiwari M, Lakhwani D, Tripathi RD, Trivedi PK (2015) Differential expression of microRNAs by arsenate and arsenite stress in natural accessions of rice. Metallomics 7:174–187

Skinner MK (2011) Environmental epigenomics and disease susceptibility. EMBO Rep 12:620–622

Song Y, Ci D, Tian M, Zhang D (2016) Stable methylation of a non-coding RNA gene regulates gene expression in response to abiotic stress in *Populus simonii*. J Exp Bot 67:1477–1492

Srivastava S, Srivastava AK, Suprasanna P, D'Souza SF (2012) Identification and profiling of arsenic stress-induced microRNAs in *Brassica juncea*. J Exp Bot 64:303–315

Stefanska B, Huang J, Bhattacharyya B, Suderman M, Hallett M, Han ZG, Szyf M (2011) Definition of the landscape of promoter DNA hypomethylation in liver cancer. Cancer Res 71:5891–5903

Steward N, Ito M, Yamaguchi Y, Koizumi N, Sano H (2002) Periodic DNA methylation in maize nucleosomes and demethylation by environmental stress. J Biol Chem 277:37741–37746

Straif K, Benbrahim-Tallaa L, Baan R, Grosse Y, Secretan B, El Ghissassi F, Bouvard V, Guha N, Freeman C, Galichet L, Cogliano V (2009) A review of human carcinogens–part C: metals, arsenic, dusts, and fibres. Lancet Oncol 10:453–454

Sun CQ, Arnold R, Fernandez-Golarz C, Parrish AB, Almekinder T, He J, Ho SM, Svoboda P, Pohl J, Marshall FF, Petros JA (2006) Human beta-defensin-1, a potential chromosome 8p tumor suppressor: control of transcription and induction of apoptosis in renal cell carcinoma. Cancer Res 66:8542–8549

Sunkar R, Kapoor A, Zhu JK (2006) Posttranscriptional induction of two Cu/Zn superoxide dismutase genes in Arabidopsis is mediated by downregulation of miR398 and important for oxidative stress tolerance. Plant Cell 18:2051–2065

Szyf M (2011) The implications of DNA methylation for toxicology: toward toxicomethylomics, the toxicology of DNA methylation. Toxicol Sci 120:235–255

Takahashi M, Barrett JC, Tsutsui T (2002) Transformation by inorganic arsenic compounds of normal Syrian hamster embryo cells into a neoplastic state in which they become anchorage-independent and cause tumors in newborn hamsters. Int J Cancer 99:629–634

Tan M (2010) Analysis of DNA methylation of maize in response to osmotic and salt stress based on methylation-sensitive amplified polymorphism. Plant Physiol Biochem 48:21–26

Thomas DJ, Li J, Waters SB, Xing W, Adair BM, Drobna Z, Devesa V, Styblo M (2007) Arsenic (+3 oxidation state) methyltransferase and the methylation of arsenicals. Exp Biol Med (Maywood) 232:3–13

Tuli R, Chakrabarty D, Trivedi PK, Tripathi RD (2010) Recent advances in arsenic accumulation and metabolism in rice. Mol Breed 26:307–323

Uthus EO, Davis C (2005) Dietary arsenic affects dimethylhydrazine-induced aberrant crypt formation and hepatic global DNA methylation and DNA methyltransferase activity in rats. Biol Trace Elem Res 103:133–145

Vahter ME (2007) Interactions between arsenic-induced toxicity and nutrition in early life. J Nutr 137:2798–2804

Waalkes MP, Liu J, Chen H, Xie Y, Achanzar WE, Zhou YS, Cheng M, Diwan BA (2004) Estrogen signaling in livers of male mice with hepatocellular carcinoma induced by exposure to arsenic in utero. J Natl Cancer Inst 96:466–474

Wang GG, Allis CD, Chi P (2007) Chromatin remodelling and cancer, part I: covalent histone modifications. Trend Mol Med 13:363–372

Wang Z, Zhao Y, Smith E, Goodall GJ, Drew PA, Brabletz T, Yang C (2011) Reversal and prevention of arsenic-induced human bronchial epithelial cell malignant transformation by microRNA-200b. Toxicol Sci 121:110–122

Weake VM, Workman JL (2010) Inducible gene expression: diverse regulatory mechanisms. Nat Rev Genet 11:426–437

Wilson AS, Power BE, Molloy PL (2007) DNA hypomethylation and human diseases. Biochim Biophys Acta 1775:138–162

World Health Organization (WHO) (2006) Guidelines for drinking water quality. WHO, Geneva, pp 306–308

Wysocka J, Allis CD, Coonrod S (2006) Histone arginine methylation and its dynamic regulation. Front Biosci 11:344–355

Xie Y, Liu J, Benbrahim-Tallaa L, Ward JM, Logsdon D, Diwan BA, Waalkes MP (2007) Aberrant DNA methylation and gene expression in livers of newborn mice transplacentally exposed to a hepatocarcinogenic dose of inorganic arsenic. Toxicology 236:7–15

Yaish MW (2013) DNA methylation-associated epigenetic changes in stress tolerance of plants. In: Rout GR, Das AB (eds) Molecular stress physiology of plants. Springer, New York, pp 427–440

Yaish MW, Colasanti J, Rothstein SJ (2011) The role of epigenetic processes in controlling flowering time in plants exposed to stress. J Exp Bot 62:3727–3735

Yong-Villalobos L, González-Morales SI, Wrobel K, Gutiérrez-Alanis D, Cervantes-Peréz SA, Hayano-Kanashiro C, Oropeza-Aburto A, Cruz-Ramírez A, Martínez O, Herrera-Estrella L (2015) Methylome analysis reveals an important role for epigenetic changes in the regulation of the *Arabidopsis* response to phosphate starvation. Proc Natl Acad Sci U S A 112:E7293–E7302

Yu LZ, Luo YF, Liao B, Xie LJ, Chen L, Xiao S, Li JT, Hu SN, Shu WS (2012) Comparative transcriptome analysis of transporters, phytohormone and lipid metabolism pathways in response to arsenic stress in rice (*Oryza sativa*). New Phytol 195:97–112

Zardo G, Ciolfi A, Vian L, Starnes LM, Billi M, Racanicchi S, Maresca C, Fazi F, Travaglini L, Noguera N, Mancini M, Nanni M, Cimino G, Lo-Coco F, Grignani F, Nervi C (2012) Polycombs and microRNA-223 regulate human granulopoiesis by transcriptional control of target gene expression. Blood 119:4034–4046

Zhang AH, Bin HH, Pan XL, Xi XG (2007) Analysis of p16 gene mutation, deletion and methylation in patients with arseniasis produced by indoor unventilated-stove coal usage in Guizhou, China. J Toxicol Environ Health A 70:970–975

Zhou X, Sun H, Ellen TP, Chen H, Costa M (2008) Arsenite alters global histone H3 methylation. Carcinogenesis 29:1831–1836

Zhou X, Li Q, Arita A, Sun H, Costa M (2009) Effects of nickel, chromate, and arsenite on histone 3 lysine methylation. Toxicol Appl Pharmacol 236:78–84

Zhou LG, Liu YH, Liu ZC, Kong DY, Duan M, Luo LJ (2010) Genome-wide identification and analysis of drought-responsive microRNAs in *Oryza sativa*. J Exp Bot 61:4157–4168

Zykova TA, Zhu F, Lu C, Higgins L, Tatsumi Y, Abe Y, Bode AM, Dong Z (2006) Lymphokine-activated killer T-cell-originated protein kinase phosphorylation of histone H2AX prevents arsenite-induced apoptosis in RPMI7951 melanoma cells. Clin Cancer Res 12:6884–6893

Prospects of Combating Arsenic: Physico-chemical Aspects

Soumya Chatterjee, Mridul Chetia, Anna Voronina, and Dharmendra K. Gupta

Contents

1 Introduction 104
2 Chemical and Physico-chemical Processes for Arsenic Removal 105
2.1 Lime Softening 106
2.2 Coagulation/Filtration 106
2.3 Sorption 107
2.3.1 Ion Exchange 107
2.3.2 Arsenic Adsorption Using Fixed Bed 108
2.3.3 Arsenic Removal by Layered Double Hydroxide 108
2.3.4 Nano Materials in Arsenic Removal 109
2.4 Reverse Osmosis/Nano-Filtration 109
2.5 Other Related Technologies 109
3 Technology Selection Criteria and Available Indigenous Methods: An Appraisal 110
4 Vulnerability and Scope of Arsenic Remediation Through Available Technologies in Countries Like Bangladesh and India 111
5 Widely Accepted Small Scale Arsenic Removal Units 112
5.1 Biosand Filter (BSF) 112
5.2 Magc-Alcan Filter (Two Bucket Filter) 113
5.3 Nirmal Filter 113
5.4 Shapla and Surokka Filter 113

S. Chatterjee
Defence Research Laboratory, Defence Research and Development Organization (DRDO), Ministry of Defence, Post Bag No. 2, Tezpur 784001, Assam, India

M. Chetia
Department of Chemistry, D.R. College, Golghat 785621, Assam, India

A. Voronina
Department of Radiochemistry and Applied Ecology, Physical Technology Institute, Ural Federal University, Mira str, 19 Ekaterinburg, Russia

D.K. Gupta (✉)
Institut für Radioökologie und Strahlenschutz (IRS), Gottfried Wilhelm Leibniz Universität Hannover, Herrenhäuser Str. 2, Hannover 30419, Germany
e-mail: guptadk1971@gmail.com

D.K. Gupta, S. Chatterjee (eds.), *Arsenic Contamination in the Environment*,
DOI 10.1007/978-3-319-54356-7_5

5.5 Bucket Treatment Unit (BTU) 114
5.6 3-Kolshi Filter 114
5.7 SONO Filter 114
5.8 Kanchan Arsenic Filter (KAF) 115
5.9 Subterranean Arsenic Removal Technology (SAR) 116
6 Conclusions 117
References 118

1 Introduction

Arsenic (As) is a Class A human carcinogen and its chronic exposure results in severe health effects including cancer (NRC 1999; 2001). Ingestion of contaminated drinking water is among the most awful environmental health challenges nowadays. According to several reports, arsenic contamination is intimidating the well-being and livelihood of more than a 137 million people in 70 different countries of the world (http://www.worstpolluted.org/projects_reports/display/76 downloaded on 11.02.16). The presence of arsenic in groundwater has been reported extensively in recent years from different parts of the world, including countries in North America and Latin America (viz., USA, Canada, Mexico, Argentina, Bolivia, Brazil, and Nicaragua), Australia, and Southeast Asia (viz., Bangladesh, China, Nepal, Vietnam, Cambodia, and India); (Acharyya et al. 2000; Smedley and Kinniburgh 2002; USEPA 2000; Bhattacharya et al. 1997, 2002; Bhattacharyya et al. 2005; Mukherjee et al. 2006). The environmental problem of arsenic toxicity in groundwater of the entire Bengal delta of the Ganga–Padma–Meghna–Brahmaputra (GPMB) river plain, covering several districts of West Bengal and Bangladesh, creates apprehension toward the scientific community and considered as the worst arsenic-affected alluvial basin (Guha Mazumder et al. 2005; Chetia et al. 2011, 2012; Jiang et al. 2013). In northeastern states of India, the presence of arsenic has been identified in Assam (21 districts out of 24 districts), Arunachal Pradesh (06 districts), Tripura (03 districts), Nagaland (02 districts), and Manipur (01 district) (Das et al. 2009; Hoque et al. 2011; Chetia et al. 2012). Where subsurface water is primarily used as a source of drinking water, the magnitude of the severity with arsenic toxicity is very high. In various districts of Bangladesh and West Bengal, India, 79.9 and 42.7 million people respectively are reported to be affected by groundwater arsenic poisoning (Sen Gupta et al. 2009; Drobna et al. 2009; Spivey 2011; Williams et al. 2011; Web info page downloaded from: http://patient.info/doctor/arsenic-poisoning). With this report of "unprecedented scale" of arsenic contamination and related arsenicosis, World Health Organization in 2002 asserted that largest mass poisoning of a population in history is now underway in Bangladesh (Naujokas et al. 2013; WHO 2001; Web info page downloaded from: http://www.who.int/features/archives/feature206/en/, downloaded on 03 March 2016). The maximum contaminant level (MCL) of arsenic is 0.010 mg L^{-1} (10 μg L^{-1}).

2 Chemical and Physico-chemical Processes for Arsenic Removal

Arsenic is a redox-sensitive element with standard reduction potential E° = 0.56 V and standard oxidation potential E° = −0.67 V. Two inorganic forms of arsenic typically occurs in water: trivalent arsenite, As(III) and pentavalent arsenate, As(V). As(III) compounds predominantly present in nature and is neutral, while As(V) species are negatively charged in the pH range of 4–10. Removal of the element from water is a challenging task altogether. There are several potential best available technologies (BATs) as reviewed by US Environment Protection Agency (US EPA 2001a, b; 2005; CAWST 2009). There are several factors for choosing appropriate chemical based arsenic removal technology. Water quality along with costs, and disposal options for residual (hazardous or nonhazardous) wastes, are the factors for the assortment of the appropriate and effective treatment systems. In most of the chemicalprocesses, optimizing arsenic removal requires pre-oxidation step and pH adjustment. Oxidation step is critical for attaining best possible performance for the unit. Oxidation is effective to convert neutral As(III) to negatively charged As(V) for better removal effectiveness. Chlorine, ozone, permanganate, and manganese dioxide components are effective oxidizing agents for this process (Table 1 representing common oxidizing agents). Further, pH adjustment is occasionally essential to optimize elimination of arsenic (AWWA 1999; CAWST 2009). Few technologies like iron-based treatment technologies have a higher removal capacity of arsenic at a lower pH (<8).

Table 1 Comparison of major arsenic oxidizing agents

Oxidant	Advantage	Disadvantage
Chlorine	• Cost: comparatively low • Oxidation fast, in <01 min • Disinfection capabilities • Also effective for iron Oxidation	• Chlorine gas: Require special storage and handling • Membrane fouling • Formation of disinfection byproducts • Not much effective for manganese oxidation
Ozone	• Oxidation fast, in <01 min • Disinfection capabilities • By-products: Nil • Also effective for iron and manganese oxidation	• Initial cost: relatively high • Interference: TOC and Sulfide that increase the required contact time and dose • Formation of disinfection byproducts
Permanganate	• Oxidation fast, in <01 min • No formation of disinfection byproducts • Unreactive with membranes • Also effective for iron and manganese oxidation	• Initial cost: relatively high • No disinfection capability • Water color changes, if over dosed • Difficulty in handling
Solid Phase Oxidants	• Oxidation, in approx • 1.5 min (absence of interfering reductants) • By-products—Nil • Storage Not required	• Require: backwashing • Require: dissolved oxygen • No disinfection capabilities • Interference: Iron, manganese, TOC, and sulfide that increase the required contact time and dose

The primary methods (Table 2) for removing arsenic from drinking waters (Clifford and Lin 1991; Chang et al. 1994; Xu et al. 1988; Clifford 1999; Fields et al. 2000a, b; Chwirka et al. 2000; CAWST 2009; Chutia et al. 2009) include:

- Lime-softening (adding lime to soften water (remove Ca and Mg) often removing substantial amounts of arsenic).
- Coagulation-assisted micro-filtration (CMF).
- Coagulation cum filtration (by adding Fe(III) or Al(III) salts to form arsenic-sorbing flocs and subsequent filtration).
- Ion exchange to remove anionic As species by using a resin.
- Fixed bed adsorption-uses adsorbent, typically a metal hydroxide such as activated alumina (AA) or ferric hydroxide.
- Oxidation/filtration (iron removal)—Oxidizing reduced iron to remove arsenic through sorption/co-precipitation/coagulation.
- Physical filtration—To remove colloidal-bound arsenic.
- Membrane processes—Membrane removal of arsenic by reverse osmosis (RO) or Nano-filtration (NF).

2.1 *Lime Softening*

Arsenic removal can be increased effectively at higher pH through precipitation using lime softening treatment (Sorg and Logsdon 1978). Lime softening is commonly used to remove magnesium and calcium cations from solution. Lime is added to increase the pH above 10.5 to form magnesium hydroxide which is then co-precipitated with As(V). The process is thought to be a combination of sorption, co-precipitation, and occlusion on the calcium and magnesium solids (McNeill and Edwards 1997). Iron oxidation through aeration or chlorination also results in arsenic removal. Therefore, arsenic removal may be targeted indirectly through Fe removal (Fields et al. 2000b). McNeill and Edwards (1997) showed that Al and Fe precipitation processes in conventional water treatment plants can be optimized for removal of arsenic at low cost.

2.2 *Coagulation/Filtration*

Iron (Fe III) or aluminum (Al III) salts are used to form flocs that sorbs and co-precipitate the dissolved arsenic (Chang et al. 1994; Fields et al. 2000a) which are subsequently filtered out. The precipitated $Fe(OH)_3$ or $Al(OH)_3$ can be removed either by granular media filtration or membrane microfiltration. If granular media filtration is used a flocculation step must be included to facilitate growth of floc particles. Use of a membrane micro filter is a better alternative to get the desired result. Arsenic removal processes using Al or Fe hydroxide floc work best at pH < 7.5 (Edwards 1994; Chen et al. 1999).

Table 2 Comparison of common arsenic removal techniques (After, Kumar et al. 2015)

Treatment process	As (III) removal efficiency	As (V) removal efficiency	Advantages	Disadvantages	Cost (relative)
Fe Coagulation	60–90%	>90%	Simple operation, Good efficiency	Pre-oxidation require for better removal of As(III)	Low
Alum Coagulation	<30%	>90%	Simple operation, Good efficiency, Low chemical cost	Low efficiency for As(III), Generation of Toxic sludge	Low
Coagulation/ co-precipitation	<30%	>90%	Good efficiency, Low chemical cost	Low efficiency for As(III), Generation of Toxic sludge Require trained manpower to operate	Low
Activated Alumina adsorption	60–90%	>90%	Simple operation, Good efficiency, Commercially available chemicals	Readjustment of pH is essential, Residue disposal hazardous	Medium
Iron based sorbents	30–60%	>90%	Simple operation, Good efficiency	Readjustment of pH is essential, Replacement of media	Medium
Reverse Osmosis	60–90%	60–90%	Good efficiency, Removes other contaminants too	Regular maintenance required, High running cost	High–very high
Nano-filtration	60–90%	60–90%	Good efficiency	High running cost, Regular maintenance required,	Very High

2.3 *Sorption*

2.3.1 Ion Exchange

It is a process of physicochemical ion exchange between a solid resin phase and a solution phase. The technology is widely used for contaminant removal form water (Clifford 1999) with familiar applications of this technique involves use of cationic exchange of Na^{+} for Ca^{2+} and Mg^{2+} in case of home water softeners. This technology is applicable for low volume drinking water applications (Xiong et al. 1987; Wang et al. 2016). Polystyrene-based resins remove anionic arsenate species but not H_3AsO_4. The selectivity sequence for strong base anion exchange resins is (Clifford 1999):

$SO_4^{-2} > HAsO_4^{2-} > CO_3^{-2}$.
$NO_3^- > Cl^- > H_2AsO_4^-$,
$HCO_3^- >> Si(OH)_4$ and $As(OH)_3^o$.

2.3.2 Arsenic Adsorption Using Fixed Bed

Adsorption of arsenic onto metal oxide or metal hydroxide surfaces has been well known for many years. Adsorption based systems usually utilize metal (hydr)oxides such as iron (hydr)oxide (HFO), ferrihydrite, zirconium dioxide, titanium dioxide, goethite, and hematite. These metal (hydr)oxides have excellent arsenic selectivity. Typically, they form inner sphere complexes with the metal atoms in the sorbent using oxygen bridges (Medpelli et al. 2015). This process is economic and having easy working conditions and accessibility of low-cost adsorbents. Further, to improve their removal affinities, adsorbents can be modified by using different treatment methods like oxidation. Early studies focused on the use of activated alumina (γ-Al_2O_3) primarily as a selective adsorbent for As and Fe (Sorg and Logsdon 1978; Xu et al. 1988; Clifford and Lin 1991; Sorg 1993), but in recent years a wide variety of other adsorbents have been used. Al and Fe(III)-based sorbents tend to work best at pH 7 or below. Activated Alumina (AA) is a good adsorbent for arsenic(V) rather than arsenic(III) (Wang et al. 2000). Its preparation involves partial dehydration $Al(OH)_3$ at high temperatures. The sequence of selectivity and adsorption of AA is: $OH^- > H_2AsO_4^- > H_3SiO_4^- > F^- > HSeO_3^- >$ TOC $> SO_4^{2-} > H_3AsO_3$ (Clifford 1999). AA sorption of arsenic works best at pH below 6.5. AA is usually regenerated using 1–4% NaOH followed by a sulfuric acid wash which causes some destruction of the media, and limits the number of times the media can be cycled. Many groundwater sources are having pH 7 or more (basic in nature). pH adjustment is therefore an important factor in technology selection. The technology is associated with initially lowering, and subsequently rising, water pH (to avoid corrosion in plumbing) which increases costs, complexity. Further, and perhaps most importantly, and this technique may require proper training or systems to handling hazardous acid and caustic solutions.

2.3.3 Arsenic Removal by Layered Double Hydroxide

Layered double hydroxide (LDH) or hydrotalcite-like compound with the general formula $[M^{2+}{}_{1-x}\, M^{3+}{}_x(OH)_2]^{x+}{}_{Ax}{}^{-}\cdot nH_2O$ can be considered as a good adsorbent for the removal of toxic As from water. Due to large surface area and high anion exchange capacity of LDH, the compound may be a good adsorbent for the removal of As from contaminated water (Chetia et al. 2012). Study on the calcination–rehydration reaction, in batch experiment (30 ± 1 °C) showed the removal of >98% As from a solution by adsorption process that follows the Freundlich-type adsorption isotherm. Factors like pH, adsorbent dose, and shaking time influenced the rate of As removal (Chetia et al. 2012).

2.3.4 Nano Materials in Arsenic Removal

With the advent of nanotechnology, the variety of adsorbents and their scope of application have been widened for the treatment of contaminated water. Nanoparticles like monocrystalline titanium dioxide, nanoscale zero valent iron (nZVI) with their large active surface area have shown promising results (Rahman et al. 2009). Further, magnetic nanoparticles as an adsorbent are much better than conventional adsorption processes. Removal of arsenic with magnetite (Fe_3O_4) nanoparticles showed the highest level adsorption capacity with small-diameter nanoparticles (10 nm) for both As(III) and As(V) (Tuutijärvi et al. 2009; Ibrahim et al. 2016; Formoso et al. 2016). Iron oxide-coated sand was used in many studies for arsenic removal (Boddu et al. 2008). Iron compounds such as hematite, goethite iron oxide-coated materials and granular ferric hydroxide (GFH) are the preferred group of nanomaterials for arsenic adsorption because they lead to low leaching of adsorbed arsenic from exhausted adsorbent (Ghanizadeh et al. 2010). Iron oxides in various forms have been also used for treatment of radioactive and heavy metals from water and wastewater solutions. The study of Lien and Wilkin (2005) showed that zero-valent iron can efficiently remove arsenite from water.

2.4 *Reverse Osmosis/Nano-Filtration*

Size exclusion process is the basis of reverse osmosis and nano-filtration (NF) technology. Subjected to a pressure gradient, a semipermeable membrane allows water to pass through while retaining certain ions. RO membranes are more selective then NF membranes. Due to higher selectivity, RO systems require higher driving pressures hence higher energy consumption. Arsenic rejection in RO and NF is relatively insensitive to pH. However, arsenite being anionic at high pH, but uncharged at lower pH, is rejected more completely at $pH > 8$ (Narasimhan et al. 2005). Both RO and NF are relatively expensive water treatment systems, consume large amounts of water and generate a considerable amount of waste brine. There are other issues that limit the membrane life, which include the presence of suspended solids, microbes, hydrogen sulfide, iron and manganese, and organics. Therefore typically pretreatment is required to keep the system more durable.

2.5 *Other Related Technologies*

Couple of new technologies like sand ballasted coagulation sedimentation, fluidized-bed in situ oxidation adsorption, coagulation-assisted ceramic media filtration, and immersed media with carrier particles has also been reported to ameliorate the contaminated water (Narasimhan et al. 2005; Bazrafshan et al. 2015). Sand ballasted coagulation sedimentation uses sand and polymer additions to coagulation to boost arsenic removal. Fluidized-bed in situ oxidation adsorption involves adsorbing ferrous iron onto a continuously generated sand surface. Oxidation of the iron leads to uptake and removal of arsenic. Coagulation–assisted ceramic media filtration is similar to CMF except that ceramic filters are used in the floc removal step. Immersed media with carrier particles is an in situ filtration process.

3 Technology Selection Criteria and Available Indigenous Methods: An Appraisal

1. First of all, as standard set by WHO for drinking water, any technology for arsenic removal must have the capacity to remove arsenic below 10 ppb.
2. Point-Of-Use (POU) Treatment; the compliance technology or devices for POU are important for eliminating arsenic from the drinking water. POU treatment has many advantages like low capital and treatment costs in comparison to centralized treatment, making more economically feasible. However, in many countries including USA, Safe Drinking Water Act needs that "devices be owned, controlled, and maintained by the public water utility or by an agency under contract with the water utility" (USEPA 2001a).
3. Again, appropriate cost-effective technology to help small community water systems (<10,000 customers) or individual households in rural areas may be helpful.
4. Further, technology should be eco-friendly, the removal techniques and applicable materials should not produce any toxic materials which in turn may create problem to health of human being and other organisms.
5. Easy handling and operation of the technology is mostly required. Simplicity of the technology helps in propagation to the larger sector including unskilled people.
6. Water quality consideration is another important driver for choosing the technology for arsenic removal. Among predominant forms in natural waters like arsenate [As(V)] and arsenite [As(III)], arsenate can be more efficiently removed either by iron coagulants, adsorptive media or precipitation of natural iron. Source water iron concentration in any regional water is the main component as it likely to play a major role for treatability of a given water source. Strong oxidant or increasing levels of iron concentration leads to conversion of As(III) to As(V) leading to improvement of arsenic removal. Since, a strong affinity is present on the iron surfaces to adsorb arsenic; it is therefore considered that iron-based treatment technologies (like adsorptive media with iron-based products) are mainly effectual at eliminating arsenic from aqueous systems along with iron. Maintenance of pH is another important criterion to augment the arsenic removal capacity. pH 7.0 (with a range of 7–8) and pH of 5.5 are the acceptable for iron-based and alumina-based adsorption media respectively. Iron oxides based coagulation/filtration processes require a pH range of 5.5–8. However,

Table 3 As per the EPA guidelines Fe/As ratio in the regional source water should drive the technology selection

Fe/As levels	Applicable technologies
High Fe levels (>0.3 mg L^{-1}). High Fe:As ratio (>20:1)	Adsorption and co-precipitation (arsenic removal by iron removal)
Moderate Fe levels (>0.3 mg L^{-1}) Low Fe:As ratio (<20:1)	Coagulation/filtration with the addition of iron salts.
Low Fe levels (<0.3 mg L^{-1})	Adsorptive media, coagulation/filtration, and ion exchange.

changes in pH do not have any significant effect on ion exchange process. As per the EPA guidelines Fe–As ratio in the regional source water should drive the technology selection (Table 3).

7. Treatment process and residuals management is important for choosing technology on arsenic removal (Kramer et al. 2009). There are many technical concerns involved with arsenic treatment technologies. Under definite situation, economic viability of a technology may be assessed; however, consideration on residual management and disposal is another important aspect. With limited discharge options, selecting an arsenic removal technology should consider handling and disposal methods in priority. High arsenic contaminated liquid media disposal is hazardous (need on-site treatment before releasing into environment), while most solid wastes (spent media) can be disposed of in the landfills.
8. Monitoring and examining of implementable technology is another vital aspect. Though treatment options and adsorptive media testing is an expensive, time-intensive process and require recommended laboratory or devices.

4 Vulnerability and Scope of Arsenic Remediation Through Available Technologies in Countries Like Bangladesh and India

Supply of arsenic-free potable water after removal of arsenic using suitable techniques appears to be viable. Problems such as sludge disposal, operation and maintenance, etc. associated with the system must be taken care for (Ahmed 2003; Shanmugapriya et al. 2015; Singh et al. 2015). Based on above mentioned working principles, a number of arsenic removal devices have been developed, and many of them have already been extended to fields. However, satisfactory performance or user's satisfaction in relation to operation and maintenance, and sludge disposal like criteria led to narrowing down the number of products available in the market. Iron co-precipitation based technologies have been reported to be the most appropriate and effective, with minimal operation and maintenance problems (Joshi and Sahu 2014). A number of opinions subsist regarding groundwater arsenic contamination in the Ganga–Brahmaputra river basins of geological origin Indian subcontinent (included Bangladesh and Nepal region). It may be geological origin and percolation of different chemicals and fertilizer residues may have played a modifying role in its further intensification. Identification of parental rocks or outcrops is yet to be recognized, including their sources, routes, transport, speciation, and occurrence in Holocene aquifers along fluvial tracks of the Ganga–Brahmaputra valley and in scattered places, adjoining to it, in their basins.

The speculation of sources in the Gangetic plains ranged from the sulfide belts of Bihar, Uttar Pradesh, and North Bengal to the coal seams of the neighboring Gondwana Basins, the basic rocks of the Rajmahal Traps, the metamorphic schists of the Lesser Himalaya (Information downloaded from: http://www.cgwb.gov.in/documents/papers/incidpapers/Paper%208%20-%20Ghosh.pdf; http://www.who.int/bulletin/volumes/90/11/11-101253/en/). The question of the possible role of excessive withdrawal of groundwater for its triggering, however, has continued to have divided opinions. The chemical processes such as redox potential, sorption, precipitation–dissolution, pH, influence of other competing ions, and biological transformation under different soil–water environmental conditions; influence the perturbation of arsenic in a system, having presence of source material and/or conditions of enrichment (Ravenscroft et al. 2009; Ly 2012). Whether the processes of physicochemical transformation were only influenced by excessive groundwater exploitation or there were other coupled actions of a number of hydrogeological and geo-environmental disturbances, over the periods, are yet to be recognized. Surfacing new arsenic affected areas, in every additional survey, is a matter of concern. It was reported that the contaminated waters are enriched in Fe, Mn, Ca, Mg, bicarbonates, and depleted in sulfate, fluoride, chloride; pH ranged from 6.5 to 8; redox condition usually in reducing; high on organic matter content; lodged mostly in sand coatings, or sorbed on clays, HFOs, and organic matters; As concentration diminishing down-depth, which brings out a generalized geochemical perception that could help develop in situ remediation of arsenic. It has been proved that arsenic has affinity with iron in groundwater both positively and negatively, depending upon the condition. There is wide-scale report of the presence of dissolved iron, in arsenic contaminated groundwater, and of co-precipitation of iron and arsenic under oxidizing condition. The relationship between As and Fe can be interpreted as a significant one that in these instances iron played the scavenger role, adsorbing arsenic from water as it precipitated out, again desorbing arsenic into water as it redissolved in response to appropriate change of Eh-pH conditions. This gives a positive hope of a plausible way of in situ remediation of the problem of As contamination by removal of Fe from groundwater before withdrawal.

5 Widely Accepted Small Scale Arsenic Removal Units

5.1 Biosand Filter (BSF)

BSF is an indigenous, traditional, simple slow sand water filtration device, particularly effective for discontinuous use. The filters are in use for water treatment at community or household level for hundreds of years. BSF consists of a container filled with sand and gravel, where physical straining of flow water helps in removal of turbidity, iron, manganese and pathogens from drinking water. The modified engineering design for the BSFs as slow intermittent filter was developed in the 1990s at University of Calgary (Canada) by Dr. David Manz. Most widely used version of BSF is usually made up of concrete (95 cm height X 36 cm in width) with a cost of establishment ranging approximately between 20 and 40 USD and with the flow rate of 20–40 L/h. The amendments of the BSF include a provision of 5 cm shallow layer of water sits atop the sand and that forms biofilm (Schmutzdecke) with sufficient oxygen to allow. This biofilm helps in reducing the pathogenic load of the water. Further, user friendly cleaning technique of sand bed, negligible operating costs, requirement of no consumables, etc. help the BSF an affordable option for mass application (Baker and Duke 2006; http://www.sswm.info/content/biosand-filter; Bera 2013).

5.2 *Magc-Alcan Filter (Two Bucket Filter)*

Activated alumina media is filled in the series of two buckets that removes arsenic by adsorption. The development of media by thermal dehydration (at 250–1150 °C) of an aluminum hydroxide has been done by MAGC Technologies and Alcan (USA). Maintaining pH level is essential for continuing better performance of arsenic removal (http://akvopedia.org/wiki/Arsenic_filter).

5.3 *Nirmal Filter*

On the same adsorption based principle using activated alumina as media, Indian Nirmal filters less costly than Magc-Alcan filter. It performs by using activated alumina and followed by filtration through a ceramic candle. With 80–90% arsenic removal efficiency, the filter needs to be regenerated in every 6 months (http://akvopedia.org/wiki/Arsenic_filter).

5.4 *Shapla and Surokka Filter*

International Development Enterprises (IDE), Bangladesh has developed adsorption based earthen Shapla household arsenic removal technology. Ferrous sulfate coated brick chips, iron coated sand is used for the adhesion of arsenic molecule from the water as the water passes through the filter. As per the reports, 20 kg media can filter 4000 L of arsenic-contaminated water reducing arsenic concentrations to undetectable levels and average supplying 25–32 L of drinking water per day (http://akvopedia.org/wiki/Arsenic_filter, www.jalmandir.com/arsenic/shapla/

shapla-arsenic-filter.html). The household Surokka arsenic filter is an economic variant made out of iron containing yellow sands, a local component that adsorbs arsenic.

5.5 Bucket Treatment Unit (BTU)

Coagulation–flocculation based bucket treatment unit is another arsenic removal device. It is based upon oxidation (forming As(V) from As(III) and coagulation processes (using potassium permanganate and aluminum sulfate respectively), subsequent filtration using cloth and sand filtration bed (http://archive.unu.edu/env/Arsenic/Tahura.pdf; http://www.sswm.info/content/arsenic-removal-technologies).

5.6 3-Kolshi Filter

Widely accepted household arsenic removal unit being implemented in Bangladesh is the three-kolshi filter unit (Ahmed 2001; Munir et al. 2001). The filter unit is based on a traditional water purification system consist of three clay pitchers, or kolshi, stacked vertically in a frame. The top kolshi contains a layer of Fe filings and a layer of coarse sand, the middle kolshi contains a layer of charcoal and a layer of fine sand, and the bottom kolshi collects the filtered water. The operation procedure is very simple, and no chemicals are required. The total cost of the system is 250–300 in Taka (US \$5–6), of which 50% is the cost of the stand (Ahmed 2001). But this system has some disadvantages, where removal capacity of As decreases rapidly after 3–4 months. Coagulation/filtration is a conventional treatment process that adds a chemical coagulant (typically iron sulfate or iron chloride) to contaminated water. The physical or chemical properties of dissolved or suspended contaminants are modified by coagulants, which then settle from solution and can be removed by filtration. As part of the coagulation process, arsenic is co-precipitated with the iron. The stirring process helps to build the flocs into larger particles.

5.7 SONO Filter

It is a three-bucket system with a composite iron matrix (CIM) as the active arsenic elimination constituent (Fig. 1). It works on the principle of surface complexation of arsenic on the CIM followed by a filtration, and does not require any chemical treatment. Further, the production of toxic wastes is negligible. The cost of the filter is about \$40, and it can produce 20–30 $L\ h^{-1}$ for daily drinking and cooking needs. SONO filters have been well deployed all over Bangladesh and continue to provide more than a billion liters of safe drinking water (Munir et al. 2001).

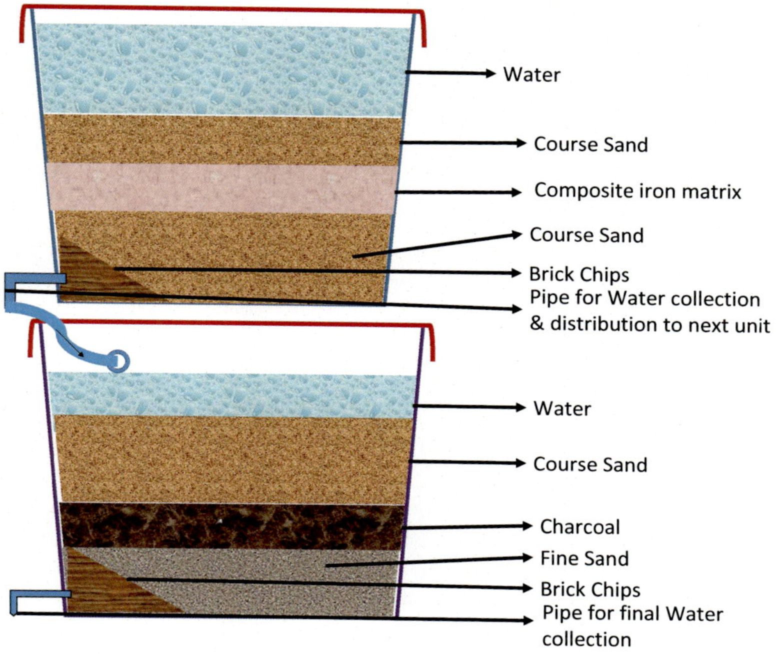

Fig. 1 Diagrammatic details of different components of SONO filter(Please change this figure with attached figure)

5.8 *Kanchan Arsenic Filter (KAF)*

The Kanchan Arsenic Filter (KAF) is a variation of the bio sand filter. It has been developed as the Kanchan™ filter (Fig. 2) under the Nepal Water Project in collaboration with Massachusetts Institute of Technology (MIT, USA) and Nepal-based nongovernmental Environment and Public Health Organization (ENPHO). The filter can remove Fe, As, turbidity, and odor and is useful for microbiological water treatment without using any chemicals (http://ceeserver3.mit.edu/~water; Bera 2013; http://resources.cawst.org/asset/kanchan-arsenic-filter-fact-sheet-simplified_en). The KAF is composed of two removal units: the As-removal unit and the pathogen removal unit. The As-removal unit consists of a plastic diffuser basin, iron nails, and some brick chips. The pathogen removal unit consists of sand and gravel layers. It works on a simple principle: when the nails are exposed to air and water, they rust quickly, producing ferric hydroxide (Fe rust) particles, which are an excellent adsorbent for As. When As-contaminated water is poured into the filter, the toxic element is rapidly adsorbed onto the surface of the ferric hydroxide particles. The As loaded Fe particles are then flushed into the sand layer below. The layer of fine sand traps the As-loaded Fe particles in the top few centimeters, thus effectively removing As. The system works at optimal pH (6.5–8.5) and water flow rate

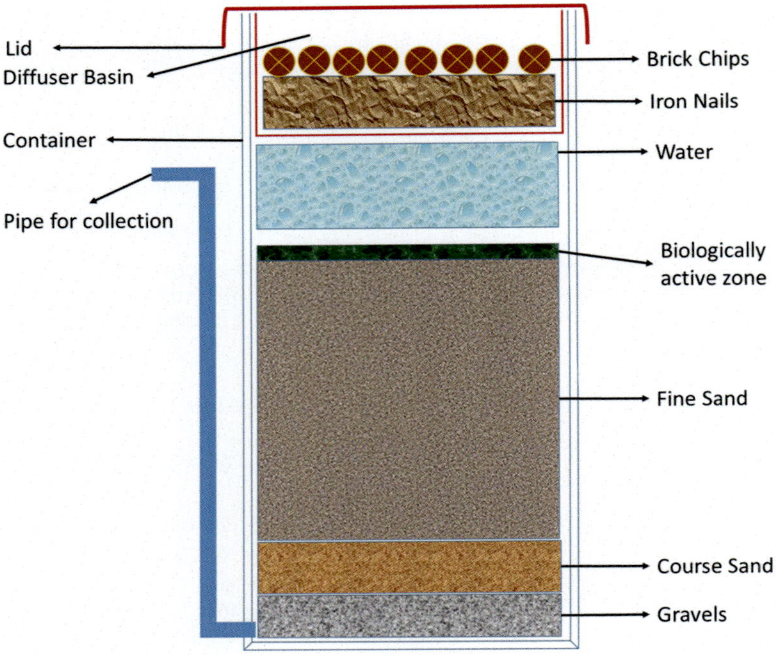

Fig. 2 Diagrammatic details of different components of Kanchan arsenic filter (Adopted from: http://web.mit.edu/watsan/img_nepal_worldbank3.html; http://chinawaterrisk.org/resources/analysis-reviews/kafbio-sand-filter-arsenic-be-gone/)

between 10 and 30 L h^{-1}. It can remove both As (III) and As (V) without any chemical treatment. Field research by MIT and ENPHO showed that As removal using the filter was in the range 85–95%. Independent field studies of the Kanchan™ As filter in Nepal by the Tribhuvan University, Kathmandu University and United States Peace Corp also confirm the rate of As removal.

5.9 *Subterranean Arsenic Removal Technology (SAR)*

It is an interesting approach to treat the contaminated aquifers itself to reduce the arsenic, (along with iron and manganese) load before distribution (Fig. 3). This in situ treatment neither requires any chemical, nor produces any toxic wastes, thereby preserving normal permeability of the aquifer. Prof. B Sengupta and his team of Belfast University conceptualized the plan and implemented the same in some places of West Bengal, India (http://www.insituarsenic.org/origin.html). The technology is based on recharging a part of the groundwater with aeration and subsequent adsorption-based phenomenon, where underground aquifer is turned into a

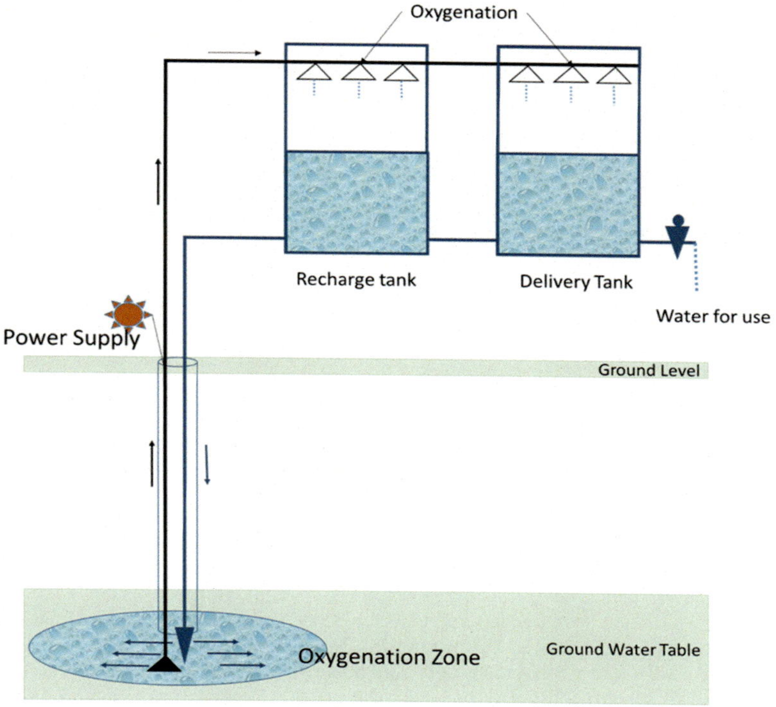

Fig. 3 Diagrammatic representation of subterranean arsenic removal technology (SAR) (Adopted from: http://www.insituarsenic.org/details.html)

natural biochemical reactor (http://www.insituarsenic.org/innovation.html). Autocatalytic effect of the oxidation products and chemoautotrophic microorganisms (bioremediation) further accelerate the oxidation processes. Oxygen incorporation causes an increment of redox potential of water. The oxidation zone (zone where the air is being mixed) of water of aquifer becomes the spot of diverse physical, chemical, and biological processes (http://www.insituarsenic.org/details.html). This method is cost-effective and ecofriendly.

6 Conclusions

Rapid growth, industrialization, and indiscriminate waste disposal are contributing to generating high levels of water and environmental pollution. Arsenic is one of the very important elements responsible for severe pollution to water, soil, foodstuffs, and all, resulting in severe health effects. As is the Group A human carcinogen. Ingestion of arsenic contaminated drinking water is affecting 137 million people of the world. For removal of this arsenic several physicochemical techniques are common. Among them, lime softening and iron co-precipitation have been testified as the maximum effective removal technologies. The systems based on these technologies are running reasonably well. A couple of simplified systems are widely used in Bangladesh and West Bengal. However, still there are operation and maintenance problems. Further, introduction of nano-based research and related systems for arsenic removal, however, is creating new hope and fresh challenges as well. The existing arsenic removal technologies, along with newer avenues, require further improvement to make them appropriate, user friendly, and sustainable for their large-scale effective uses.

References

Acharyya SK, Lahiri S, Raymahashay BC, Bhowmik A (2000) Arsenic toxicity of groundwater in parts of the Bengal basin in India and Bangladesh: the role of quaternary stratigraphy and Holocene Sea-level fluctuation. Environ Geol 39:1127–1137

Ahmed MF (2001) An overview of arsenic removal technologies in Bangladesh and India. http://archive.unu.edu/env/Arsenic/Ahmed.pdf

Ahmed MF (2003) Arsenic contamination: Regional and global scenario. In: Arsenic contamination: Bangladesh perspective, TN-Bangladesh, BUET, Dhaka, Bangladesh

AWWA-American Water Works Association (1999) Water quality and treatment: a handbook of community water supply, 5th edn. McGraw-Hill, New York

Baker DL, Duke WF (2006) Intermittent slow sand filters for household use–A field study in Haiti. IWA Publishing, London. (Downloaded from: https://wiki.umn.edu/pub/EWB/Uganda_Groundwater_Supply/Slow_Sand_Filter_Study.pdf)

Bazrafshan E, Mohammadi L, Ansari-Moghaddam A, Mahvi AH (2015) Heavy metals removal from aqueous environments by electrocoagulation process—a systematic review. J Environ Health Sci Eng 13:74

Bera S (2013) Humble filter makes a comeback. Down to Earth magazine. http://www.downtoearth.org.in/coverage/humble-filter-makes-a-comeback-40066

Bhattacharya P, Chatterjee D, Jacks G (1997) Occurrence of As contaminated groundwater in alluvial aquifers from the Delta Plains, eastern India: option for safe drinking water supply. Int J Water Resour Dev 13:79–92

Bhattacharya P, Jacks G, Ahmed KM, Khan AA, Routh J (2002) Arsenic in groundwater of the Bengal Delta plain aquifers in Bangladesh. Bull Environ Contam Toxicol 69:538–545

Bhattacharyya R, Jana J, Nath B, Sahu SJ, Chatterjee D, Jacks G (2005) Groundwater As mobilization in the Bengal Delta plain. The use of ferralite as a possible remedial measure: a case study. Appl Geochem 18:1435–1451

Boddu VM, Abburi K, Talbott JL, Smith ED, Haasch R (2008) Removal of arsenic (III) and arsenic (V) from aqueous medium using chitosan-coated biosorbent. Water Res 42:633–642

CAWST-Center for Affordable Water and Sanitation Technology (2009) Household water treatment for arsenic removal fact sheet. (Data compiled from: http://www.cawst.org/files/governance/Annual%20Reports/2009_CAWST_Annual_Report.pdf; http://www.sswm.info/sites/default/files/reference_attachments/CAWST%202009%20Household%20Water%20Treatment%20and%20Safe%20Storage%20Fact%20Sheet%20Academic.pdf)

Chang SD, Ruiz H, Bellamy WD, Spangenberg CW, Clark DL (1994) Removal of arsenic by enhanced coagulation and membrane technology. In: Proceedings of the national conference on environmental engineering, ASCE, New York

Chen H, Frey M, Clifford D, McNeill L, Edwards M (1999) Arsenic treatment considerations. J Am Water Works Assoc 91:74–85

Chetia M, Chatterjee S, Banerjee S, Nath MJ, Singh L, Srivastava RB, Sarma HP (2011) Groundwater arsenic contamination in Brahmaputra river basin: a water quality assessment in Golaghat (Assam), India. Environ Monit Assess 173:371–385

Chetia M, Goswamee RL, Banerjee S, Chatterjee S, Singh L, Srivastava RB, Sarma HP (2012) Arsenic removal from water using calcined Mg–Al layered double hydroxide. Clean Technol Environ Policy 14:21–27

Chutia P, Kato S, Kojima T, Satokawa S (2009) Arsenic adsorption from aqueous solution on synthetic zeolites. J Hazard Mater 162:440–447

Chwirka J, Thomson B, Stomp JMI (2000) Removing arsenic from groundwater. J Amer Water Works Assoc 92:79–88

Clifford D (1999) Ion exchange and inorganic adsorption. In: Letterman RD (ed) Water quality and treatment, 5th edn. McGraw-Hill, New York

Clifford D, Lin CC (1991) Arsenic III and arsenic V removal from drinking water in San Ysidro, New Mexico. EPA/600/2-91/011, USEPA, Cincinnati, OH

Das B, Rahman MM, Nayak B, Pal A, Chowdhury UK, Mukherjee SC, Saha KC, Pati S, Quamruzzaman Q, Chakraborti D (2009) Groundwater arsenic contamination, its health effects and approach for mitigation in West Bengal, India and Bangladesh. Water Qual Expo Health 1:5–21

Drobna Z, Styblo M, Thomas DJ (2009) An overview of arsenic metabolism and toxicity. Curr Protoc Toxicol 42:4311–4316

Edwards M (1994) Chemistry of arsenic removal during coagulation and Fe-Mn oxidation. J Amer Water Works Assoc 86:64–78

Fields K, Chen A, Wang L (2000a) Arsenic removal from drinking water by coagulation/filtration and lime softening plants. EPA/600/R-00/063, USEPA, Cincinnati, OH

Fields K, Chen A, Wang L (2000b) Arsenic removal from drinking water by iron removal plants. EPA/600/R-00/086, USEPA, Cincinnati, OH

Formoso P, Muzzalupo R, Tavano L, Filpo GD, Nicoletta FP (2016) Nanotechnology for the environment and medicine. Mini Rev Med Chem 16:668–675

Ghanizadeh G, Ehrampoush MH, Ghaneian MT (2010) Application of iron impregnated activated carbon for removal of arsenic from water. Iran J Environ Health Sci Eng 7:145–156

Guha Mazumder DN, Steinmaus C, Bhattacharya P, von Ehrenstein OS, Ghosh N, Gotway M, Sil A, Balmes JR, Haque R, Hira-Smith MM, Smith AH (2005) Bronchiectasis in persons with skin lesions resulting from arsenic in drinking water. Epidemiology 16:760–765

Hoque MA, Burgess WG, Shamsudduha M, Ahmed KM (2011) Delineating low-arsenic groundwater environments in the Bengal aquifer system, Bangladesh. Appl Geochem 26:614–623

Ibrahim RK, Hayyan M, AlSaadi MA, Hayyan A, Ibrahim S (2016) Environmental application of nanotechnology: air, soil, and water. Environ Sci Pollut Res Int 23:13754–13788.

Jiang JQ, Ashekuzzaman SM, Jiang A, Sharifuzzaman SM, Chowdhury SR (2013) Arsenic contaminated groundwater and its treatment options in Bangladesh. Int J Environ Res Public Health 10:18–46

Joshi J, Sahu O (2014) Protection of human health by low cost treatment in rural area. J Biomed Eng Technol 2:5–9

Kramer T, Loeppert R, Wee H (2009) Assessment of arsenic treatment residuals: analysis and stabilization techniques. IWA Publishing, London

Kumar A, Namdeo M, Mehta R, Agrawala V (2015) Effect of arsenic contamination in potable water and its removal techniques. Int J Water Wastewater Treat 1:2

Lien HL, Wilkin RT (2005) High-level arsenite removal from groundwater by zero-valent iron. Chemosphere 59:377–386

Ly TM (2012) Arsenic contamination in groundwater in Vietnam: An overview and analysis of the historical, cultural, economic, and political parameters in the success of various mitigation options. Pomona Senior Theses. Paper 41 http://scholarship.claremont.edu/pomona_theses/41

McNeill LS, Edwards M (1997) Predicting As removal during metal hydroxide precipitation. J Amer Water Works Assoc 89:75–86

Medpelli D, Sandoval R, Sherrill L, Hristovski K, Seo DK (2015) Iron oxide-modified nonporous geopolymers for arsenic removal from ground water. Resour Efficient Technol 1:19–27
Mukherjee A, Sengupta MK, Hossain MA, Ahamed S, Das B, Nayak B, Lodh D, Rahman MM, Chakraborti D (2006) Arsenic contamination in groundwater: a global perspective with emphasis on the Asian scenario. J Health Popul Nutr 24:142–163
Munir AKM, Rasul SB, Habibuddowla M, Alauddin M, Hussam A, Khan AH (2001) Evaluation of performance of Sono 3-Kolshi filter for arsenic removal from groundwater using zero valent iron through laboratory and field studies. http://archive.unu.edu/env/Arsenic/Munir.pdf
Narasimhan R, Thomson B, Chwirka J, Lowry J (2005) Chemistry and treatment of arsenic in drinking water. American Water Works Association Research Foundation, Denver
Naujokas MF, Anderson B, Ahsan H, Aposhian HV, Graziano JH, Thompson C, Suk WA (2013) The broad scope of health effects from chronic arsenic exposure: update on a worldwide public health problem. Environ Health Perspect 121:295–302
NRC-National Research Council (1999) Arsenic in drinking water. National Academy Press, Washington, DC
NRC-National Research Council (2001) Arsenic in drinking water: update. National Academy Press, Washington, DC
Rahman MM, Naidu R, Bhattacharya P (2009) Arsenic contamination in groundwater in the Southeast Asia region. Environ Geochem Health 31:9–21
Ravenscroft P, Brammer H, Richards K (2009) Arsenic pollution: a global synthesis. Wiley-Blackwell, Chichester; Malden
Sen Gupta B, Chatterjee S, Rott U, Kauffman H, Bandopadhyay A, DeGroot W, Nag NK, Carbonell-Barrachina AA, Mukherjee S (2009) A simple chemical free arsenic removal method for community water supply—a case study from West Bengal, India. Environ Pollut 157:3351–3353
Shanmugapriya SP, Rohan J, Alagiyameenal D (2015) Arsenic pollution in India–an overview. J Chem Pharma Res 7:174–177
Singh R, Singh S, Parihar P, Singh VP, Mohan PS (2015) Arsenic contamination, consequences and remediation techniques: a review. Ecotoxicol Environ Saf 112:247–270
Smedley PL, Kinniburgh DG (2002) A review of the source, behaviour and distribution of arsenic in natural waters. Appl Geochem 17:517–568
Sorg T (1993) Removal of arsenic from drinking water by conventional treatment methods. Proceedings of the 1993 AWWA WQTC
Sorg T, Logsdon TJ (1978) Treatment technology to meet the interim primary drinking water regulations for inorganics: part 2. J Am Water Works Assoc 70:379–393
Spivey A (2011) Arsenic and infectious disease: a potential factor in morbidity among Bangladeshi children. Environ Health Perspect 119:218
Tuutijärvi T, Lu J, Sillanpää M, Chen G (2009) As(V) adsorption on maghemite nanoparticles. J Hazard Mater 166:1415–1420
US EPA-US Department of Health and Human Services (2005) Public health service agency for toxic substances and disease registry. Division of toxicology and environmental medicine. http://www.baltimorehealth.org/info/ATSDR%20fact%20sheet.pdf
US EPA-US Environmental Protection Agency (2000) Arsenic occurrence in drinking water, Office of Health and Environment Assessment. U.S. Environmental Protection Agency, Washington, DC
US EPA-US Environmental Protection Agency (2001a) Method 200.9. Trace elements in water, solids, and biosolids by stabilization temperature graphite furnace atomic adsorption spectrometry, environmental monitoring systems laboratory, office of research and development. Downloaded from: http://www.cromlab.es/Articulos/Metodos/EPA/200/200_9-bio.pdf. Downloaded on 29.04.2016
US EPA-US Environmental Protection Agency (2001b) National primary drinking water regulations; arsenic and clarifications to compliance and new source contaminants monitoring: Final rule, Federall Register. Downloaded from: https://www.federalregister.gov/arti-

cles/2001/01/22/01-1668/national-primary-drinking-water-regulations-arsenic-and-clarifications-to-compliance-and-new-source. Downloaded on 29.04.2016

Wang GQ, Huang YZ, Gang JM, Wang SZ, Xiao BY, Yao H, Hu Y, Gu YL, Zhang C, Liu KT (2000) Endemic arsenism, fluorosis and arsenic-fluoride poisoning caused by drinking water in Kuitun, Xinjiang. Chin Med J 113:524–524

Wang X, Liu Y, Zheng J (2016) Removal of As(III) and As(V) from water by chitosan and chitosan derivatives: a review. Environ Sci Pollut Res Int 23:13789–13801

WHO-World Health Organization (2001) Arsenic compounds, environmental health criteria 224, 2nd edn. World Health Organization, Geneva

Williams PN, Zhang H, DavisonW MAA, Hossain M, Norton GJ, Islam MR, Brammer H (2011) Organic matter-solid phase interactions are critical for predicting arsenic release and plant uptake in Bangladesh paddy soils. Environ Sci Technol 45:6080–6087

Xiong XZ, Li PJ, Wang YS, Ten H, Wang LP, Song LP (1987) Environmental capacity of arsenic in soil and mathematical model. Chin J Environ Sci 8:8–14

Xu H, Allard B, Grimvall A (1988) Influence of pH and organic-substance on the adsorption of As(V) on geologic materials. Water Air Soil Pollut 40:295–305

Further consultation for the preparation of MS was done form the web information as mentioned below:

1. Treatment technologies for arsenic removal: http://nepis.epa.gov/Exe/ZyPDF.cgi/20017IDW.PDF?Dockey=20017IDW.PDF
2. Arsenic treatment technologies for soil, waste, and water https://clu-in.org/download/remed/542r02004/arsenic_report.pdf; http://www.niehs.nih.gov/news/assets/docs_a_e/arsenic_treatment_technologies_for_soil_waste_and_water_508.pdf
3. Arsenic mitigation strategies (https://www.epa.gov/sites/production/files/2015-09/documents/train5-mitigation.pdf)
4. Treatment options: (http://www.epa.gov/sites/production/files/2015-09/documents/treatment_options_tom_sorg.pdf)
5. Guidelines for Arsenic Removal Treatment for Small Public Drinking Water Systems. http://epa.ohio.gov/portals/28/documents/engineering/ArsenicManual.pdf
6. A simple solution for arsenic problem Kanchan Filter: http://web.mit.edu/watsan/Docs/Other%20Documents/KAF/KAF%20booklet%20final%20Jun05.pdf

Arsenic and Its Effect on Major Crop Plants: Stationary Awareness to Paradigm with Special Reference to Rice Crop

Soumya Chatterjee, Sonika Sharma, and Dharmendra K. Gupta

Contents

1 Arsenic and Associated Problem: Global Scenario 123
2 Rice and Arsenic: Entwined Aspects of Arsenic in Staple Food/Rice as Staple Food 124
3 General Contributing Factors for Arsenic Uptake and Transport 125
4 Bioaccumulation of Arsenic in Crop Plants 126
5 As-Induced Stress and Physiological Response of Plants 127
6 Mechanism of Arsenic Transportation and Removal from Crop Plants 128
7 Rice Crop Model for Arsenic Toxicity 130
8 Arsenic Bioavailability in Rice Rhizosphere 133
9 Arsenic Transport and Speciation in Rice: Role of Silicon 134
10 Arsenic Species Unloading at Rice Grain 135
11 Conclusion 135
References 136

1 Arsenic and Associated Problem: Global Scenario

Certain pollutants remain in the environment for extensive periods and affect our health. The situation becomes more critical in the case of heavy metal pollution. In the present scenario, the world is facing a serious issue of arsenic contamination as high levels of As contaminate groundwater, geothermal spring water, downstream wetlands, and estuary ecosystems (Lièvremont et al. 2009; Liu et al. 2016). It is

S. Chatterjee (✉) • S. Sharma
Defence Research Laboratory, Defence Research and Development Organization (DRDO), Ministry of Defence, Post Bag No. 2, Tezpur 784001, Assam, India
e-mail: drlsoumya@gmail.com

D.K. Gupta
Institut für Radioökologie und Strahlenschutz (IRS), Gottfried Wilhelm Leibniz Universität Hannover, Herrenhäuser Str. 2, Hannover 30419, Germany

D.K. Gupta, S. Chatterjee (eds.), *Arsenic Contamination in the Environment*,
DOI 10.1007/978-3-319-54356-7_6

ubiquitous metalloid (atomic number 33) with different allotropes (Mohan and Pittman 2007; Tangahu et al. 2011). Being a reactive metal, it can mix with other reactive metals to form organic and inorganic arsenic compounds. In organic form (organoarsenics) As is less toxic, found predominantly in fish and shellfish like prawns, and also added in pesticides to protect cotton plants from damage (http://www.baltimorehealth.org/info/ATSDR%20fact%20sheet.pdf.U.S). However, arsenite, the inorganic form is toxic to living organisms due to greater solubility and mobility than organic forms of arsenic, as it can readily be absorbed by gastrointestinal track (National Ground Water Association 2001; Chutia et al. 2009; Ampiah-Bonney et al. 2007; Vaclavikova et al. 2008). Commonly, arsenic exists in its various oxidation forms like trivalent As trioxide, As trichloride, As trisulfide, sodium arsenite, and pentavalent form including As pentafluoride. Organic form includes arsenobetaine, arsenocholine, tetramethylarsonium salts, arsenosugars, and As containing lipids. Among these states, the pentavalent form (arsenates) contaminates underground water (Ampiah-Bonney et al. 2007; Vaclavikova et al. 2008; Kumar et al. 2015).

Agriculture and crop production needs water. However, the crop water need mainly depends on: the climate and weather (sunshine, temperature, humidity, wind speed) during the production, crop type (some crops require more water like rice, maize or sugarcane), and growth stage of the crop (http://www.fao.org/docrep/s2022e/s2022e07.htm). For example, rice requires about 1200–2500 mm of water per unit delta of production with variability in requirement at its different growth phase (http://nptel.ac.in/courses/Webcourse-contents/IIT%20Kharagpur/Water%20Resource%20Engg/pdf/m3l03.pdf). And mostly, groundwater is the source of irrigation water. However, groundwater contamination with arsenic has drawn the attention of the world. Use of arsenic-contaminated groundwater for various purposes including crop production led to an exposure of arsenic poisoning in more than 100 million people (Rahman et al. 2009a, b) and according to World Bank policy report 2005 (http://siteresources.worldbank.org/INTANNREP2K5/Resources/51563_English.pdf). Southeast Asian countries are mostly affected by As-related diseases. This worldwide problem is extensively found in different countries like Bangladesh, India, China, Taiwan, Vietnam, the USA, Argentina, Chile, and Mexico, Nepal, Myanmar, Pakistan, Lao People's Democratic Republic, Cambodia, and has recently been reported in lowlands of Sumatra in Indonesia etc. (Mandal and Suzuki 2002; Chakraborti et al. 2004; Bhattacharya et al. 2012; Mukherjee et al. 2006; Winkel et al. 2008; Rahman MA et al. 2009; Rahman MM et al. 2009; Mahanta et al. 2015).

2 Rice and Arsenic: Entwined Aspects of Arsenic in Staple Food/Rice as Staple Food

Rice is the staple food for more than half of the global population. Rice is most popular especially in the Southeast Asian countries. As per the available data (FAOSTAT data 2014—http://faostat3.fao.org/; https://top5ofanything.com/list/47d811ad/

Rice-Producing-Countries; http://www.countryranker.com/top-10-largest-producers-of-rice-in-the-world/ downloaded on 21 April 2016), China is the world's largest rice producer (204.3 million metric tonnes (mmt), 28.4% of world production[W.P.]) followed by India (152.6 mmt, 21.2% of W.P.), Indonesia (69.0 mmt, 9.6% of W.P.), Vietnam (43.7 mmt, 6% of W.P.), Thailand (37.8 mmt, 5.2% of W.P.). Among cereal crops, rice (*Oryza sativa*) is well known for its efficiency in arsenic (As) uptake and accumulation, which occur through the soil–water–rice pathway (Wang et al. 2015; Li et al. 2016). Ironically, these regions of the world is worst contaminated with geochemical occurrence of As in groundwater. Zhao et al. (2010a) reported that, in Bangladesh, approximately 1000 tonnes of As has been introduced into the rice cultivated land through irrigation with As-contaminated groundwater of shallow underground aquifers (Zhao et al. 2010a). On the other hand, anthropogenic activities and improper arsenic rich waste dumping are causing serious concern in the Hunan province (China) over the production of rice (as grains are reported to contain up to 723 ng g^{-1} with maximum contaminant level: 150 ng g^{-1} in China) (Zhu et al. 2008a, b; Okkenhaug et al. 2012; Wang et al. 2015). In India, states like West Bengal, Bihar, Uttar Pradesh, Jharkhand, Assam, Chhattisgarh, and Manipur have been detected with maximum occurrence of high As concentrations in groundwater (Mukherjee et al. 2006). The study reveals arsenic accumulation in major crops like rice, pulses, and vegetables due to arsenic-contaminated irrigated water (0.318–0.643 mg L^{-1}) and in soil (5.70–9.71 mg kg^{-1}). The fractionation of As varies in rice grain in the following order: rice straw > bran > whole grain > polished rice > husk (Carey et al. 2012).

With rice being the staple food, incipient data has suggested that it is an important dietary source of As exposure and may augment to adverse health outcomes. A study on 18,470 persons of Bangladesh has shown a substantial relationship of rice intake with both urinary As and prevalent skin lesions (Melkonian et al. 2013). A similar report from USA showed a positive association between rice consumption and urinary arsenic (Gilbert-Diamond et al. 2011). The processes of bioavailability, root uptake, rhizosphere, transport, accumulation, and grain unloading of arsenic are imperative aspects of research to alleviate arsenic in rice (Ma et al. 2008; Meharg et al. 2008; Zhao et al. 2013; Wang et al. 2015).

3 General Contributing Factors for Arsenic Uptake and Transport

Since the last century, anthropogenic activities like industrialization, mining operations, use of different As-based pesticides, and irrigation with As-contaminated groundwater have resulted in pollution globally including soil pollution (Adriano 2001; Meharg et al. 2009). In soil, a dynamic equilibrium exists in between different metals in various chemical forms and this is govern by other contributing factor like soil physical, chemical, biological properties and metal– soil particle affinity (Jiang and Singh 1994; Delgado and Go'mez 2016) (Fig. 1). Soil acts as natural

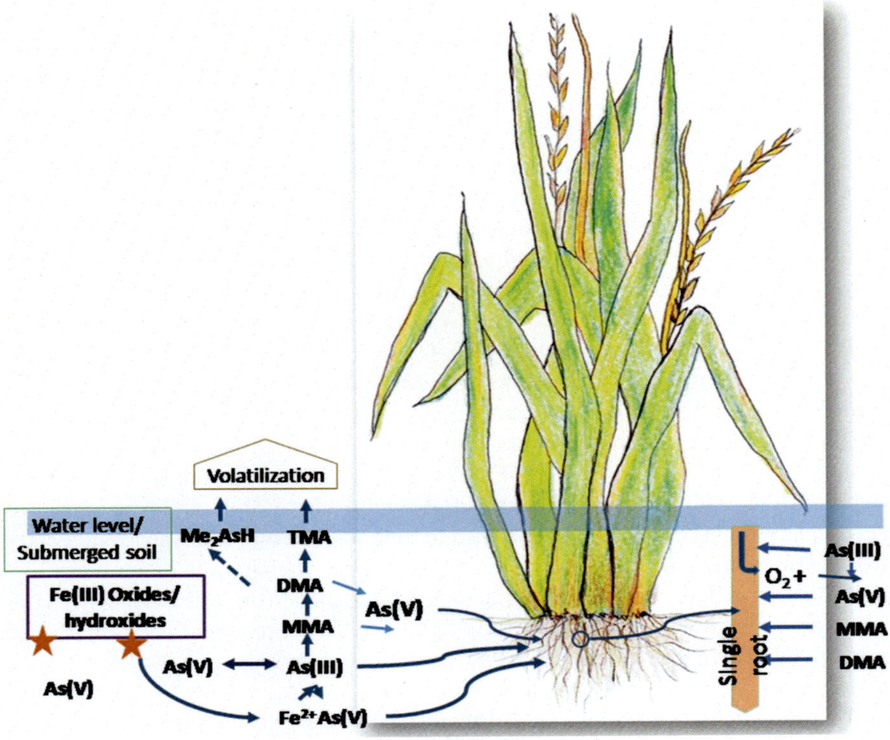

Fig. 1 Schematic represesntation of arsenic in agriculture soil and its speciation (Adopted from Zhao et al. 2010a, b)

buffer and is also responsible for the exchange of chemical elements and substances within the biosphere. The lithology of parent rock decides the concentration of As in soil, like igneous rock and its subtype contains As. The causative root of soil contamination is irrigation with underground water carrying high amounts of arsenic. A small percent of soluble form of metal is readily available for plant uptake and transported through roots. Microclimate present in root rhizosphere, i.e., association of microbes with root and root exudates, also contributes to absorption of metal ions from soil (Wenzel et al. 2009). In *Populus nigra*, arsenic accumulation was found at a concentration of 0.2 mg As g^{-1} dry weight of plant root (Tangahu et al. 2011).

4 Bioaccumulation of Arsenic in Crop Plants

Crops that grow under submerged flooded conditions are vulnerable to toxicity especially in case of arsenic, due to its high bioavailability in soil. In plants, root is the primary part which get exposed to As which in turn effects the root elongation and proliferation. For example, in Guandu wetland of Taiwan, As concentration was determined in aqueous, solid, and plant phases and reported that *Kandelia obovata* (23.69 mg kg^{-1}) can accumulate highest As than surrounding water (0.0018 mg L^{-1}) followed by soils (17.24 mg kg^{-1}). Also with in plants parts of *Kandelia obovata* was decreased from the roots (19.74 mg kg^{-1}) to the stems (1.76 mg kg^{-1}), leaves (1.71 mg kg^{-1}), and seedlings (0.48 mg kg^{-1}) (Liu CW et al. 2014; Liu WJ et al. 2014). Similarly, in *O. sativa* (Aman rice, Ratna variety) and *C. capsularis* (jute) order of bioaccumulation of arsenic was found in decreasing order root > basal stem > median stem > apical stem > leaves > grains (Bhattacharya et al. 2014). It has also been studied that in rice plants, accumulation of arsenic is highest in the root zones and reduce notably in the upper parts of the plant (Liu ZJ et al. 2004). About 40% of the total translocated As is in As^V form determined by As speciation in various species (Zhao et al. 2009; Rascio and Navari-Izzo 2011). Plants also reported to contain methylated As. A study by Raab et al. (2005) showed the absorption of 14 forms of As in the roots of *Helianths annuus* grown in As-rich soil. In root of submerged plants, oxidation occurs either through aerial roots or bacterial community present in rhizosphere (Rahman and Hassler 2014). In case of rice plants, downward movement or supply of oxygen from leaves to the roots also results in aeration of rhizosphere. Arsenic uptake in such conditions depends on the oxidation of As-contained FeS_2 in the aerial roots and/or adsorption of As on root surface. Fe-plaque (iron oxide) formation was reported around the roots of *Kandelia obovata*. The role of Fe-plaque (iron oxide) at molecular level and its relation with arsenic is still unknown (Liu WJ et al. 2004; Chen et al. 2005); however, it has been reported that arsenic has high affinity for iron plaques and can restrict the translocation to the remaining part of plant especially in edible parts like straw, husk, and grain (Garnier et al. 2010; Norra et al. 2006). This phenomenon was further confirmed in the roots of *Kandelia obovata*, where the iron plaques around the root can prevent the uptake of As and protect (Meharg 2004; Fitz and Wenzel 2002; Fitz et al. 2003; Liu WJ et al. 2004; Mirza et al. 2014). Nevertheless, some microbial community residents of rhizosphere in rice plants can increase the bioavailability of arsenic by solubilizing ferric iron in the rhizosphere by exuding siderophores to the root–plaque interface (Kraemer 2004; Bhattacharya et al. 2012), and due to this the accumulated arsenic concentration can reach up to 160 mg kg^{-1} (Bhattacharya et al. 2007).

5 As-Induced Stress and Physiological Response of Plants

In plants exposure to heavy metal stress generally provokes generation of free radical species or reactive oxygen species (ROS) and to counterbalance it antioxidant defence system activates. Recently, role of As in altering sensing and signalling has been studied extensively (Verbruggen et al. 2009) along with contaminated water and agricultural soil, arsenic enters in to food chain and with time it gets accumulates inside human body and acts as a slow poison. Plants face an extra burden due to arsenic accumulation; roots absorb arsenic from soil and form metalloid that hamper the plant growth by inhibiting root extension and proliferation, and it also effects biomass accumulation and results in loss of reproductive capacity, hence reducing yield and fruit production and in turn reducing the economic growth of a country (Garg and Singla 2011). From both biological and toxicological points of view, As accumulation has importance, as it can attain 0.7 mg As g^{-1} dry weight of plants or rise more than this limit depending upon the plant species, pH, redox conditions, surrounding mineral composition, and microbial activities influence the oxidation state of arsenic and hence decide the extent of its absorption. For example, decrease in the pH enhances the accessibility of water soluble As that can be taken up by the roots (Fitz and Wenzel 2002; Gonzaga et al. 2008). Arsenic accumulation is phytotoxic and express some physiological changes in plants like reduced root elongation, loss of root branching, chlorosis in leaves, and shrinking or necrosis in aerial plant parts (Carbonell-Barrachina et al. 1998). Impaired growth of plant species was found during hydroponic experiment with arsenate and arsenite, reported in *Holcus lanatus*, *Lupinus albus*, and *Triticum aestivum* (Hartley-Whitaker et al. 2001; Geng et al. 2006). Biochemical changes in response to arsenic concentration results in decrement in chlorophyll (chl) and carotenoid concentrations along with some antioxidant like polyamines (PAs) in plants (Mascher et al. 2002). Also, by consuming these edible parts, health faces an inevitable exposure to arsenic, with resultant accumulation and toxicity. In leguminous crops like soybean, it inhibits nitrogen fixation by reducing the nodule number (Vázquez et al. 2008). At higher concentrations As toxicity leads to impaired metabolic processes, ultimately results in plant death. Unbalancing of ions due to As assimilation in cells causes leakages of cellular membranes (Singh et al. 2006). Moreover, in response to oxidative stress due to high arsenic concentration, membrane damage caused by lipid peroxidation forms a by-product malondialdehyde. Stoeva and Bineva (2003) reported that transpiration rate reduces on arsenic exposure in plants.

6 Mechanism of Arsenic Transportation and Removal from Crop Plants

Plants absorb a range of heavy metals from soil. Some are essential micronutrients such as Cu, Zn, Mn, Fe, Ni and Co while some are nonessential (Cd, Hg, As, and Pb) that is required in ultra-trace amount for animal and plant but creates toxicity when supplied in supra-optimal values. Mobility of metal ions across root cellular membrane and loading in to xylem allows access of metals into plant tissues. A small portion of total amount of metal can reach up to the cells, while rest portion are transported further into the apoplast and some are bound to the cell wall substances (Gregor 1999). Metal ions bind to high affinity chemical receptors present on the root surface (Salt et al. 1995; Salt and Kramer 2000; Dushenkov et al. 1997). Route of heavy metal intake in vivo is through apoplastic diffusion from root epidermis to cortex, crossing barrier for apoplastic diffusion, i.e., casparian strip act thereafter reach up to plant tissue. Moreover, transpiration pump facilitates, root symplast ion movement into xylem apoplast (Salt et al. 1995; Marschner 2003). Physiological concentrations of heavy metals in the plant cell are regulated by tonoplast as well as metal transporters on the plasma membrane. Generally metals are transported through transport systems that include CDF family, Nramps, ZIP family, ABC transporters, heavy metal ATPases (HMAs) and CAX family antiporters, Cyclic Nucleotide Gated Channels (CNGC), and copper transporter. These transporters help in the acquisition, distribution, and homeostasis of heavy metals in plants. The potentiality of As accumulation and resistance are governed by genetic differences and diversity in detoxification processes in different plant species (Meharg and Hartley-Whitaker 2002; Meharg et al. 2008, 2009). Ma et al. (2001) first reported that the Chinese brake fern (*Pteris vittata*) is able to hyperaccumulate up to 22,630 mg As kg^{-1} in shoot. Thereafter, Zhao et al. (2002) suggested hyperaccumulation capacity of other species of *Pteris* like *P. vittata*, *P. cretica*, *P. longifolia* and *P. umbrosa* to a similar extent. Arsenic and phosphorous both belongs to Group 15 elements of periodic table with different atomic number, however arsenic in its pentavalent form mimic the chemical analog of phosphate and interfere with plant important metabolic processes by replacing phosphorus by arsenic. It also impairs the balance of phosphate supply that hampers the vital process of oxidative phosphorylation and ATP generation to carried out various metabolic processes (Finnegan and Chen 2012; Tripathi et al. 2007; Gupta et al. 2013; Chao et al. 2014; Meadows 2014).

For better understanding of As effect at cellular level, it is important to understand the form of As in soil/water matrix, mechanism of intake of this form in plants, chemical conversion of As inside cell and various transporter specific to As. Inorganic arsenic (arsenate) absorbed by roots then within root cell, As(V) is reduced to As(III) and translocated through the xylem with water through phosphate transporters and minerals as an As(III) -S compound, and then it is stored as As(III) in the fronds (Ma et al. 2001), this arsenic-laced water in the xylem reach in different part of shoots, hence accumulates in whole plant (Tripathi et al. 2007; Gupta et al. 2013). Within plant tissue, solubility of Arsenic increased through chelation with cyclohexylenedinitrotetraacetic acid, nitrilotriacetic acid, or As–sulfur complexes and thus facilitates mobilization. Arsenic translocation to the shoots and vascular tissues in hyperaccumulator plants as predominant arsenite form probably occurs through a different transporter like aqua glyceroporins of the NIP (Nodulin 26-like intrinsic protein) family members (Ma et al. 2008; Kamiya et al. 2009; Zhao et al. 2009). At higher concentrations As accumulation create problems and thus for normal plant growth mechanisms must exist that assure the needs of cellular metabolism by regulating the uptake and distribution of these metal ions within different cells and organelles, thereby combating the hazardous effect of toxic metals. A study by Zheng et al. (2012) observed variations in As level during plant growth with decrement in methylated As while inorganic As remains constant.

Recently, Dai-Yin Chao and colleagues in 2014 have discovered an arsenic reducing enzyme in the plant *Arabidopsis thaliana* by genome-wide association mapping of 349 types of *A. thaliana* collected from around the world and demonstrated the role of this enzyme in arsenic elimination. Plant converts arsenate in arsenite, by reduction process and the arsenite is ejected back into the soil from the roots. The study revealed that plants with higher arsenic concentration (Kr-0) have a cytosine at a specific nucleotide whereas plants with average arsenic levels (Col-0) have a thymine on loci controlling natural variation in arsenic accumulation of chromosome 2 and the gene responsible for the expression of arsenate reductases is High Arsenic Content 1 (HAC1). This protein accumulates in the root epidermis and in the pericycle cells surrounding the central vascular tissue. Loss in function of HAC1 causes more loading into the xylem and hyperaccumulation of arsenic in shoot, hence resulting in arsenic toxicity (Chao et al. 2014; Meadows 2014). Besides HAC1, nonenzymatic antioxidant and its isomers like glutathione, glutaredoxin, dithiothreitol (DTT) and triosephosphate isomerase help in reduction of arsenate, as in presence of a reductant, some enzymes catalyze the phosphorolytic cleavage of their substrates, reducing arsenate (As^{V}) to arsenite (As^{III}) (Gregus et al. 2009; Rathinasabapathi et al. 2006; Sundaram et al. 2008). A study by Ma et al. (2008), reported that in the rice cultivars Oochikara, T-65, and Koshihikari, As^{III} is transported in the form of arsenous acid $As(OH)_3$ through silicic acid transporters (Lsi1 and Lsi2). Furthermore, As^{V} biotransformed by arsenate reductase to less toxic organic compounds such as DMA (dimethylarsinic acid), MMA (monomethylarsonic acid), or as inorganic As(III) complexed with thiol groups (Pickering et al. 2000; Dhankher et al. 2006). As accumulation promotes the synthesis of antioxidants combating the ill effects of arsenic; ascorbate, γ-Glu-Cys-Gly tripeptide glutathione (GSH), and GSH oligomer ([γ-Glu-Cys]n-Gly) phytochelatin (PC) were reported from the root zones (Schmöger et al. 2000; Gupta et al. 2004; Geng et al. 2006; Singh et al. 2006; Khan et al. 2009; Li et al. 2014) while, anthocyanin was predominant in the leaf tissues (Catarecha et al. 2007). In addition, in non-accumulator or "excluder" plants, root exudates help in preventing metal uptake inside the cells by chelating metals (Marschner 2003).

7 Rice Crop Model for Arsenic Toxicity

Rice is reported as an accumulator of As because of dominance of As in agricultural soil and the plant has the ability to uptake and transport (Kumar et al. 2015). In Bangladesh, researchers quantified arsenic content in 150 paddy rice with respect to seasonal variations and found in the range of 10–420 μg kg^{-1} (Williams et al. 2007; Zhu et al. 2008a, b; Meharg et al. 2009). However, process like parboiling and milling reduces arsenic content in rice by 19% in 21 collected samples (Duxbury et al. 2003). Arsenic toxicity in rice plant decreases rate of photosynthesis, reducing the chlorophyll content, and in turn affecting the growth and yield of rice (Rahman et al. 2007). In 1987, arsenic-contaminated soil (25 mg kg^{-1}) resulted in drop yield by 10% (Xiong et al. 1987). In variety T-Aman rice, the supplementation of arsenic in soil significantly reduced yield of grain and straw (Azad et al. 2009). The permissible limit of arsenic in rice gains is 1.0 mg kg^{-1} according to WHO recommendation and the tolerable daily intake (TDI) should not exceed 2.1 μg inorganic As day^{-1} kg^{-1} body weight. Rice grain reports presence of both organic and inorganic forms of As such as As(III), As(V), monomethylarsenic acid (MMA), dimethyl arsenic acid, and tetramethylarsonate (Juskelis et al. 2013; Sinha and Bhattacharyya 2015; Yu et al. 2017). Elevated levels of arsenic in soil reduce percent seed germination and plantlet growth in rice. However, the toxicity effect on rice seedling was found more prominent in the case of arsenite over arsenate (Abedin and Meharg 2002). Some of the dry season varieties like Purbachi of Bangladesh are better performers in arsenic-contaminated soil as they can tolerate up to 4 mg L^{-1} arsenite and 8 mg L^{-1} arsenate with no effect on seed germination (Abedin and Meharg 2002). A study by Williams and group in 2005 with rice varieties collected from all over the world to decipher the role of genetic variation in arsenic uptake and speciation found that USA long grain rice has a mean value of 0.26 μg As g^{-1} and Bangladeshi rice 0.13 μg As g^{-1} As. In rice of Europe, the USA, Bangladesh, and India, arsenic is present mainly in Arsenite (As^{III}); Dimethyl arsenic (DMA^{V}) and Arsenate (As^{V}) forms (Williams et al. 2005). In two rice varieties from Bangladesh the mean values for arsenic accumulation in boro and aman season rice were 183 and 117 μg kg^{-1}, respectively (Duxbury et al. 2003), while in the other rice varieties, BRRI dhan28 and BRRI hybrid dhan1, the values for BRRI dhan28 were 0.8 ± 0.1 (parboiled) and 0.5 ± 0.0 mg kg^{-1} dry weight (non-parboiled), respectively, and for BRRI hybrid dhan1 the values were 0.8 ± 0.2 (parboiled) and 0.6 ± 0.2 mg kg^{-1} dry weight (non-parboiled), respectively (Rahman et al. 2007). Rice plant can accumulate 28- and 75-fold arsenic in its different parts like roots, stem and in its grains (Rahman et al. 2007).The maximum concentrations of As were in the root (248 ± 65 mg kg^{-1}) and the lowest in the grain (1.25 ± 0.23 mg kg^{-1}) (Abedin and Meharg 2002). Some other varieties of rice have potential to accumulate arsenic in following order: Faridpur (boro) 0.51 > Satkhira (boro) 0.38 > Satkhira (aman) 0.36 > Chuadanga (boro) 0.32 > Meherpur (boro) 0.29 μg g^{-1} As (Williams et al. 2006). According to FDA test in 2012 Basmati rice from India, Pakistan, or California, and sushi rice from the USA had the lowest levels of total inorganic arsenic and can be a better choice to avoid arsenic toxicity. Basmati rice contains 63 μg kg^{-1} of As, and in jasmine rice the average arsenic content is 69 μg kg^{-1} (a-survey-of-inorganic-arsenic-in-rice-and-rice-products-on-the-swedish-market-2015---part-1.pdf10/56). Moreover, high-yielding varieties of rice in India include Ratna, IR 50, Ganga kaveri, and Tulsa (a West Bengal local rice variety). IR 50 and Ganga kaveri varieties can accumulate up to 1.99 and 2.43 mg kg^{-1} d.w. in soil containing arsenic 40.0 mg kg^{-1}, whereas Tulsa rice variety accumulates 0.50 ± 0.08 mg kg^{-1} dry weight of arsenic (Bhattacharya et al. 2011). Another example is from rice sold in Spanish local markets; the uncooked rice showed total As =0.188 ± 0.078 μg g^{-1} and inorganic As was measured as 0.114 ± 0.046 μg g^{-1} d.w. and even after cooking rice retained between 45 and 107% of the As(V) (Torres-Escribano et al. 2008). Figure 2 schematically represent of diverse components involved in As uptake, transportation, and detoxification in the plants.

Arsenic in its trivalent form adversely effects overall growth of rice seedling by reducing photosynthetic efficiency which further activates MAP kinase cascade,

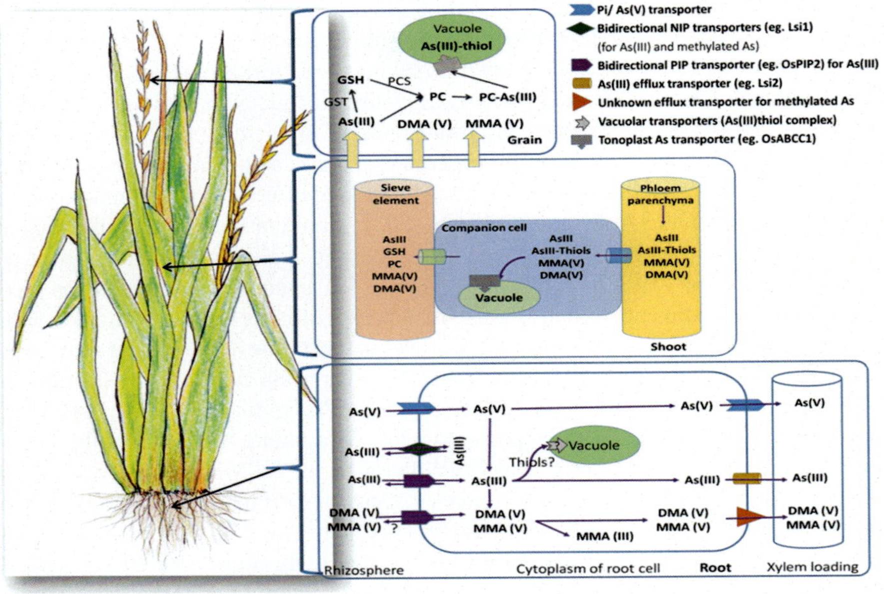

Fig. 2 Schematic representation of diverse components involved in As uptake, transport, and detoxification in the plants (adopted from Kumar et al. 2015; Zhao et al. 2010a, b)

ROS and NO production in arsenic-induced stress signal transduction in plants (Rao et al. 2011). Arsenate at different concentrations was supplemented in sandy soil Na_2HAsO_4 (5, 10, and 50 mg kg^{-1} As soil) along with a heavy metal mixture (5 mg Cd, 300 mg Zn, and 10 mg kg^{-1} As soil). The accumulation of arsenic in the clove shoot resulted in activation of antioxidants and antioxidant enzymes to combat the stress. It was reported that expression of superoxide dismutase (SOD) activity, along with isoenzymes, Mn-SOD, and Cu/Zn-SOD increased in response to arsenic accumulation. In addition to it, peroxidase activity (POD) was also enhanced (Mascher et al. 2002). According to a study conducted by Bhattacharya et al. (2010) the highest and lowest mean arsenic concentrations (milligrams per kilogram) in different important crops irrigated through the water channels of Ganga–Meghna–Bramhaputra basin in Nadia district, West Bengal were found in potato (0.654) and in turmeric (0.003), respectively. Higher mean arsenic concentrations (milligrams per kilogram) were observed in Boro rice grain (0.451), arum (0.407), amaranth (0.372), radish (0.344), Aman rice grain (0.334), lady's finger (0.301), cauliflower (0.293), and brinjal (0.279) which was in the permissive limit. Another case study revealed that vegetables grown at Chapai-Nawabgonj of Bangladesh showed As accumulation in the order of tomato > potato > red amaranth > katua data (Farid et al. 2015).

8 Arsenic Bioavailability in Rice Rhizosphere

Rice is usually grown under flooded conditions. In this anaerobic situation, mostly As(III) species dominate. As readily desorbed from soil (solid phase), As(III) bioavailability increases greatly (Stroud et al. 2011). Experiment with 14 paddy soils from Bangladesh, China, and the UK, Khan et al. (2010) showed that under flooded condition, As(III) accounted for >95% of total soluble arsenic (Khan et al. 2010; Wang et al. 2015). Anoxic conditions also promote reductive dissolution of iron (Fe) oxyhydroxides and subsequent arsenic mobilization, along with higher abundance of Fe-reducing bacteria (Somenahally et al. 2011). In comparison with constantly flooded areas, intermittent flooded situations may reduce 86% of arsenic in pore water, 55% in the root plaque, and 41% in rice grains (Somenahally et al. 2011; Wang et al. 2015).

Soil mineralogy, especially presence of Fe (hydr) oxides, signifies the major pool of arsenic in soils and thereby its solubility (Miretzky and Cirelli 2010). In anaerobic conditions, from aqueous phase, rapid co-precipitation of exogenous As(V) takes place with dissolved Fe^{III}/Fe^{II}, preventing As(V) reduction to As(III) (Fan et al. 2014). Fe plaque usually forms on mature rice plant's root surfaces gradually. The plaque comprises amorphous or crystalline Fe(oxyhydr)oxides (e.g., γ-FeOOH, α-FeOOH, $Fe(OH)_3$-nH_2O). Aerenchyma tissue present in the rice root rhizosphere supplies oxygen in its vicinity, promoting growth of Fe plaques, which can impede arsenic movement towards roots by sorbing it (Liu et al. 2006; Wu et al. 2011, 2013; Syu et al. 2013; Wang et al. 2015). However, younger roots are vulnerable to arsenic transport (both trivalent and pentavalent) as little or no formations of Fe plaques and decreased soil redox potential (Eh) at deeper layer (Yamaguchi et al. 2014). Both As(V) and As(III), due to the deficiency of Fe plaque formation, could move capably into rice through the young roots. Complete drainage, on the other hand, leads to a huge increment of arsenic (V). Therefore, intermittent drainage of paddy fields particularly during grain filling could be effective (Nakamura and Katou 2012; Yamaguchi et al. 2014; Wang et al. 2015). Further, root aerenchyma contributes towards the oxygen diffusion into rhizosphere zone. This loss of oxygen of plants (termed radial oxygen loss—ROL, which varies among different genotypes) put forth a substantial effect on Fe plaque formation and subsequent arsenic sink and bioavailability in an inverse relation (Lee et al. 2013; Wu et al. 2015; Mei et al. 2012; Pan et al. 2014). Therefore, rice genotypes with higher ROL rate are likely to have higher Fe plaque formations (Fe and As accumulation), leading to decreased arsenic mobilization in rhizospheric zone and subsequent translocation into shoots and grains (Mei et al. 2012; Pan et al. 2014). This property of plants is important to select appropriate rice cultivars for breeding in arsenic contaminated areas (Wang et al. 2015).

Arsenic mobilization also positively correlated with dissolved organic carbon (DOC), which is usually abundant in paddy fields. It is reported that though DOC enhances arsenic mobility in soil, the subsequent high accumulation of arsenic in rice grains does not occur, which may be due to sequestration of arsenic (forming

As-DOC complexes) that reduces its bioavailability and subsequent transportation into plant (Buschmann et al. 2006; Williams et al. 2011; Wang et al. 2015). Thus, DOC concentration and composition may have a significant role on the bioavailability of arsenic in paddy soils.

On the contrary, though biochar (a carbon-rich, porous, and fine granular pyrolysis product) have the capacity to reduce the bioavailability of different heavy metals (e.g., Cd, Pb, Cu, and Zn), arsenic mobility in soils tends to be enhanced probably due to the raised soil pH (Manyà 2012; Zheng et al. 2012). Thus addition of simple biochar is not advisable in arsenic rich soil that enhances its mobility, transport, and accumulation in rice. However, as proposed by Wang et al. (2015), biochar with Fe oxide mixture (e.g., physical mix, impregnation of biochar by Fe oxyhydroxide) may be a potentially useful component for reducing a range of heavy metal cations in soil. pH, clay content, phosphorus and sulfate status, etc. could also influence the arsenic bioavailability to paddy plants (Yang et al. 2012; Sahoo and Kim 2013; Li et al. 2014; Sahoo and Mukherjee 2014; Sun et al. 2014).

9 Arsenic Transport and Speciation in Rice: Role of Silicon

As(III) (uncharged arsenious acid—$H_3AsO_3{}^0$) translocation is high in anoxic environment of flooded paddy soils at pH < 8 (Su et al. 2010). Transfer of the As(III) occurs through nodulin26-like intrinsic proteins (NIPs) for silicon (Si) uptake (OsLsi1—Si influx transporter) and translocation (OsLsi2—Si efflux transporter) (Ma et al. 2006, 2007, 2008; Wang et al. 2015). Chen et al. (2012) reported the protective role of arbuscular mycorrhizal fungi (AMF) in arsenic translocation that suppresses mRNA expression of OsLsi1 and OsLsi2. Thus, AMF helps in biomass and grain yield without escalating accumulation of arsenic in grain under As stress (Chen et al. 2012; Wu et al. 2013), which might be an interesting approach to develop a cost-effective mitigation strategy. Further, improvement of soil with Si (silicon) has been reported useful in abating arsenic level in rice grain, as both Si and As(III) compete for the same transporter in root cell (Fleck et al. 2013; Tripathi et al. 2013; Liu CW et al. 2014; Liu WJ et al. 2014). Upon silicon application under flooded condition, approximately 33% reduction of As(III) concentration in polished rice was found as a consequence of ~50% reduced vascular transportation of arsenic (Fleck et al. 2013; Wang et al. 2015). However, some silicon amendments (e.g., sodium silicate, diatomaceous earth, calcium silicate) can increase soil pH resulting in elevated arsenic desorption into soil aqueous phase and entry of As(III), MMA, DMA into rice. Although As^0 and Si share transport pathways, arsenic is sequestered in vacuoles of the endodermis, pericycle, and xylem parenchyma cells, while Si in the cell walls of the endodermal cells (Seyfferth and Fendorf 2012; Wang et al. 2015).

Another subfamily of plant aquaporins (rice plasma membrane intrinsic proteins—OsPIP) is involved with AsIII tolerance and permeability in rice (Mosa et al. 2012). Significant downregulation of expressions of identified OsPIPs was found when they are exposed to As(III); however, bidirectional AsIII transport by OsPIPs (both influx and efflux) are found during heterogeneous overexpression of OsPIPs in Arabidopsis (Wang et al. 2015). Interestingly, yeast plasma membrane transporter ScAcr3p can be expressed in rice as it has As(III) extruding capacity, which could be a approach to decrease arsenic in rice grains by elevating As(III) efflux from roots (Duan et al. 2012, 2013; Wang et al. 2015). Rice root has the efficiency to reduce As(V) to As(III) and subsequent transport to xylem sap (Su et al. 2010). Kopittke et al. (2014) using X-ray absorption near-edge spectroscopy demonstrated that As(V) reduction takes place in all tissues of root apex. Although As(III) is highly mobile, formation of AsIII-thiol complex (with the involvement of phytochelatin or GSH) in rice root restrict its translocation from root to shoot (Zhao et al. 2009).

10 Arsenic Species Unloading at Rice Grain

Arsenic accumulation in grain is the major aspect related to dietary exposure to arsenic. Researchers suggest active phloem pathway is the major contributor for unloading arsenic in the grain (Carey et al. 2010, 2011; Zhao et al. 2012). Introduction of 73As-labelled As(III) at grain filling stage through root suggest the similar opinion. Further, Song et al. (2014) demonstrated a significant decrease in arsenic translocation into rice grains by facilitating As sequestration in vacuoles by a rice tonoplast transporter (OsABCC1) localized to phloem companion cells. Therefore, overexpression of OsABCC1 may be a practical measure to reduce arsenic accumulation in grains (Wang et al. 2015).

Among different As species, DMA exhibits higher mobility, greater dispersal rate than iAs (Carey et al. 2010, 2011). DMA can be concentrated into nucellar epidermis, tissues (endosperm, aleurone, and embryo) causing significant damage to seed setting rate (spikelet sterility) and yield loss (Zheng MZ et al. 2013). Interestingly, DMA is derived from microbial methylation catalyzed by the methyltransferase enzyme (ArsM) mostly in acidic paddy soils (Jia et al. 2012, 2013; Lomax et al. 2012; Marapakala et al. 2012; Ye et al. 2012; Zheng RL et al. 2013). Development of low-DMA cultivars is the option to reduce methylated arsenic in rice grain. Chromosome 6 and 8 harbor three quantitative trait loci (QTLs) that control DMA content in rice grain and account for ~73% of total phenotypic variance in DMA (Kuramata et al. 2013). This might be an effective option through selecting cultivars; however, further studies are essential to get a clear picture on rice QTLs (Wang et al. 2016; Mahender et al. 2016).

11 Conclusion

Arsenic bioavailability in soil is high which may influence the production and yield of major crops, mainly cereal crops like rice which grows in submerged flooded condition. Some practices like rain water harvesting for crop irrigation, bioremediation by microbes resistant to arsenic, use of natural arsenic chelators, hyperaccumulating plants, and the aerobic farming of paddy crops are effective to mitigate arsenic contamination. Further, arsenic methylation may be linked with various influencing factors like rhizospheric microorganisms and their activities, water regime, soil and mineral composition, organic matter, nutritional status, and biochar. Again, parboiling of rice grains can also reduce its arsenic content. Protein modulation and other associated factors can be beneficial to reduce the extent of bioaccumulation in economically important crop or may produce low arsenic content crops. Detailed multidimensional integrated investigations are required to know more about arsenic mobility, transport, accumulation, and resistance in crop plants grown in arsenic-contaminated areas, which will help further to mitigate one of the biggest environmental health adversities of chronic arsenic exposure.

References

Abedin MJ, Meharg AA (2002) Relative toxicity of arsenite and arsenate on germination and early seedling growth of rice (*Oryza sativa* L). Plant Soil 243:57–66

Adriano DC (2001) Trace elements in the terrestrial environment biogeochemistry, bioavailability, and risks of metals. Springer, New York

Azad MAK, Islam MN, Alam A, Mahmud H, Islam MA, Karim MR, Rahman M (2009) Arsenic uptake and phytotoxicity of T-aman rice (*Oryza sativa* L.) grown in the as-amended soil of Bangladesh. Environmentalist 29:436–440

Ampiah-Bonney RJ, Tyson JF, Lanza GR (2007) Phytoextraction of arsenic from soil by *Leersia oryzoides*. Int J Phytorem 9:31–40

Bhattacharya P, Mukherjee AB, Bundschuh J, Zevenhoven R, Loeppert R (2007) Arsenic in soil and groundwater environment: biogeochemical interactions, health effects and remediation. Elsevier, Amsterdam

Bhattacharya P, Samal AC, Majumdar J, Santra SC (2010) Arsenic contamination in Rice, wheat, pulses, and vegetables: a study in an arsenic affected area of West Bengal, India. Water Air Soil Pollut 213:3–13

Bhattacharya P, Samal AC, Majumdar J, Banerjee S, Santra SC (2011) Arsenic toxicity in four different varieties of Rice (*Oryza sativa* L.) of West Bengal, Proceedings of UGC Sponsored National Seminar on Advances in Environmental Science and Technology, Vivekanandan College, Kolkata, India, 5–6 Feb 2011

Bhattacharya S, Gupta K, Debnath S, Ghosh UC, Chattopadhyay DJ, Mukhopadhyay A (2012) Arsenic bioaccumulation in rice and edible plants and subsequent transmission through food chain in Bengal basin: a review of the perspectives for environmental health. Toxicol Environ Chem 94:429–441

Bhattacharya P, Jovanovic D, Polya D (2014) Best practice guide on the control of arsenic in drinking water. IWA Publishing, London

Buschmann J, Kappeler A, Lindauer U, Kistler D, Berg M, Sigg L (2006) Arsenite and arsenate binding to dissolved humic acids: influence of pH, type of humic acid, and aluminum. Environ Sci Technol 40:6015–6020

Carbonell-Barrachina A, Aarabi MA, Delaune RD, Gambrell RP, Patrick WHJ (1998) Bioavailability and uptake of arsenic by wetland vegetation: effects on plant growth and nutrition. J Environ Sci Health 33:45–66

Carey AM, Scheckel KG, Lombi E, Newville M, Choi Y, Norton GJ, Charnock JM, Feldmann J, Meharg AA, Price AH (2010) Grain unloading of arsenic species in rice. Plant Physiol 152:309–319

Carey AM, Norton GJ, Deacon C, Scheckel KG, Lombi E, Punshon T, Guerinot ML, Lanzirotti A, Choi Y, Newville M, Meharg AA, Price AH (2011) Phloem transport of arsenic species from flag leaf to grain during grain filling. New Phytol 192:87–98

Carey AM, Lombi E, Donner E, de Jonge MD, Punshon T, Jackson BP, Guerinot ML, Price AH, Meharg AA (2012) A review of recent developments in the speciation and locationof arsenic and selenium in rice grain. Anal Bioanal Chem 402:3275–3286

Catarecha P, Segura MD, Franco-Zorrilla JM, García-Ponce B, Lanza M, Solano R, Paz-Ares J, Leyva A (2007) A mutant of the Arabidopsis phosphate trans-porter PHT1;1 displays enhanced arsenic accumulation. Plant Cell 19:1123–1133

Chao DY, Chen Y, Chen J, Shi S, Chen Z, Wang CC, Danku JM, Zhao FJ, Salt DE (2014) Genome-wide association mapping identifies a new arsenate reductase enzyme critical for limiting arsenic accumulation in plants. PLoS Biol 12:e1002009

Chakraborti D, Sengupta MK, Rahman MM, Ahamed S, Chowdhury UK, Hossain MA, Mukherjee SC, Pati S, Saha KC, Dutta RN, Quamruzzaman Q (2004) Groundwater arsenic contamination and its health effects in the Ganga–Meghna–Brahmaputra plain. J Environ Monit 6:74–83

Chen Z, Zhu YG, Liu WJ, Meharg AA (2005) Direct evidence showing the effect of root surface iron plaque on arsenite and arsenate uptake into rice (*Oryza sativa*). New Phytol 165:91–97

Chen X, Li H, Chan WF, Wu C, Wu F, Wu S, Wong MH (2012) Arsenite transporters expression in rice (*Oryza sativa* L.) associated with arbuscular mycorrhizal fungi (AMF) colonization under different levels of arsenite stress. Chemosphere 89:1248–1254

Chutia P, Kato S, Kojima T, Satokawa S (2009) Arsenic adsorption from aqueous solution on synthetic zeolites. J Hazard Mater 162:440–447

Delgado A, Go'mez JA (2016) The soil. Physical, chemical and biological properties. In: Villalobos FJ, Fereres E (eds) Principles of agronomy for sustainable agriculture. Springer International Publishing AG, Switzerland. doi:10.1007/978-3-319-46116-8_2

Dhankher OP, Rosen BP, McKinney EC, Meagher RB (2006) Hyperaccumulation of arsenic in the shoots of *Arabidopsis* silenced for arsenate reductase (ACR2). Proc Natl Acad Sci U S A 103:5413–5418

Duan G, Kamiya T, Ishikawa S, Fujiwara T, Arao T (2012) Expressing ScACR3 in rice enhanced arsenite efflux and reduced arsenic accumulation in rice grains. Plant Cell Physiol 53:154–163

Duan G, Liu W, Chen X, Hu Y, Zhu Y (2013) Association of arsenic with nutrient elements in rice plants. Metallomics 5:784–792

Dushenkov S, Vasudev D, Gleba D, Fleisher D, Ting KC, Ensley B (1997) Removal of uranium from water using terrestrial plants. Environ Sci Technol 31:3468–3474

Duxbury JM, Mayer AB, Lauren JG, Hassan N (2003) Food chain aspects of arsenic contamination in Bangladesh: effects on quality and productivity of rice. J Environ Sci Health A Tox Hazard Subs Environ Eng 38:61–69

Fan JX, Wang YJ, Liu C, Wang LH, Yang K, Zhou DM, Li W, Sparks DL (2014) Effect of iron oxide reductive dissolution on the transformation and immobilization of arsenic in soils: new insights from X-ray photoelectron and X-ray absorption spectroscopy. J Hazard Mater 279:212–219

Farid ATM, Roy KC, Hossain KM, Sen R (2015) Effect of arsenic contaminated irrigation water on vegetable (file:///F:/Arsenic%20Book/my%20dnld/introduction/guidelines-drinking-water-quality-management-for-new-zealand-2015-oct15%20(1).pdf)

Fitz WJ, Wenzel WW (2002) Arsenic transformations in the soil-rhizosphere-plant system: fundamentals and potential application to phytoremediation. J Biotechnol 99:259–278

Fitz WJ, Wenzel WW, Zhang H, Nurmi J, Stipek K, Fischerova Z, Schweiger P, Köllensperger G, Ma LQ, Stingeder G (2003) Rhizosphere characteristics of the arsenic hyperaccumulator *Pteris vittata* L. and monitoring of phytoremoval efficiency. Environ Sci Technol 37:5008–5014

Finnegan PM, Chen W (2012) Arsenic toxicity: the effects on plant metabolism. Front Physiol 3:182

Fleck AT, Mattusch J, Schenk MK (2013) Silicon decreases the arsenic level in rice grain by limiting arsenite transport. J Plant Nutr Soil Sci 176:785–794

Garg N, Singla P (2011) Arsenic toxicity in crop plants: physiological effects and tolerance mechanisms. Environ Chem Lett 9:303–321

Garnier JM, Travassac F, Lenoble V, Rose J, Zheng Y, Hossain MS, Chowdhury SH, Biswas AK, Ahmed KM, Cheng Z, van Geen A (2010) Temporal variations in arsenic uptake by rice plants in Bangladesh: the role of iron plaque in paddy fields irrigated with groundwater. Sci Total Environ 408:4185–4193

Geng CN, Zhu YG, Hu Y, Williams P, Meharg AA (2006) Arsenate causes differential acute toxicity to two P-deprived genotypes of rice seedlings (*Oryza sativa* L.) Plant Soil 279:297–306

Gilbert-Diamond D, Cottingham KL, Gruber JF, Punshon T, Sayarath V, Gandolfi AJ, Baker ER, Jackson BP, Folt CL, Karagas MR (2011) Rice consumption contributes to arsenic exposure in US women. Proc Natl Acad Sci 108:20656–20660

Gonzaga MIS, Santos JAG, Ma LQ (2008) Phytoextraction by arsenic hyperaccumulator *Pteris vittata* L. from six arsenic-contaminated soils: repeated harvests and arsenic redistribution. Environ Pollut 154:212–218

Gregor M (1999) Metal availability and bioconcentration in plants. In: Prasad MNV, Hagemeyer J (eds) Heavy metal stress in plants. Springer-Verlag, Berlin, Heidelberg, pp 1–27

Gregus Z, Roos G, Geerlings P, Ne'meti B (2009) Mechanism of thiol-supported arsenate reduction mediated by phosphorolytic-arsenolytic enzymes: II. Enzymatic formation of arsenylated products susceptible for reduction to arsenite by thiols. Toxicol Sci 110:282–292

Gupta DK, Tohoyama H, Joho M, Inouhe M (2004) Changes in the levels of phytochelatins and related metal binding peptides in chickpea seedlings exposed to arsenic and different heavy metal ions. J Plant Res 117:253–256

Gupta DK, Inouhe M, Rodríguez-Serrano M, Romero-Puerta MC, Sandalio LM (2013) Oxidative stress and arsenic toxicity: Role of NADPH oxidases. Chemosphere 90:1987–1996

Hartley-Whitaker J, Ainsworth G, Vooijs R, Ten Bookum W, Schat H, Meharg AA (2001) Phytochelatins are involved in differential arsenate tolerance in Holcus lanatus. Plant Physiol 126:299–306

Jia Y, Huang H, Sun GX, Zhao FJ, Zhu YG (2012) Pathways and relative contributions to arsenic volatilization from rice plants and paddy soil. Environ Sci Technol 46:8090–8096

Jia Y, Huang H, Zhong M, Wang FH, Zhang LM, Zhu YG (2013) Microbial arsenic methylation in soil and rice rhizosphere. Environ Sci Technol 47:3141–3148

Jiang QQ, Singh BR (1994) Effect of different forms and sources of arsenic on crop yield and arsenic concentration. Water Air Soil Pollut 74:321–343

Juskelis R, Li W, Nelson J, Cappozzo JC (2013) Arsenic speciation in rice cereals for infants. J Agric Food Chem 61:10670–10676

Kamiya T, Tanaka M, Mitani N, Ma JF, Maeshima M, Fujiwara T (2009) $NIP_{1;1,}$ an aquaporin homolog, determines the arsenite sensitivity of *Arabidopsis thaliana*. J Biol Chem 284:2114–2120

Khan I, Ahmad A, Iqbal M (2009) Modulation of antioxidant defence system for arsenic detoxification in Indian mustard. Ecotoxicol Environ Saf 72:626–634

Khan MA, Stroud JL, Zhu YG, McGrath SP, Zhao FJ (2010) Arsenic bioavailability to rice is elevated in Bangladeshi paddy soils. Environ Sci Technol 44:8515–8521

Kopittke PM, de Jonge MD, Wang P, McKenna B, Lombi E, Paterson D, Howard D, James S, Spiers K, Ryan C, Johnson AAT, Menzies NW (2014) Laterally resolved speciation of arsenic in roots of wheat and rice using fluorescence-XANES imaging. New Phytol 201:1251–1262

Kraemer S (2004) Iron oxide dissolution and solubility in the presence of siderophores. Aquat Sci Res 66:3–18

Kumar S, Dubey RS, Tripathi RD, Chakrabarty D, Trivedi PK (2015) Omics and biotechnology of arsenic stress and detoxification in plants: current updates and prospective. Environ Int 74:221–230

Kuramata M, Abe T, Akira K, Kaworu E, Taeko S, Masahiro Y, Satoru I (2013) Genetic diversity of arsenic accumulation in rice and QTL analysis of methylated arsenic in rice grains. Rice 6:1–10

Lee CH, Hsieh YC, Lin TH, Lee DY (2013) Iron plaque formation and its effect on arsenic uptake by different genotypes of paddy rice. Plant Soil 363:231–241

Li J, Dong F, Lu Y, Yan Q, Shim H (2014) Mechanisms controlling arsenic uptake in rice grown in mining impacted regions in South China. PLoS One 9:e108300

Li N, Wang J, Song W-Y (2016) Arsenic uptake and translocation in plants. Plant Cell Physiol 57:4–13

Lièvremont D, Bertin PN, Lett MC (2009) Arsenic in contaminated waters: biogeochemical cycle, microbial metabolism and biotreatment processes. Biochimie 91:1229–1237

Liu CW, Wang CJ, Kao YH (2016) Assessing and simulating the major pathway and hydrogeochemical transport of arsenic in the Beitou–Guandu area. Taiwan Environ Geochem Health 38:219

Liu WJ, Zhu YG, Smith FA, Smith SE (2004) Do phosphorus nutrition and iron plaque alter arsenate (As) uptake by rice seedlings in hydroponic culture? New Phytol 162:481–488

Liu ZJ, Boles E, Rosen BP (2004) Arsenic trioxide uptake by hexosepermeases in *Saccharomyces cerevisiae*. J Biol Chem 279:17312–17318

Liu WJ, Zhu YG, Hu Y, Williams PN, Gault AG, Meharg AA, Charnock JM, Smith FA (2006) Arsenic sequestration in iron plaque, its accumulation and speciation in mature rice plants (*Oryza sativa* L.) Environ Sci Technol 40:5730–5736

Liu CW, Chen YY, Kao YH, Maji SK (2014) Bioaccumulation and translocation of arsenic in the ecosystem of the Guandu Wetland, Taiwan. Wetlands 34:129–140

Liu WJ, McGrath SP, Zhao FJ (2014) Silicon has opposite effects on the accumulation of inorganic and methylated arsenic species in rice. Plant Soil 376:423–431

Lomax C, Liu WJ, Wu LY, Xue K, Xiong JB, Zhou JZ, McGrath SP, Meharg AA, Miller AJ (2012) Methylated arsenic species in plants originate from soil microorganisms. New Phytol 193:665–672

Ma LQ, Komar KM, Tu C, Zhang WH, Cai Y, Kennelley ED (2001) A fern that hyperaccumulates arsenic-a hardy, versatile, fast-growing plant helps to remove arsenic from contaminated soils. Nature 409:579

Ma JF, Tamai K, Yamaji N, Tamai K, Konishi S, Fujiwara T, Katsuhara M, Yano M (2006) A silicon transporter in rice. Nature 440:688–691

Ma JF, Yamaji N, Mitani N, Tamai K, Konishi S, Fujiwara T, Katsuhara M, Yano M (2007) An efflux transporter of silicon in rice. Nature 448:209–212

Ma JF, Yamaji N, Mitani N, Xu XY, Su YH, McGrath SP, Zhao FJ (2008) Transporters of arsenite in rice and their role in arsenic accumulation in rice grain. Proc Natl Acad Sci U S A 105:9931–9935

Mahanta C, Enmark G, Nordborg D, Sracek O, Nath B, Nickson RT, Herbert R, Jacks G, Mukherjee A, Ramanathan AL, Choudhury R, Bhattacharya P (2015) Hydrogeochemical controls on mobilization of arsenic in groundwater of a part of Brahmaputra river floodplain, India. J Hydrol: Reg Stud 4:154–171

Mahender A, Anandan A, Pradhan SK, Pandit E (2016) Rice grain nutritional traits and their enhancement using relevant genes and QTLs through advanced approaches. Springerplus 5:2086

Mandal BK, Suzuki KT (2002) Arsenic round the world: a review. Talanta 58:201–235

Manyà JJ (2012) Pyrolysis for biochar purposes: a review to establish current knowledge gaps and research needs. Environ Sci Technol 46:7939–7954

Marapakala K, Qin J, Rosen BP (2012) Identification of catalytic residues in the As(III) S-adenosylmethionine methyltransferase. Biochemistry 51:944–951

Marschner H (2003) Mineral nutrition of higher plants. Academic, New York

Mascher R, Lippmann B, Holzinger S, Bergmann H (2002) Arsenate toxicity: effects on oxidative stress response molecules and enzymes in red clover plants. Plant Sci 163:961–969

Meadows R (2014) How plants control arsenic accumulation. PLoS Biol 12:e1002008

Meharg AA (2004) Arsenic in rice-understanding a new disaster for South-East Asia. Trends Plant Sci 9:415–417

Meharg AA, Hartley-Whitaker J (2002) Arsenic uptake and metabolism in arsenic resistant and nonresistant plant species. New Phytol 154:29–43

Meharg AA, Lombi E, Williams PN, Scheckel KG, Feldmann J, Raab A, Zhu YG, Islam R (2008) Speciation and localization of arsenic in white and brown rice grains. Environ Sci Technol 42:1051–1057

Meharg AA, Williams PN, Adomako E, Lawgali YY, Deacon C, Villada A (2009) Geographical variation in total and inorganic arsenic content of polished (white) rice. Environ Sci Technol 43:1612–1617

Mei XQ, Wong MH, Yang Y, Dong HY, Qiu RL, Ye ZH (2012) The effects of radial oxygen loss on arsenic tolerance and uptake in rice and on its rhizosphere. Environ Pollut 165:109–117

Melkonian S, Argos M, Hall MN, Chen Y, Parvez F, Pierce B, Cao H, Aschebrook-Kilfo B, Ahmed A, Islam T, Slavcovich V, Gamble M, Haris PI, Graziano JH, Ahsan H (2013) Urinary and dietary analysis of 18,470 Bangladeshis reveal a correlation of rice consumption with arsenic exposure and toxicity. PLoS One 8:e80691

Miretzky P, Cirelli AF (2010) Remediation of arsenic-contaminated soils by iron amendments: a review. Crit Rev Environ Sci Technol 40:93–115

Mirza N, Mahmood Q, Maroof Shah M, Pervez A, Sultan S (2014) Plants as useful vectors to reduce environmental toxic arsenic content. Scient World J 2014:921581

Mohan D, Pittman CU Jr (2007) Arsenic removal from water/waste water using adsorbents-a critical review. J Hazard Mater 142:1–53

Mosa KA, Kumar K, Chhikara S, Mcdermott J, Liu Z, Musante C, White JC, Dhankher OP (2012) Members of rice plasma membrane intrinsic proteins subfamily are involved in arsenite permeability and tolerance in plants. Transgen Res 21:1265–1277

Mukherjee A, Sengupta MK, Hossain MA, Ahamed S, Das B, Nayak B, Lodh D, Rahman MM, Chakraborti D (2006) Arsenic contamination in groundwater: a global perspective with emphasis on the Asian scenario. J Health Popul Nutr 24:142–163

Nakamura K, Katou H (2012) Arsenic and cadmium solubilisation and immobilization in paddy soils in response to alternate submergence and drainage. In: Selim HM (ed) Competitive sorption and transport of heavy metals in soils and geological media. CRC Press, Boca Raton pp 373–397

National Ground Water Association (2001) Arsenic what you need to know. http://www.ngwa.org/ASSETS/A0DD107452D74B33AE9D5114EE6647ED/Arsenic.pdf

Norra S, Berner ZA, Agarwala P, Wagner F, Chandresekharam D, Stüben D (2006) Impact of irrigation with as rich groundwater on soil and crops: a geochemical case study in west Bengal Delta plain, India. Appl Geochem 20:1890–1906

Okkenhaug G, Zhu YG, He J, Li X, Luo L, Mulder J (2012) Antimony (Sb) and arsenic (As) in Sb mining impacted paddy soil from Xikuangshan, China: differences in mechanisms controlling soil sequestration and uptake in rice. Environ Sci Technol 46:3155–3162

Pan W, Wu C, Xue S, Hartley W (2014) Arsenic dynamics in the rhizosphere and its sequestration on rice roots as affected by root oxidation. J Environ Sci 26:892–899

Pickering IJ, Prince RC, George MJ, Smith RD, George GN, Salt DE (2000) Reduction andcoordination of arsenic in Indian mustard. Plant Physiol 122:1171–1177

Raab A, Schat H, Meharg AA, Feldmann J (2005) Uptake, translocation and transformation of arsenate and arsenite in sunflower (*Helianthus annuus*): formation of arsenic-phytochelatin complexes during exposure to high arsenic concentrations. New Phytol 168:551–558

Rahman MA, Hassler C (2014) Is arsenic biotransformation a detoxification mechanism for microorganisms? Aquat Toxicol 146:212–219

Rahman MA, Hasegawaa H, Rahman MM, Rahman MA, Miah MAM (2007) Accumulation of arsenic in tissues of rice plant (*Oryza sativa* L.) and its distribution in fractions of rice grain. Chemosphere 69:942–948

Rahman MA, Hasegawa H, Kadohashi K, Maki T, Ueda K (2009) Hydroxyiminodisuccinic acid (HIDS): a novel biodegradable chelating ligand for the increase of iron bioavailability and arsenic phytoextraction. Chemosphere 77:207–213

Rahman MM, Naidu R, Bhattacharya P (2009) Arsenic contamination in groundwater in the Southeast Asia region. Environ Geochem Health 31:9–21

Rao KP, Vani G, Kumar K, Wankhede DP, Misra M, Gupta M, Sinha AK (2011) Arsenic stress activates MAP kinase in rice roots and leaves. Arch Biochem Biophys 506:73–82

Rascio N, Navari-Izzo F (2011) Heavy metal hyperaccumulating plants: how and why do they do it? And what makes them so interesting? Plant Sci 180:169–181

Rathinasabapathi B, Wu S, Sundaram S, Rivoal J, Srivastava M, Ma LQ (2006) Arsenic resistance in *Pteris vittata* L.: identification of a cytosolic triosephosphate isomerase based on cDNA expression cloning in *Escherichia coli*. Plant Mol Biol 62:845–857

Sahoo PK, Kim K (2013) A review of the arsenic concentration in paddy rice from the perspective of geoscience. Geosci J 17:107–122

Sahoo PK, Mukherjee A (2014) Arsenic fate and transport in the groundwater-soil-plant system: an understanding of suitable rice paddy cultivation in arsenic enriched areas. In: Sengupta D (ed) Recent trends in modelling of environmental contaminants. Springer, India, pp 28–30

Salt DE, Kramer U (2000) Mechanisms of metal hyperaccumulation in plants. In: Raskin I, Ensley BD (eds) Phytoremediation of toxic metals: using plants to clean up the environment. Wiley, New York, pp 231–245

Salt DE, Prince RC, Pickering IJ, Raskin I (1995) Mechanism of cadmium mobility and accumulation in Indian mustard. Plant Physiol 109:1426–1433

Schmöger M, Oven M, Grill E (2000) Detoxification of arsenic by phytochelatins in plants. Plant Physiol 122:793–801

Seyfferth AL, Fendorf S (2012) Silicate mineral impacts on the uptake and storage of arsenic and plant nutrients in rice (*Oryza sativa* L.) Environ Sci Technol 46:13176–13183

Singh N, Ma LQ, Srivastava M, Rathinasabapathi B (2006) Meta- bolic adaptations to arsenic induced oxidative stress in *Pteris vittata* L. and *Pteris ensiformis* L. Plant Sci 170:274–282

Sinha B, Bhattacharyya K (2015) Arsenic toxicity in rice with special reference to speciation in Indian grain and its implication on human health. J Sci Food Agric 95:1435–1444

Somenahally AC, Hollister EB, Yan W, Gentry TJ, Loepper RH (2011) Water management impacts on arsenic speciation and iron-reducing bacteria in contrasting rice-rhizosphere compartments. Environ Sci Technol 45:8328–8335

Song WY, Yamaki T, Yamaji N, Ko D, Jung KH, Fujii-Kashino M, An G, Martinoia E, Lee Y, Ma JF (2014) A rice ABC transporter, OsABCC1, reduces arsenic accumulation in the grain. Proc Natl Acad Sci U S A 111:15699–15704

Stoeva N, Bineva T (2003) Oxidative changes and photosynthesis in oat plants grown in As- contaminated soil. Bulg J Plant Physiol 29:87–95

Stroud JL, Khan MA, Norton GJ, Islam MR, Dasgupta T, Zhu YG, Price AH, Meharg AA, McGrath SP, Zhao FJ (2011) Assessing the labile arsenic pool in contaminated paddy soils by isotopic dilution techniques and simple extractions. Environ Sci Technol 45:4262–4269

Su YH, McGrath SP, Zhao FJ (2010) Rice is more efficient in arsenite uptake and translocation than wheat and barley. Plant Soil 328:27–34

Sun HJ, Rathinasabapathi B, Wu B, Luo J, Pu LP, Ma LQ (2014) Arsenic and selenium toxicity and their interactive effects in humans. Environ Int 69:148–158

Sundaram S, Rathinasabapathi B, Ma LQ, Rosen BP (2008) An arsenate activated glutaredoxin from the arsenic hyperaccumulator fern *Pteris vittata* L. regulates intracellular arsenite. J Biol Chem 283:6095–6101

Syu CH, Jiang PY, Huang HH, Chen WT, Lin TH, Lee DY (2013) Arsenic sequestration in iron plaque and its effect on As uptake by rice plants grown in paddy soils with high contents of As, iron oxides, and organic matter. Soil Sci Plant Nutr 59:463–471

Tangahu BV, Sheikh Abdullah SR, Basri H, Idris M, Anuar N, Mukhlisin M (2011) A review on heavy metals (As, Pb, and Hg) uptake by plants through phytoremediation. Int J Chem Eng 2011:31

Torres-Escribano S, Leal M, Vélez D, Montoro R (2008) Total and inorganic arsenic concentrations in rice sold in Spain, effect of cooking, and risk assessments. Environ Sci Technol 42:3867–3872

Tripathi RD, Srivastava S, Mishra S, Singh N, Tuli R, Gupta DK, Frans J, Maathuis M (2007) Arsenic hazards: strategies for tolerance and remediation by plants. Trends Biotechnol 25:158–165

Tripathi P, Tripathi RD, Singh RP, Dwivedi S, Goutam D, Shri M, Trivedi PK, Chakrabarty D (2013) Silicon mediates arsenic tolerance in rice (*Oryza sativa* L.) through lowering of arsenic uptake and improved antioxidant defence system. Ecol Eng 52:96–103

Vaclavikova M, Gallios GP, Hredzak S, Jakabsky S (2008) Removal of arsenic from water streams: an overview of available techniques. Clean Technol Environ Policy 10:89–95

Vázquez S, Esteban E, Carpena RO (2008) Evolution of arsenate toxicity in nodulated white lupine in a long-term culture. J Agric Food Chem 56:8580–8587

Verbruggen N, Hermans C, Schat H (2009) Mechanisms to cope with arsenic or cadmium excess in plants. Curr Opin Plant Biol 12:364–372

Wang X, Peng B, Tan C, Ma L, Rathinasabapathi B (2015) Recent advances in arsenic bioavailability, transport, and speciation in rice. Environ Sci Pollut Res 22:5742–5750

Wang H, Xu X, Vieira FG, Xiao Y, Li Z, Wang J, Nielsen R, Chu C (2016) The power of inbreeding: NGS-based GWAS of rice reveals convergent evolution during rice domestication. Mol Plant 9:975–985

Wenzel WW (2009) Rhizosphere processes and management in plant-assisted bioremediation (phytoremediation) of soils. Plant and Soil 321:385–408

Williams PN, Price AH, Raab A, Hossain SA, Feldmann J, Meharg AA (2005) Variation in arsenic speciation and concentration in paddy rice related to dietary exposure. Environ Sci Technol 39:5531–5540

Williams PN, Islam MR, Adomako EE, Raab A, Hossain SA, Zhu YG, Feldmann J, Meharg AA (2006) Increase in rice grain arsenic for regions of Bangladesh irrigating paddies with elevated arsenic in ground waters. Environ Sci Technol 40:4903–4908

Williams PN, Villada A, Deacon C, Raab A, Figuerola J, Green AJ, Feldmann J, Meharg AA (2007) Greatly enhanced arsenic shoot assimilation in rice leads to elevated grain levels compared to wheat and barley. Environ Sci Technol 41:6854–6859

Williams PN, Zhang H, Davison W, Meharg AA, Hossain M, Norton GJ, Islam MR, Brammer H (2011) Organic matter-solid phase interactions are critical for predicting arsenic release and plant uptake in Bangladesh paddy soils. Environ Sci Technol 45:6080–6087

Winkel L, Berg M, Stengel C, Rosenberg T (2008) Hydrogeological survey assessing arsenic and other groundwater contaminants in the lowlands of Sumatra, Indonesia. Appl Geochem 23:3019–3028

World Bank Policy Report (2005) Towards a more effective operational response: Arsenic contamination of groundwater in South and East Asian Countries Vol I and II, Washington, DC. http://hdl.handle.net/10986/8524; http://hdl.handle.net/10986/8526

Wu C, Ye Z, Shu WS, Zhu YG, Wong MH (2011) Arsenic accumulation and speciation in rice are affected by root aeration and variation of genotypes. J Exp Bot 62:2889–2898

Wu C, Li H, Ye Z, Wu F, Wong MH (2013) Effects of As levels on radial oxygen loss and As speciation in rice. Environ Sci Pollut Res 20:8334–8341

Wu F, Hu J, Wu S, Wong MH (2015) Grain yield and arsenic uptake of upland rice inoculated with arbuscular mycorrhizal fungi in As spiked soils. Environ Sci Pollut Res 22:8919–8926

Xiong XZ, Li PJ, Wang YS, Ten H, Wang LP, Song LP (1987) Environmental capacity of arsenic in soil and mathematical model. Chin J Environ Sci 8:8–14

Yamaguchi N, Ohkura T, Takahashi Y, Maejima Y, Arao T (2014) Arsenic distribution and speciation near rice roots influenced by iron plaques and redox conditions of the soil matrix. Environ Sci Technol 48:1549–1556

Yang X, Hou Q, Yang Z, Zhang X, Hou Y (2012) Solid-solution partitioning of arsenic (As) in the paddy soil profiles in Chengdu plain, Southwest China. Geosci Front 3:901–909

Ye J, Rensing C, Rosen BP, Zhu YG (2012) Arsenic biomethylation by photosynthetic organisms. Trends Plant Sci 17:155–162

Yu Z, Qiu W, Wang F, Lei M, Wang D, Song Z (2017) Effects of manganese oxide-modified biochar composites on arsenic speciation and accumulation in an indica rice (*Oryza sativa* L.) cultivar. Chemosphere 168:341–349

Zhao FJ, Dunham SJ, McGrath SP (2002) Arsenic hyperaccumulation by different fern species. New Phytol 156:27–31

Zhao FJ, Ma JF, McGrath SP, Meharg AA (2009) Arsenic uptake and metabolism in plants. New Phytol 181:777–794

Zhao FJ, Ago Y, Mitani N, Li RY, Su YH, Yamaji N, McGrath SP, Ma JF (2010a) The role of the rice aquaporin Lsi1 in arsenite efflux from roots. New Phytol 186:392–399

Zhao FJ, Zhu YG, Meharg AA (2010b) Arsenic as a food chain contaminant: mechanisms of plant uptake and metabolism and mitigation strategies. Annu Rev Plant Biol 61:535–559

Zhao FJ, Stroud J, Khan M, McGrath SP (2012) Arsenic translocation in rice investigated using radioactive 73As tracer. Plant Soil 350:413–420

Zhao FJ, Zhu YG, Meharg AA (2013) Methylated arsenic species in rice: geographical variation, origin, and uptake mechanisms. Environ Sci Technol 47:3957–3966

Zheng RL, Cai C, Liang JH, Huang Q, Chen Z, Huang YZ, Arp HPH, Sun GX (2012) The effects of biochars from rice residue on the formation of iron plaque and the accumulation of Cd, Zn, Pb, As in rice (*Oryza sativa* L.) seedlings. Chemosphere 89:856–862

Zheng MZ, Li G, Sun GX, Shim H, Cai C (2013) Differential toxicity and accumulation of inorganic and methylated arsenic in rice. Plant Soil 365:227–238

Zheng RL, Sun GX, Zhu YG (2013) Effects of microbial processes on the fate of arsenic in paddy soil. Chin Sci Bull 58:186–193

Zhu YG, Sun GX, Lei M, Teng M, Liu YX, Chen NC, Wang LH, Carey AM, Deacon C, Raab A, Meharg AA, Williams PN (2008a) High percentage inorganic arsenic content of mining impacted and non-impacted Chinese rice. Environ Sci Technol 42:5008–5013

Zhu YG, Williams PN, Meharg AA (2008b) Exposure to inorganic arsenic from rice: a global health issue? Environ Pollut 154:169–171

Uptake, Transport, and Remediation of Arsenic by Algae and Higher Plants

Anindita Mitra, Soumya Chatterjee, and Dharmendra K. Gupta

Contents

1 Introduction 146
2 Mechanism of Uptake of Inorganic and Organic Species of Arsenic by Algae and Plants 148
2.1 Arsenic Uptake in Algal Cells 148
2.2 Arsenate Uptake in Plants 148
2.3 Arsenite Uptake in Plants 150
2.4 Uptake of Organic Species of As by Plants 150
3 Biochemical Fate of As in Algal Cells 151
3.1 As^{III} Oxidation 151
3.2 As^{V} Reduction 151
3.3 Arsenic Methylation 152
3.4 Formation of Arsenosugars and Arsenolipids 152
3.5 Detoxification of Arsenic by Metal Chelators 153
4 Biochemical Fate of As in Non-hyperaccumulator and Hyperaccumulator Plants 153
5 Factors Affecting As Availability in Plants 155
6 Phytoremediation of Arsenic 155
6.1 Algae 155
6.2 Terrestrial Plants 156
6.3 Aquatic Plants 156

A. Mitra (✉)
Bankura Christian College, Bankura 722101, West Bengal, India
e-mail: aninditajashmitra@rediffmail.com

S. Chatterjee
Defence Research Laboratory, Defence Research and Development Organization (DRDO), Ministry of Defence, Post Bag No. 2, Tezpur 784001, Assam, India

D.K. Gupta
Institut für Radioökologie und Strahlenschutz (IRS), Gottfried Wilhelm Leibniz Universität Hannover, Herrenhäuser Str. 2, Hannover 30419, Germany

D.K. Gupta, S. Chatterjee (eds.), *Arsenic Contamination in the Environment*,
DOI 10.1007/978-3-319-54356-7_7

7 Biotechnological Approaches to Improve As Phytoremediation 157
7.1 Augmented Synthesis of As Chelators 157
7.2 By Producing Double Transgenic Plants Co-Expressing Bacterial Arsenate Reductase and γ-ECS Genes 159
7.3 Regulating Arsenate Reductase Activity 159
8 Prevention of As Uptake by Food Crops 159
8.1 Restriction of As at the Underground Level 160
8.2 Limiting Uptake of As in Root 160
8.3 Increasing As^{III} Efflux Rate 160
8.4 Manipulation of Lsi1 and Lsi2 Genes 160
8.5 Volatilization of As 160
9 Conclusion 161
References 161

1 Introduction

Arsenic (As) is a toxic metalloid of global concern (Roy et al. 2014). The natural source of arsenic in the environment is geochemical weathering of rocks, volcanic emission, but the level is raised mostly by anthropogenic activities such as application of pesticides, wood preservatives, mining and smelting of ores, and combustion of fossil fuel (Wang and Mulligan 2006). Due to severe surface water shortage, progressively more farmers are using recycled water and sewage sludge that further contributes to As accumulation in agricultural lands. Arsenic is also released into the environment as a by-product from chemical manufacturing sites (Dhankher et al. 2002).

As is often described as a metalloid as it belongs to Group VA of the periodic table and its electronegativity is not sufficient to give As a metallic character. The oxidation states and electron orbitals of As are similar to those of phosphorus. It is commonly bound to carbon, iron, oxygen, and sulfur, forming inorganic and organic arsenicals in various oxidation states: −III, 0, +III, and +V (Hughes 2006; Gupta et al. 2011). In nature, As exists as inorganic or organic species but is normally not encountered in its elemental state. Between the two inorganic forms, the more highly oxidized pentavalent arsenate (As^V, occurss as $H_2AsO_4^-$ and $HAsSO_4^{2-}$ in most environments) is prevalent in aerobic environments, while the more highly reduced trivalent arsenite As^{III} is the predominant form in anaerobic environments. The organic forms of As are mostly methylated and exist as monomethylarsinic acid (MMA^V), dimethylarsinic acid (DMA^V), and trimethylarsine oxide ($TMAO^V$) found at low concentrations in most soils (Finnegan and Chen 2012). Methylated organic forms of As are higher in concentrations in anaerobic soils than in aerobic soils (Abedin et al. 2002). Arsines are volatile compounds of mono-, di-, and trimethylated derivatives of As^{III} (MMA^{III}, DMA^{III}, TMA^{III}).

As is a nonessential metalloid to algae and plants and causes toxic response after gaining entry into the cell (Ullrich-Eberius et al. 1989; Garg and Singla 2011). Under severe As stress, in microalgae As toxicity may be enhanced as it subdues several detoxification mechanisms, resulting in oxidative stress and inhibition of cell division (Levy et al. 2005). As toxicity may be elicited due to disruption of phosphorus metabolism by incorporating As^{V} into phosphorylated compounds vital for ATP synthesis or by blocking sulfhydryl (−SH) groups, interfering with the activity of important cellular enzymes such as pyruvate dehydrogenase and 2-oxoglutarate dehydrogenase eventually resulting in membrane leakage and cell death by producing reactive oxygen species (Lloyd and Oremland 2006).

Like arsenate (As^{V}) and arsenite (As^{III}), the methylated forms of As are also phytotoxic (Zhao et al. 2010), plants have developed a range of strategies to combat As toxicity including chelation and sub-sequestration of complexes in vacuoles (Gupta et al. 2011, 2013a, b; Li et al. 2016). The first tissue of the plants to be exposed to As is the root. Most plants have the capacity to load much of the As within root but it depends on the plant species whether the metalloid will be translocated to shoot or other tissues. After absorption through the roots the metalloid inhibits root expansion and proliferation (Finnegan and Chen 2012). Upon translocation to the shoots As can severely impede growth of the plants by slowing or arresting accumulation of biomass, as well as induce loss of fertility, yield, and fruit production (Garg and Singla 2011). Several reports are there indicating that the elevated concentration of As in the soil causes a significant reduction in crop yield (Marin et al. 1993; Meharg 2004; Zhu et al. 2008). As toxicity at cellular level damages membrane which leads to electrolyte leakage (Singh et al. 2006) and is often accompanied by oxidative stress due to increased production of malondialdehyde, a by-product of lipid peroxidation (Finnegan and Chen 2012). In plants, As exposure induces antioxidant defense mechanism. In the root As intoxication triggers the synthesis of ascorbate, γ-Glu-Cys-Gly-tripeptide glutathione (GSH) and the GSH oligomer, phytochelatin particularly in root (Schmoger et al. 2000; Li et al. 2004; Geng et al. 2006; Singh et al. 2006; Khan et al. 2009) and accumulation of anthocyanin in the leaves (Catarecha et al. 2007).

The detrimental effects of arsenicals on human is an increasing menace chiefly due to contaminated drinking water and food as the levels of arsenic have been elevated in soil and groundwater across the globe (Dopp et al. 2010). The maximum concentration limit of As for drinking water recommended by World Health Organization is 10 $\mu g\ L^{-1}$. In some countries such as Bangladesh and India groundwater arsenic concentration has exceeded the WHO safety limits, i.e., 2000 $\mu g\ L^{-1}$ (Hossain 2006). As contamination in human occurs through consumption of cereals, vegetables, and fruits irrigated with arsenic contaminated water (Zhao et al. 2006). The consequence is a global epidemic of arsenic poisoning, leading to skin lesion, cancer of bladder, lung, and kidney and other symptoms (Mondal et al. 2006).

Remediation of As-contaminated soil and groundwater therefore is an urgent need for providing safe drinking water and food. Phytoremediation is an efficient bioremediation strategy that involves green algae and suitable plants including arsenic hyperaccumulating ferns and some aquatic or terrestrial angiosperms that efficiently remove the metalloid from highly contaminated soil without exhibiting any sign of intoxication (Singh et al. 2015; Wang et al. 2015). Arsenic cannot be degraded from the contaminated sites but can be transformed into less toxic products either by reduction/oxidation, can be extracted within plants or in some cases volatilized. The phytoremediation method of arsenic includes phytostabilization, phytoextraction, rhizofiltration, and phytovolatilization (Chatterjee et al. 2013). As photosynthetic organisms besides plants, algae also play an important role in controlling metal concentrations in lakes and ocean (Sigg 1985, 1987). Algae have been reported to have the ability to absorb metals from the environment and can accumulate higher concentration of metals than the surrounding water (Chang et al. 2005; Shamsuddoha et al. 2006). Mitra et al. (2012) reported that algae belonging to the group Chlorophyta and Cyanophyta absorbed and accumulated arsenic and boron from environment and thus act as hyper-phytoremediators, reducing the arsenic and boron levels in contaminated water.

For the betterment of the As phytoremediation, a general scheme of arsenic uptake, transport, the biochemical pathway for metabolism and detoxification of the metalloid by the algae as well as plants is necessary. The objective of this review is, therefore, to provide an overview about the uptake of the inorganic and organic species of As, their translocation and biochemical fate in algae and plants and to explore the current concepts of phytoremediation along with their limitations and challenges associated with the developed processes.

2 Mechanism of Uptake of Inorganic and Organic Species of Arsenic by Algae and Plants

2.1 Arsenic Uptake in Algal Cells

In prokaryotic (e.g., cyanobacteria) and eukaryotic algae (e.g., *Chlorella* sp.), arsenate (As^V) enters into cells via phosphate transporters, while arsenite (As^{III}) is absorbed through plasma membrane via aquaglyceroporins and hexose permeases (Levy et al. 2005; Zhang et al. 2014). The arsenate is a competitive inhibitor of phosphate due to chemical similarity between PO_4^{3-} and AsO_4^3 (Maeda et al. 1988; Xue et al. 2014). As reported by several researchers addition of phosphate significantly decreased the As^V uptake by the microalgae *Chlorella salina* (Karadjova et al. 2008), *Chlamydomonas reinhardtii* (Planas and Healey 1978), and *Skeletonema costatum* (Sanders 1979). Therefore, arsenate uptake and resulting toxicity in algae could be reduced by increasing PO_4^{3-} concentration in the environment (Guo et al. 2011).

2.2 Arsenate Uptake in Plants

The mechanisms of As uptake in plants have been reviewed in depth (Tripathi et al. 2007; Zhao et al. 2008, 2010). Arsenate (As^V) is an analog of inorganic phosphate (Pi) and therefore shares the common transport pathway across the plasma membrane in higher plants (Fig. 1) as evidenced from physiological and

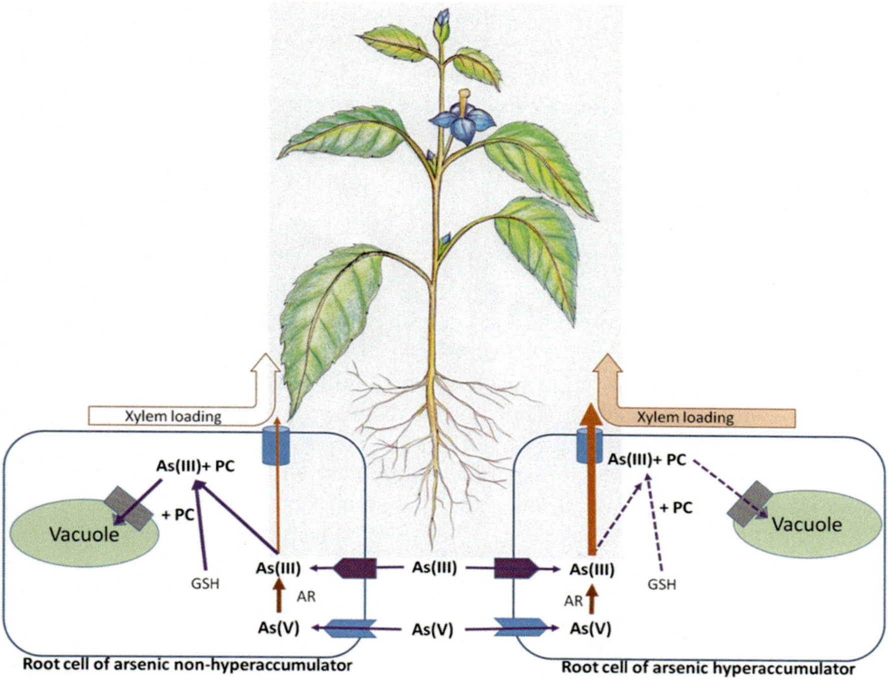

Fig. 1 Schematic representation of arsenic uptake and metabolism in roots of non-hyperaccumulator and hyperaccumulator plants (Arsenic influx is performed by aquaporin like membrane transporters, while efflux via arsenite efflux carrier) (Modified from Zhao et al. 2008)

electrophysiological studies (Ullrich-Eberius et al. 1989; Meharg et al. 1994; Gupta et al. 2011). The uptake mechanism involves cotransport of each phosphate ($H_2PO_4^-$)/arsenate ($H_2AsO_4^-$) molecule with two protons ($2H^+$) (Ullrich-Eberius et al. 1989). Among 100 of phosphate transporters the Pi transporter belonging to the PHT1 family, most of which are present in the roots is likely to be involved in As^V transport (Ullrich-Eberius et al. 1989; Bucher 2007; Wu et al. 2011). Pht protein transports As^V in As hyperaccumulators (Wang et al. 2002; Tu and Ma 2003), As-tolerant non-hyperaccumulators (Meharg and Macnair 1992; Bleeker et al. 2003), and also in As-sensitive non-accumulators (Abedin et al. 2002; Esteban et al. 2003) although the affinity of different phosphate transporters for As^V was greater in hyperaccumulator plants in comparison to non-accumulator plant species (Poynton et al. 2004; Wang et al. 2002). *Arabidopsis thaliana* double mutant for Pht 1;1 and Pht 1;4 was found to much more resistant to arsenate than wild type plants supporting the role of Pht 1;1 and Pht 1;4 in arsenate transport (Shin et al. 2004). Phosphates in the soil compete with arsenate in the uptake by plants as the transporters have higher affinity for phosphate than arsenate (Meharg et al. 1994). Under low levels of Pi, As^V may outcompete Pi for entry through the root and thereby increasing uptake and phytotoxicity, while larger amounts of phosphates compete with arsenate at root surfaces to decrease uptake and phytotoxicity (Peterson and Girling 1981; Tu and Ma 2003). As-tolerant plants such as *Holcus lanatus* and *Cytisus striatus*, adapted to grow in soil with higher As concentration, avoid As toxicity by restricting the influx of As by constitutive suppression of high-affinity phosphate/As^{V} transport (Meharg and Macnair 1992; Bleeker et al. 2003).

2.3 Arsenite Uptake in Plants

Arsenite (As^{III}) is the predominant As species in reducing environment such as submerged soils (Marin et al. 1993). As^{III} is able to enter root cells through nodulin 26-like intrinsic proteins (NIPs) (Meharg and Jardine 2003; Bienert et al. 2008; Isayenkov and Maathuis 2008; Ma et al. 2008; Mitra et al. 2014). NIP belongs to the plant major intrinsic proteins family (MIPs), commonly known as aquaporin's (Fig. 1) the other three plant aquaporins include tonoplast intrinsic protein (TIPs), plasma membrane intrinsic protein PIP), and small basic intrinsic protein (SIPs) (Chaumont et al. 2005; Maurel et al. 2008). NIPs, also called aquaglyceroporins (Wallace et al. 2006), because they are able to transport multiple uncharged solutes including glycerol, urea, ammonia, boric acid, silicic acid (Wallace et al. 2006), and arsenite (Ma et al. 2008) but impervious to water (Bienert et al. 2008). Unlike arsenate, arsenite uptake is inhibited by glycerol and antimonite but not by phosphate (Zhao et al. 2008). The uptake of As species in aquatic macrophytes is regulated by three mechanisms: (1) active uptake by phosphate transporter, (2) passive uptake with the help of aquaglyceroporins and (3) physicochemical adsorption in the root (Rahman and Hasegawa 2011).

Ma et al. (2008) have identified a major pathway for the entry of arsenite in the root of rice cultivars Oochikara T-65, through the *OsNIP* 2;1 transporter also called *Lsi1*, primarily involved in silicon transport. Another arsenite transporter *Lsi2* was also reported by Ma et al. (2008) which mediate arsenite efflux from root toward the direction of xylem. It has been observed that arsenite concentration in the xylem sap from two different rice species mutant for *Lsi2* gene, was much lower than those xylem sap of wild species of rice because in mutants As^{III} accumulation was much lower in the shoots in comparison to wild type (Ma et al. 2008). Rice is a hyperaccumulator of As, as this plant efficiently transports Si in the shoots, thereby involuntarily co-transporting arsenic along with silicon which ultimately accumulates in higher concentration in the straw and grain (Zhao et al. 2008).

2.4 Uptake of Organic Species of As by Plants

Organic forms of arsenic are methylated arsenic species occur usually in a small proportion in soil, such as MMA^{V} and DMA^{V}. They may be derived from previous application of arsenical pesticides, herbicides or may be synthesized by microorganisms. Plants uptake these compounds less efficiently than that of inorganic As species (Raab et al. 2007; Carbonell-Barrachina et al. 1998). The mechanism involved in the uptake of methylated arsenic species by plants is not clearly known. The aquatic plants uptake As^{III} in the form of dimethylarsinic acid (DMAA) and monomethyl arsinic acid (MMAA) passively through aquaglyceroporin channel (Rahman and Hasegawa 2011). It has been reported in rice that aquaporin *OsLsi1* was involved in the uptake of pentavalent methylated As species and the loss of function in rice *OsLsi1* led to an 80% reduction in MMA^V uptake and 50% for DMA^V compared to wild species (Li et al. 2009). However, MMA^V and DMA^V are taken up by plant roots at a slower rate than that of arsenate or arsenite (Abedin et al. 2002; Abbas and Meharg 2008). On the contrary, the mobility of MMA^V and DMA^V within the plant tissue appears to be substantially greater than that of inorganic arsenic species (Li et al. 2009; Carey et al. 2010, 2011). Although *OsLsi1* is involved in the uptake of organic arsenic species, *OsLsi2* in plants has no role in the efflux of the MMA^V and DMA^V (Li et al. 2009).

3 Biochemical Fate of As in Algal Cells

3.1 As^{III} Oxidation

Following uptake, microalgae are able to oxidize As^{III} to As^V as evidenced from *Synechocystis* and *Cyanidiales* (Qin et al. 2009; Zhang et al. 2011, 2014). A detail study of uptake, accumulation, and transformation of arsenic species in microalgae *Synechocystis* sp. PCC6803 was studied by Zhang et al. (2011). According to their study when As^{III} entered into the *Synechocystis* cells, the oxidation process is stimulated, transforming As^{III} to As^V for the purpose of detoxification. However, contrasting report of Qin et al. (2009) showed that As^{III} oxidation occurred outside of the cytoplasmic compartment of the algae and was related to the extracellular enzymes like carbonic anhydrase and extracellular phosphatase. A number of studies showed that in cystosolic fractions of microalgae cells 99% of the total As was in the form of As^V (Duncan et al. 2015; Levy et al. 2005; Wang ZH et al. 2013; Wang Y et al. 2013). To reduce the toxic effect of the arsenite a kind of survival mechanism has been attained in microalgae by restricting the oxidation process either in the periplasm or near the outer membrane (Zhang et al. 2014). However, an As^{III} oxidase in microalgae (*Synechocystis*) has yet to be identified (Zhang et al. 2014).

3.2 As^V Reduction

Although As^{III} has higher toxicity than As^V, a number of microalgae such as *Dunaliella sp., C. salina, Closterium aciculare* reduce As^V to As^{III}, either for methylation, or for efflux out of the cell (Hasegawa et al. 2001; Hellweger et al. 2003; Karadjova et al. 2008; Takimura et al. 1996). In *C. aciculare* and *C. salina* most of the As^{V} was converted into As^{III} after uptake from the growth medium (Hasegawa et al. 2001; Karadjova et al. 2008). Arsenate reduction is essential for further sequestration into vacuole (Zhang et al. 2013), in the form of stable complexes with glutathione (GSH) or phytochelatins (PCs) (Bleeker et al. 2006) and it generally occurs within 1–2 days (Levy et al. 2005).

3.3 Arsenic Methylation

One of the As detoxification process found in microalgae is the methylation of As^{III} to less toxic MMA^{V} and DMA^{V}, catalyzing by the enzyme arsenite methyltransferases (ArsM) (Thomas et al. 2007; Ye et al. 2012). The production of MMA^{V} and DMA^{V} species by marine algae and fresh water algae has been reported by a number of workers (Sanders and Riedel 1993; Maeda et al. 1992; Hasegawa et al. 2001). An extremophilic algae *Cyanidioschyzon* sp. isolate 5508 was reported to oxidize As^{III} to As^{V}, reduced As^{V} to As^{III} and also methylated As^{III} to DMA^{V} and TMAO (trimethylarsine oxide) (Qin et al. 2009). Two genes CmarsM7 and CmarsM8 encoding methyltransferase, isolated from this species and it was found that arsenite hypersensitive strain of *E. coli* harboring these genes became resistant to As (Qin et al. 2009). In addition to mono and dimethylarsenic products As^{III} methylation also includes less toxic arsenic species such as TMAO and volatile TMA (Yin et al. 2011).

3.4 Formation of Arsenosugars and Arsenolipids

Formation of arsenoribosides as the metabolic products of arsenic has been detected in diatom *Chaetoceros concavicornis* (Edmonds et al. 1997) and also in freshwater algae *Chlorella vulgaris, Monoraphidium arcuatum*, *Synechocystis, Nostoc, Chlamydomonas*. In these algae further transformation of DMA^{V} occurs to form more complex As compounds like arsenosugars, although in a meager amount (0–12%) (Murray et al. 2003; Levy et al. 2005; Miyashita et al. 2011, 2012). Certain marine microalgae such as *Dunaliella, Phaeodactylum,* and *Thalassiosira* also produced arsenoribosides even in greater amount (more than 12% in water-soluble and 6–100% in lipid soluble fractions) as reported by Duncan et al. (2013a, b, 2015). According to various researchers arsenic species is sequestered in the lipid fraction of the cells of microalgae (Irgolic et al. 1977; Wrench and Addison 1981) incorporated predominantly as OH-ribose, As^{V} and DMA moieties (Foster et al. 2008). The presence of lipid-soluble As compounds has been identified in a number of algae such as *C. vulgaris, Chlorella ovalis, Chlorella pyrenoidosa, D. tertiolecta;* bluegreen algae *Oscillatoria rubescens, Synechocystis*, diatoms *Phaeodactylum tricornutum, S. costatum, T. pseudonana* (Wang et al. 2015).

3.5 Detoxification of Arsenic by Metal Chelators

The detoxification of As, by metal chelating polypeptides such as glutathione (GSH), phytochelatins (PCs) and its derivatives that readily bind with As species in the cell, was probably achieved by shielding As^{III} with SH-groups of these biomolecules (Maeda et al. 1992). In microalgae, under metal stress, phytochelatin synthase (PCS) catalyzes the synthesis of PCs (Wesenberg et al. 2011). According to Morelli et al. (2005) substantial amount of inorganic As sequestered in vacuoles are likely in the form of As-PCs while others suggest that residue fraction of inorganic arsenic is likely to form complexes with intracellular structural elements of the cells (Levy et al. 2005; Pohl 2007).

4 Biochemical Fate of As in Non-hyperaccumulator and Hyperaccumulator Plants

Analysis of the plants exposed to arsenic shows that more than 90% of the As in the roots and shoots present in As^{III} form (Zhao et al. 2008). It is shown that following uptake by the roots, via the phosphate transporter, 96–100% of the metalloid is retained as As^{III} in the roots and shoot in *Brassica juncea* (Pickering et al. 2000), 97–100% in the leaves of *A. thaliana* (Dhankher et al. 2002), and 92–99% in the roots of tomato and rice (Xu et al. 2007). From the existing reports it is very clear that arsenate is reduced efficiently to arsenite in plant cells, and this reduction occurs either enzymatically and nonenzymatically. Arsenate to arsenite reduction carried out nonenzymatically by two molecules of glutathiones (GSH) in which, GSH is oxidized by the formation of a disulfide bond, producing a GSH dimer (GSSG) (Delnomdedieu et al. 1994). The GSSG is then rapidly recycled to two GSH molecules by GSH reductase (Foyer et al. 2011). Enzymatic reduction of arsenate occurs at much higher rate in comparison to nonenzymatic conversion as part of As detoxification process (Duan et al. 2005). Therefore, majority of As^{V} can be directly reduced to As^{III} by the enzyme arsenate reductase (AR), first isolated from bacteria and yeast (Mukhopadhyay et al. 2000). Based on sequence homology of yeast AR gene, several homologous sequences have been cloned from plants including *Arabidopsis* (*AtAsr/AtACR2*), *Holcus lanatus* (*HlAsr*), *Pteris vittata* (*PuACR2*) and rice (*OsACR2.1/OsACR2.2*) which have shown to possess AR activity (Bleeker et al. 2006; Dhankher et al. 2006; Ellis et al. 2006; Duan et al. 2007). It was found that if AR activity is increased in the plant *H. lanatus*, the plant became more tolerant to As (Bleeker et al. 2006).

Arsenite has a high affinity to the thiol (−SH) groups of peptides which justifies the presence of higher concentration of As^{III}–tristhiolate complex in the shoot (Patra et al. 2004). Binding of arsenite with the thiol group of enzymes or other proteins, alter the structure or interfere with the catalytic site of enzymes, thereby exerting its toxic effect (Meharg and Whitaker 2002). One important pathway of As detoxification in As-non hyperaccumulating plants is to substantially increase the synthesis and accumulation of glutathione (GSH) and phytochelatins (PC) after As exposure, thus become As tolerant (Grill et al. 2006; Schat et al. 2002). As^{III}-GSH or As^{III}-PC complexes has been identified in various plants such as *Rauvolfia serpentina*, *H. lanatus*, *Pteris cretica*, *Helianthus annuus*, and *B. juncea* (Pickering et al. 2000; Schmoger et al. 2000; Montes-Bayon et al. 2004; Raab et al. 2004). A variety of As conjugates can exist as identified by Raab et al. (2005). They have reported 14 different As complexes in *H. annuus* after As exposure including As^{III}-PC_3, GS-As^{III}-PC_2, As^{III}-GS_3, AS^{III}-$(PC_2)_2$ and MMA-PC_2. The predictable role of PC in As detoxification was confirmed from the study of *A. thaliana* mutant *cad1–3*, lacking the functional enzyme for PC synthesis; the mutant was found 10- to 20-fold more sensitive to arsenate than the wild type plants (Ha et al. 1999).

Final step of arsenic detoxification is the sequestration of As^{III}-PC complexes in the vacuoles of root and shoot tissues (Tripathi et al. 2007) thus preventing its interference with the cellular metabolism. In *Arabidopsis* it has been reported that As^{III}-PC conjugates are transported via ABC transporter *MRP1/ABCC1* and *MRP2/ABCC2* (Song et al. 2010). Homologues *ABCC2* transporter was reported to be up-regulated at transcription level in the rice plant when exposed to arsenic (Chakrabarty et al. 2009). Presence of ABC transporter homologues to *Arabidopsis* ABCC1 and 2 are found in many plant species (Mendoza-Cozatl et al. 2011). In *P. vittata* gametophyte, a gene, *PvACR3*, encoding arsenite effluxer protein, with little homology to the yeast *ACR3*, was found to be localized to the vacuolar membrane indicating that it likely effluxes As^{III} into the vacuole for sequestration (Indriolo et al. 2010).

In non-hyperaccumulator plants following uptake, transport of arsenic is restricted from the root to shoot due to rapid complexation with PC and sequestration within vacuoles, where acidic pH (−5.5) is favorable to the stability of the complex and thus reducing its toxic effect (Liu et al. 2010; Mendoza-Cozatl et al. 2011). The form in which As is transported from root to shoot is debatable. Studies on sporophytes of *P. vittata* have shown that As is translocated to the shoot mainly as As^{V} and stored in the fronds as As^{III} (Zhao et al. 2003). In contrast, Duan et al. (2005) reported that most of the As is translocated in its reduced form As^{III} as AR activity was exclusively restricted in the roots. Unlike non-hyperaccumulators, which rely on PC complexation for As detoxification, in hyperaccumulators *P. vittata* and *P. cretica* 60–90% As stored as inorganic arsenite (As^{III}) in the vacuole of fronds (Pickering et al. 2006; Su et al. 2008) and very little is complexed with PC in the roots and fronds (Zhao et al. 2008). Another crucial difference is that As hyperaccumulators efficiently transport As from root to shoot, whereas non-hyperaccumulators has low mobility of As translocation. The As concentration in the xylem sap of *P. vittata* was found to about twofold of magnification higher than that in the non-hyperaccumulator as reported by Zhao et al. (2008).

5 Factors Affecting As Availability in Plants

Plants collect metals from their roots and the amounts of metal absorbed by a plant depend on the concentrations of the metal in the soil; its mobility from the bulk soils to the root surface; passage from the root surface into the root; and its translocation from the root to the shoot (Wild 1988). Some primary factors like metal content of soil, dissolved organic matter, soil pH, and soil characteristics like clay, oxides, and cation exchange capacity regulate the potential mobilization of metals in soil (Kalbitz and Wenrich 1998). A "soil–plant barrier" resists excessive transfer of metal ions from contaminated soil to the food chain but in the case of certain metals like arsenic, lead, or mercury, this barrier fails to work which ultimately leads to more than 50% human intake through vegetables and cereals (Dudka and Miller 1999). The phytoavailability of As is primarily determined by soil texture, As speciation near rhizosphere, and also association of mycorrhizal fungi (Burlo et al. 1999; Carbonell-Barrachina et al. 1999; Zhao et al. 2008). Biochemical processes in the rhizosphere may influence the speciation of As and its phytoavailability (Zhao et al. 2008). As discussed before arsenate predominates in aerobic soil, but in the rhizosphere due to microbial activity anaerobic microenvironment may be created resulting transient formation of arsenite. Due to the coexistence of arsenate and arsenite in the vicinity of the roots, plants may absorb both arsenate and arsenite although growing in aerobic soils (Vetterlein et al. 2007). A number of studies reported that symbiotic relation between mycorrhizal fungi and plants confer As resistance to the host plants. As resistance by mycorrhizal colony is implicated by several methods either by suppressing the high affinity phosphate transporter in the plant roots, thus leading to less As uptake (Gonzalez-Chavez et al. 2002) or by effluxing As to the external medium as reported by Sharples et al. (2000). From the study of Ultra et al. (2007) and Chen et al. (2007) it was observed that mycorrhizal fungi restrict As translocation from root to shoot and thus reduce toxic effect, although the mechanism still remains unexplored.

6 Phytoremediation of Arsenic

6.1 *Algae*

Different functional groups present in the algal cell wall, namely carboxyl, hydroxyl, carbonyl, sulfydryl, and amino groups, are responsible for adsorption of metals and metalloids (Ting et al. 1991; Xue and Sigg 1990). Adsorption is a relatively rapid and reversible process and plays an important role in the As detoxification in different microalgal species (Wang et al. 2015). In As-contaminated water nearly 60% of total amount of this element can be removed by microalgae through adsorption (Wang ZH et al. 2013; Wang Y et al. 2013; Zhang et al. 2013). Therefore, algae can be applied for bioremediation of As in contaminated water as part of a clean technology (Brinza et al. 2007). Some species of microalgae, e.g., *Chlamydomonas*,

Chlorella vulgaris, *Scenedesmus* sp., *Synechocysis* sp. PCC 6803 were reported to remove high percentage of inorganic As (51–80%) (Bahar et al. 2013; Gunaratna et al. 2006; Jiang et al. 2011; Taboada-dela Calzada et al. 1999; Yin et al. 2012). In a study of Zhao et al. (2012) two species of macroalgae *Porphyra yezoensis* and *Laminaria japonica* were found to have the ability to metabolize inorganic arsenic to organic forms and both macroalgae have evolved arsenic resistance. Several factors regulate the As removal by microalgae such as binding capacity and growth stage, N and P levels of As-polluted water (Munoz 2014). The microalgae *Chlorella*, *Oscillatoria*, *Scenedesmus*, *Spirogyra*, and *Pandorina* were identified as As tolerant.

6.2 Terrestrial Plants

The Chinese brake fern (*P. vittata)* has an exceptional ability to hyperaccumulate very high levels of As (Ma et al. 2001), and thrives in tropical and subtropical places. Thus, *P. vittata* could be highly useful for phytoremediation of As in those regions. This fern has been shown to remove As from contaminated groundwater as well as contaminated soil as reported by a number of researchers (Natarajan et al. 2008; Mandal et al. 2012). The successful reports obtained from the field trial in 2001, where Chinese brake ferns were introduced to treat 1.5 acre site contaminated with arsenic (As) in New Jersey, North Carolina, the fern phytoextracted more than 200-fold of As in aboveground biomass (Singh et al. 2006). There are some plants that exhibit high tolerances to As in the soil without exhibiting any growth defect even though As concentration reached 500 mg kg^{-1}(Singh et al. 2007). Species of fern such as *Dryopteris filixmas*, three species of herbs *Blumea lacera*, *Mikania cordata*, *Ageratum conyzoides*, and two shrubs *Clerodendrum trichotomum* and *Ricinus communis* were reported to be suitable for phytoremediation of As. *Thelypteris palustris* has been reported to phytoextract As from a contaminated soil (Anderson et al. 2011). Shrub of willow (*Salix* sp.) is an As-tolerant plant and has good uptake capability (Dwivedi 2013). King et al. (2008) reported that *Eucalyptus cladocalyx* is a good candidate for phytostabilization of As in mine tailings. According to Tripathi et al. (2012) the fern *Marsilea* sp. and the aquatic plants *Eichhornia crassipes* and *Cyperus difformis* are very much promising for the treatment of As-contaminated paddy field. There is a report of phytostabilization of As by hybrid poplar tree (*Populus* sp.) during mine tailings at Superfund site at South Dakota in the USA containing up to 1000 mg kg^{-1} of arsenic and thus decreasing vertical migration of leachate to groundwater level (Hse 1996).

6.3 Aquatic Plants

As accumulation in plants is greatly reflected by habitat type and density and submerged plants have enhanced accumulation capacity of As in comparison to emergent and terrestrial plants (Bergqvist and Greger 2012). The floating plants

Eichhornia crassipes, *Spirodela polyrhiza*, and *Azolla pinnata*, and common wetland weed *Monochoria vaginalis* can be cited in this regard (Mahmud et al. 2008). Wetland aquatic plants *Colocasia esculenta* is the fastest As remover (Jomjun et al. 2011). Reports also exist about some aquatic plants such as duckweeds (*L. gibba*, *L.minor*, *S. polyrhiza*), water spinach (*Ipomea aquatic*a), water ferns (*A. caroliniana*, *A. filiculoides*, and *A. pinnata*), water cabbage (*Pistia stratiotes*), hydrilla (*Hydrilla verticillata*), and watercress (*Lepidium sativum*), *Portulaca tuberose*, *Portulaca oleracea*, *Eclipta alba*, and *Limnanthes* accumulating high level of As from contaminated water and therefore, extensively studied to investigate their As uptake ability and suggested to be potential for As phytoremediation (Rahman and Hasegawa 2011; Tuli et al. 2010). The aquatic plant *Hydrilla* showed 72% accumulation of total As supplied and exhibiting little toxic effect (Srivastava et al. 2011).

An environment friendly, low cost technique called phytofiltration is gaining interest to conventional cleanup of As (Singh et al. 2015). Phytofiltration techniques utilize the aquatic plants which are As resistant and accumulator. Phytofiltration involves selecting promising plants to remove As from the contaminated water and retain it in the roots. These plants are then transplanted into a constructed wetland for the purpose of removing As from the polluted water. Although most of the As is concentrated within the roots, a very low amount may be translocated to the aerial parts (Salt et al. 1998).

7 Biotechnological Approaches to Improve As Phytoremediation

Biotechnological approaches that are implied for improvement of phytoremediation process have focused mainly on the plants tolerance and accumulation of the metalloid. For this purpose transgenic plants are produced harboring the genes controlling traits such as As uptake in roots, higher tolerance, extensive translocation from root to shoot, thiol complexation of As^{III} by increasing chelators (Sharma et al. 2014), methylation of As species and subsequent volatilization (Fig. 2).

7.1 Augmented Synthesis of As Chelators

As phytoremediation can be improved by inducing the increased synthesis of chelators such as glutathione (GSH) and phytochelatins (PC) in plants. In plants the overexpression of phytochelatin synthase (PCS) gene showed promising result (Grill et al. 2006). Recently PCS gene from *C. demersum* (CdPCS1) was expressed in transgenic lines such as tobacco (Shukla et al. 2012) and Arabidopsis (Shukla et al. 2013) showed enhanced PC content along with As resistance. Although the constitutive synthesis of PCs resulted in increased As resistance and tolerance to the target plants, failure of significant enhancement of As accumulation (Li et al. 2005) and

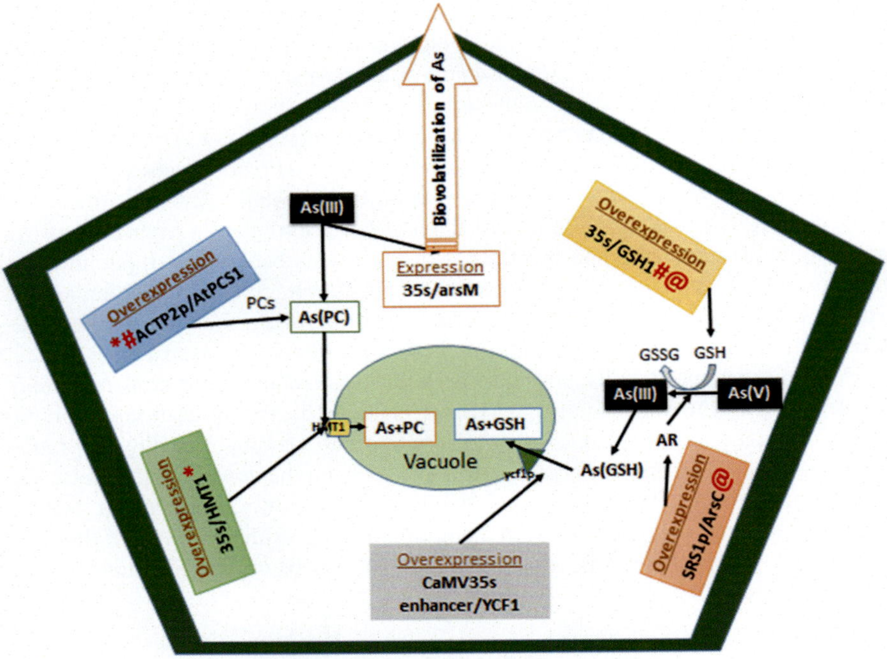

Fig. 2 Schematic presentation to As remediation by bioengineering of plant cell (Adopted from Rahman et al. 2014) [*Symbols* *#@ represents synergistic genes activities, *HMT1* human metallothionein1; *AtPCS Arabidopsis thaliana* phytochelatin synthase, *GSH* glutathione, *arsM* arsenite S-adenosyl methionine methyl transferase gene, *YCF* vacuolar membrane protein member of ABC transporter superfamily involved in the ATP-dependent transport of a wide range of GSH-conjugated substrates, including arsenic]

unwanted toxicity (Tripathi et al. 2007) have been reported. The reason might be due to the limiting supply of the essential metabolites required for PC synthesis such as cysteine, GSH, and γ-glutamylcysteine (γ-EC) (Picault et al. 2006). Reports from biochemical studies have demonstrated that induced PC synthesis after As exposure results in the depletion of GSH (Hartley-Whitaker et al. 2001). As a consequence, it leads to oxidative stress (Singh et al. 2006; Gupta et al. 2013b) and toxicity. A multigenic approach by inducing co-expression of the genes involved in GSH biosynthesis, such as γ-glutamyl cysteine synthetase (γ-ECS) and glutathione synthetase (GS) in *Arabidopsis* produced a greater effect on As tolerance and accumulation (Guo et al. 2008). Therefore, altering the levels of GSH and PCs in plants is an effective approach for increasing the As tolerance and accumulation in plants, and thus producing novel plants with strong phytoremediation potential.

7.2 *By Producing Double Transgenic Plants Co-Expressing Bacterial Arsenate Reductase and γ-ECS Genes*

Double transgenic plants were developed by Dhankher et al. (2002) overexpressing two *E. coli* genes; arsenate reductase, *arsC*, expressing in leaves and γ-glutamyl cysteine synthatase (γ-ECS) expressing in both roots and shoots. This plant was found to be highly tolerant to As in comparison to plant expressing γ-ECS alone. In transgenic *A. thaliana* expressing bacteria arsenate reductase and glutamyl cysteine synthase was found to enhance its ability to accumulate two or three times more arsenic per gram of tissue than the non-transgenic plants Dhankher et al. (2002). Another advantage of this approach is that the targeted plants achieved 17-fold higher biomass and hyperaccumulated threefold more As in the shoots than wild type plants (Dhankher et al. 2002). Overexpression of *arsC* in leaves enhances arsenate reduction to arsenite which can readily conjugates with abundant thiol-rich chelators due to overexpression of γ-ECS gene and thus the plant become manifold As tolerant which ultimately leads to more As accumulation in the aboveground biomass. Therefore, the strategy of producing transgenic plants with multiple transgenes can be used to get a synergistic effect that can surpass what either gene could accomplish on its own (Dhankher et al. 2011).

7.3 *Regulating Arsenate Reductase Activity*

In the angiosperms those are As hyperaccumulator, very meager amount of As was found to be translocated from root to shoot (Pickering et al. 2000; Dhankher et al. 2002) because these plants encode high level of endogenous arsenate reductase responsible for arsenate to arsenite conversion and thus immobilizing toxic As^{III} underground (Dhankher et al. 2002; Dhankher 2005). To achieve the goal of enhancing translocation and hyperaccumulation of As in the aboveground harvestable part, Dhankher et al. (2006) cloned the AR gene from *A. thaliana* (*AtACR2*) and knockdown the *AtACR2* by RNAi. They observed 10- to 16-folds higher As in shoot and very little amount of As retained in the roots as compared with the wild type plants.

8 Prevention of As Uptake by Food Crops

Accumulation of As in the food crops and vegetables is at alarming situation (Meharg 2004). Several strategies manipulated at genetic level can be effective to restrict As translocation in the edible plant parts such as grains or leaves.

8.1 Restriction of As at the Underground Level

As buildup in edible parts such as seeds is reduced by enhancing reduction of arsenate to arsenite in roots and increased level of arsenite-thiol complexation by co-expression of the genes encoding AR and PCS in the roots only by root specific promoter. Thus, the metalloid is phytostabilized in underground biomass or inedible parts (Dhankher et al. 2011). Very recently, heterologous expression of CdPCS1 in rice enhanced As accumulation in the roots and decreased As accumulation in aerial part including the rice grain (Shri et al. 2014).

8.2 Limiting Uptake of As in Root

Absorption of As through root can be achieved by manipulating the PHT transporter that has higher affinity for As^{V} than phosphate (Dhankher et al. 2011). High level of phosphate in the soil improved tolerance to arsenate and low accumulation of As in the grains and leaves of rice varieties as reported by Dwivedi et al. (2006).

8.3 Increasing As^{III} Efflux Rate

Transgenic rice plants expressing the *S. cerevisiae* ACR3 gene increased As^{III} efflux and also lowered As accumulation in rice grain. Transgenic plants exhibited 30% lower As concentration in root and shoot in comparison to wild type plants having similar As translocation factor (Duan et al. 2012).

8.4 Manipulation of Lsi1 and Lsi2 Genes

Low accumulation of As^{III} in crops such as rice can be promoted by knocking down the expression of *Lsi1* and *Lsi2* genes in the roots that mediate arsenite uptake in roots and translocation of the same in the shoot respectively (Dhankher et al. 2011).

8.5 Volatilization of As

Unloading of the toxic As species can be achieved by extrusion of As from cells in volatile form. This can be attained by conversion of inorganic As to methylated organic species like MMA and DMA and further volatilization to the gaseous trimethylarsine (TMA) by producing transgenic plant harboring the bacterial gene

As^{III}-S-adenosyl methyl transferase *Ars*M (Qin et al. 2006, 2009). A report from Meng et al. (2011) suggested that transgenic rice plant expressing *Ars*M produced tenfold higher volatile arsenical maintaining low As level in rice seed along with organic As species MMA^V and DMA^V in the root and shoot of transgenic rice.

9 Conclusion

As toxicity has become an increasing menace to the environment. Plants growing in As-contaminated soil exhibit reduced growth, productivity, and yield. Arsenic and its metabolites induce phytotoxicity, resulting in the alteration of metabolic pathways, affecting various physiological pathways, involved in water–nutrient balance and are often accompanied by oxidative stress. Producing transgenic plants expressing numerous defense-related genes and transporters for developing As resistance and also maximizing the plants' As removal potential would help to alleviate this problem. In most of the cases, genetical approaches involve reduced As uptake, chelation with GSH or PC, vacuolar sequestration, efflux from the roots, and biotransformation into volatile organic species of arsenic (Kumar et al. 2015). However, as reported by Hartley-Whitaker et al. (2001) overexpression of PC synthase gene leads to depletion of GSH, resulting in oxidative stress and toxicity in the targeted plants (Wojas et al. 2010). However, this situation can be overcome by the use of synthetic PCs which do not require GSH and thus can become an alternative approach for phytoremediation of heavy metals (Shukla et al. 2012) as well as As. Another important approach for effective phytoremediation is to explore the gene from the plant source for converting inorganic arsenic to organic/volatile species of arsenic. Detailed knowledge on the metabolism and detoxification pathway of As in plants is needed for this purpose and further research is necessary.

Acknowledgements The authors sincerely acknowledge Principal BCC, Directors DRL and IRS for their kind support. Sincere thanks are also due to Mrs. Swagata Chatterjee for her contribution in making handmade figures. Further, a sincere apology is rendered to many colleagues whose works could not be referred to in this review due to space limitations.

References

Abbas MHH, Meharg AA (2008) Arsenate, arsenite and dimethyl arsenic acid (DMA) uptake and tolerance in maize (*Zea mays* L.) Plant Soil 304:277–289

Abedin MJ, Cresser MS, Meharg AA, Feldmann J, Cotter-Howells J (2002) Arsenic accumulation and metabolism in rice (*Oryza sativa* L.) Environ Sci Technol 36:962–968

Anderson LL, Walsh M, Roy A, Bianchetti CM, Merchan G (2011) The potential of *Thelypteris palustris* and *Asparagus sprengeri* in phytoremediation of arsenic contamination. Int J Phytorem 13:177–184

Bahar MM, Megharaj M, Naidu R (2013) Toxicity, transformation and accumulation of inorganic arsenic species in a microalga *Scenedesmus* sp. isolated from soil. J Appl Phycol 25:913–917

Bergqvist C, Greger M (2012) Arsenic accumulation and speciation in plants from different habitats. Appl Geochem 27:615–622

Bienert GP, Schuessler MD, Jahn TP (2008) Metalloids: essential, beneficial or toxic? Major intrinsic proteins sort it out. Trend Biochem Sci 33:20–26

Bleeker PM, Hakvoort HWJ, Bliek M, Souer E, Schat H (2006) Enhanced arsenate reduction by a CDC25-like tyrosine phosphatase explains increased phytochelatin accumulation in arsenate-tolerant *Holcus lanatus*. Plant J 45:917–929

Bleeker PM, Schat H, Vooijs R, Verkleij JAC, Ernst WHO (2003) Mechanisms of arsenate tolerance in *Cytisus striatus*. New Phytol 157:33–38

Brinza L, Dring MJ, Gavrilescu M (2007) Marine micro- and macro-algal species as biosorbents for heavy metals. Environ Eng Manag J 6:237–251

Bucher M (2007) Functional biology of plant phosphate uptake at root and mycorrhiza interfaces. New Phytol 173:11–26

Burlo F, Guijarro I, Carbonell-Barrachina AA, Valero D, Martinez-Sánchez F (1999) Arsenic species: effects on and accumulation by tomato plants. J Agric Food Chem 47:1247–1253

Carbonell-Barrachina AA, Aarabi MA, DeLaune RD, Gambrell RP, Patrick WH (1998) The influence of arsenic chemical form and concentration on *Spartina patens* and *Spartina alterniflora* growth and tissue arsenic concentration. Plant Soil 198:33–43

Carbonell-Barrachina AA, Burlo F, Valero D, Lopez E, Martinez-Romero D, Martinez-Sanchez F (1999) Arsenic toxicity and accumulation in turnip as affected by arsenic chemical speciation. J Agric Food Chem 47:2288–2294

Carey AM, Norton GJ, Deacon C, Scheckel KG, Lombi E, Punshon T, Guerinot ML, Lanzirotti A, Newville M, Choi Y, Price AH, Meharg AA (2011) Phloem transport of arsenic species from flag leaf to grain during grain filling. New Phytol 192:87–98

Carey AM, Scheckel KG, Lombi E, Newville M, Choi Y, Norton GJ, Charnock JM, Feldmann J, Price AH, Meharg AA (2010) Grain unloading of arsenic species in rice (*Oryza sativa* L.) Plant Physiol 152:309–319

Catarecha P, Segura MD, Franco-Zorrilla JM, García-Ponce B, Lanza M, Solano R, Paz-Ares J, Leyva A (2007) A mutant of the Arabidopsis phosphate trans-porter PHT1;1displays enhanced arsenic accumulation. Plant Cell 19:1123–1133

Chakrabarty D, Trivedi PK, Misra P, Tiwari M, Shri M, Shukla D, Kumar S, Rai A, Pandey A, Nigam D, Tripathi RD, Tuli R (2009) Comparative transcriptome analysis of arsenate and arsenite stresses in rice seedlings. Chemosphere 74:688–702

Chang P, Kim JY, Kim KW (2005) Concentrations of arsenic and heavy metals in vegetation at two abandoned mine tailings in South Korea. Environ Geochem Health 27:109–119

Chatterjee S, Mitra A, Datta S, Veer V (2013) Phytoremediation protocol: An overview. In: Gupta DK (ed) Plant based remediation process. Springer, New York, NY, pp 1–18

Chaumont F, Moshelion M, Daniels MJ (2005) Regulation of plant aquaporin activity. Biol Cell 97:749–764

Chen BD, Xiao XY, Zhu YG, Smith FA, Xie ZM, Smith SE (2007) The arbuscular mycorrhizal fungus *Glomus mosseae* gives contradictory effects on phosphorus and arsenic acquisition by *Medicago sativa* Linn. Sci Total Environ 379:226–234

Delnomdedieu M, Basti MM, Otvos JD, Thomas DJ (1994) Reduction and binding of arsenate and dimethyl arsenate by glutathione–a magnetic resonance study. Chem Biol Interact 90:139–155

Dhankher OP, Elizabeth AH, Pilon-Smit MRH, Doty S (2011) Biotechnological approaches for phytoremediation. In: Altman A, Hasegawa PM (eds) Plant biotechnology and agriculture. Elsevier, Amsterdam

Dhankher OP, Rosen BP, Mc Kinney EC, Meagher RB (2006) Enhanced arsenic uptake in *Arabidopsis* plants by suppressing endogenous arsenate reductase AtACR2 gene. Proc Natl Acad Sci U S A 103:5413–5418

Dhankher OP (2005) Arsenic metabolism in plants: an inside story. New Phytol 168:503–505

Dhankher OP, Li Y, Rosen BP, Shi J, Salt D, Senecoff JF, Sashti NA, Meagher RB (2002) Engineering tolerance and hyperaccumulation of arsenic in plants by combining arsenate reductase and gamma glutamyl cysteine synthetase expression. Nat Biotechnol 20:1140–1145

Dopp E, von Recklinghausen U, Diaz-Bone R, Hirner AV, Rettenmeier AW (2010) Cellular uptake, subcellular distribution and toxicity of arsenic compounds in methylating and non-methylating cells. Environ Res 110:435–442

Duan G, Kamiya T, Ishikawa S, Arao T, Fujiwara T (2012) Expressing ScACR3 in rice enhanced arsenite efflux and reduced arsenic accumulation in rice grains. Plant Cell Physiol 53:154–163

Duan GL, Zhou Y, Tong YP, Mukhopadhyay R, Rosen BP, Zhu YG (2007) A CDC25 homologue from rice functions as an arsenate reductase. New Phytol 174:311–321

Duan GL, Zhu YG, Tong YP, Cai C, Kneer R (2005) Characterization of arsenate reductase in the extract of roots and fronds of Chinese brake fern, an arsenic hyperaccumulator. Plant Physiol 138:461–469

Dudka S, Miller WP (1999) Accumulation of potentially toxic elements in plants and their transfer to human food chain. J Environ Sci Health B 34:681–708

Duncan EG, Maher WA, Foster SD (2015) Contribution of arsenic species in unicellular algae to the cycling of arsenic in marine ecosystems. Environ Sci Technol 49:33–50

Duncan EG, Maher WA, Foster SD, Krikowa F (2013a) Influence of culture regime on arsenic cycling by the marine phytoplankton *Dunaliella tertiolecta* and *Thalassiosira pseudonana*. Environ Chem 10:91–101

Duncan EG, Maher WA, Foster SD, Krikowa F (2013b) The influence of arsenate and phosphate exposure on arsenic uptake, metabolism and species formation in the marine phytoplankton *Dunaliella tertiolecta*. Mar Chem 157:78–85

Dwivedi AK (2013) Arsenic in groundwater: an issue beyond boundary. In: Rajkumar D, Lal JK (eds) Biodiversity conservation & sustainable development. Biodiversity Conservation & Sustainable Development Centre for Biological Research, Puthalam, Tamil Nadu, pp 30–43

Dwivedi S, Tripathi RD, Srivastava S, Mishra S, Shukla MK, Tiwari KK, Singh R, Rai UN (2006) Growth performance and biochemical responses of three rice (*Oryza sativa* L.) cultivars grown in fly-ash amended soil. Chemosphere 67:140–151

Edmonds JS, Shibata Y, Francesconi KA, Rippington RJ, Morita M (1997) Arsenic transformations in short marine food chains studied by HPLC-ICP MS. Appl Organomet Chem 11:281–287

Ellis DR, Gumaelius L, Indriolo E, Pickering IJ, Banks JA, Salt DE (2006) A novel arsenate reductase from the arsenic hyperaccumulating *Pteris vittata*. Plant Physiol 141:1544–1554

Esteban E, Carpena RO, Meharg AA (2003) High-affinity phosphate/arsenate transport in white lupin (*Lupinus albus*) is relatively in sensitive to phosphate status. New Phytol 158:165–173

Finnegan PM, Chen W (2012) Arsenic toxicity: the effects on plant metabolism. Front Physiol 3:1–18

Foster S, Thomson D, Maher W (2008) Uptake and metabolism of arsenate by anexic cultures of the microalgae *Dunaliella tertiolecta* and *Phaeodactylum tricornutum*. Mar Chem 108:172–183

Foyer CH, Noctor G, Hodges M (2011) Respiration and nitrogen assimilation: targeting mitochondria-associated metabolism as a means to enhance nitrogen use efficiency. J Exp Bot 62:1467–1482

Garg N, Singla P (2011) Arsenic toxicity in crop plants: physiological effects and tolerance mechanisms. Environ Chem Lett 9:303–321

Geng CN, Zhu YG, Hu Y, Williams P, Meharg AA (2006) Arsenate causes differential acute toxicity to two P-deprived genotypes of rice seedlings (*Oryza sativa* L.) Plant Soil 279:297–306

Gonzalez-Chavez C, Harris PJ, Dodd J, Meharg AA (2002) Arbuscular mycorrhizal fungi confer enhanced arsenate resistance on *Holcus lanatus*. New Phytol 155:163–171

Grill E, Mishra S, Srivastava S, Tripathi RD (2006) Role of phytochelatins in phytoremediation of heavy metals. In: Singh SN, Tripathi RD (eds) Environmental bioremediation technologies, Springer, New York, pp 101–146

Gunaratna KR, Bulbul A, Imamul Huq SM, Bhattacharya P (2006) Arsenic uptake by fresh water green alga, *Chlamydomonas* species. Presented at the Philadelphia annual meeting of GSA, 22–25 Oct

Guo P, Gong Y, Wang C, Liu X, Liu J (2011) Arsenic speciation and effect of arsenate inhibition in a *Microcystis aeruginosa* culture medium under different phosphate regimes. Environ Toxicol Chem 30:1754–1759

Guo JB, Dai XJ, Xu WZ, Ma M (2008) Overexpressing *GSH1* and *AsPCS1* simultaneously increases the tolerance and accumulation of cadmium and arsenic in *Arabidopsis thaliana*. Chemosphere 72:1020–1026

Gupta DK, Inouhe M, Rodríguez-Serrano M, Romero-Puerta MC, Sandalio LM (2013a) Oxidative stress and arsenic toxicity: Role of NADPH oxidases. Chemosphere 90:1987–1996

Gupta DK, Huang HG, Nicoloso FT, Schetinger MRC, Farias JG, Li TQ, Razafindrabe BHN, Aryal N, Inouhe M (2013b) Effect of Hg, As and Pb on biomass production, photosynthetic rate, nutrients uptake and phytochelatin induction in *Pfaffia glomerata*. Ecotoxicology 22:1403–1412

Gupta DK, Srivastava S, Huang H, Romero-Puertas MC, Sandalio LM (2011) Arsenic tolerance and detoxification mechanisms in plants. In: Sherameti I, Varma A (eds) Detoxification of heavy metals (book series: soil biology). Springer, New York, NY, pp 169–180

Ha SB, Smith AP, Howden R, Dietrich WM, Bugg S, O'Connell MJ, Goldsbrough PB, Cobbett CS (1999) Phytochelatin synthase genes from *Arabidopsis* and the yeast *Schizosaccharomyces pombe*. Plant Cell 11:1153–1163

Hartley-Whitaker J, Ainsworth G, Vooijs R, Ten Bookum W, Schat H, Meharg AA (2001) Phytochelatins are involved in differential arsenate tolerance in *Holcus lanatus*. Plant Physiol 126:299–306

Hasegawa H, Sohrin Y, Seki K, Sato M, Norisuye K, Naito K, Matsui M (2001) Biosynthesis and release of methylarsenic compounds during the growth of freshwater algae. Chemosphere 43:265–272

Hellweger FL, Farley KJ, Lall U, Di Toro DM (2003) Greedy algae reduce arsenate. Limnol Oceanogr 48:2275–2288

Hossain MF (2006) Arsenic contamination in Bangladesh – a review. Agric Ecosyst Environ 113:1–16

Hse W (1996) Metal soil pollution and vegetative remediation by using poplar trees at two heavy metal contaminated sites. University of Iowa, MS thesis

Hughes MF (2006) Biomarkers of exposure: a case study with inorganic arsenic. Environ Health Persp 114:1790–1796

Indriolo E, Na GN, Ellis D, Salt DE, Banks JA (2010) A vacuolar arsenite transporter necessary for arsenic tolerance in the arsenic hyper accumulating fern *Pteris vittata* is missing in flowering plants. Plant Cell 22:2045–2057

Irgolic KJ, Woolson EA, Stockton RA, Newman RD, Bottino NR, Zingaro RA, Kearney PC, Pyles RA, Maeda S, McShane WJ, Cox ER (1977) Characterization of arsenic compounds formed by *Daphnia magna* and *Tetraselmis chuii* from inorganic arsenate. Environ Health Persp 19:61–66

Isayenkov SV, Maathuis FJM (2008) The *Arabidopsis thaliana* aquaglyceroporin AtNIP7;1 is a pathway for arsenite uptake. FEBS Lett 582:1625–1628

Jomjun N, Siripen T, Maliwan S, Jintapat N, Prasak T, Sompom C, Petch P (2011) Phytoremediation of arsenic in submerged soil by wetland plants. Int J Phytorem 13:35–46

Jiang Y, Purchase D, Jones H, Garelick H (2011) Technical note: effects of arsenate (As5?) on growth and production of glutathione (GSH) and phytochelatins (PCs) in *Chlorella vulgaris*. Int J Phytorem 13:834–844

Kalbitz K, Wenrich R (1998) Mobilization of heavy metals and arsenic in polluted wetland soils and its dependence on dissolved organic matter. Sci Total Environ 209:27–39

Karadjova IB, Slaveykova VI, Tsalev DL (2008) The bio uptake and toxicity of arsenic species on the green microalga *Chlorella salina* in seawater. Aquat Toxicol 87:264–271

Khan I, Ahmad A, Iqbal M (2009) Modulation of antioxidant defence system for arsenic detoxification in Indian mustard. Ecotoxicol Environ Saf 72:626–634

King DJ, Doronila AI, Feenstra C, Baker AJ, Woodrow IE (2008) Phytostabilisation of arsenical gold mine tailings using four eucalyptus species: growth, arsenic uptake and availability after five years. Sci Total Environ 406:35–42

Kumar S, Dubey RS, Tripathi RD, Chakrabarty D, Trivedi PK (2015) Omics and biotechnology of arsenic stress and detoxification in plants: current updates and prospective. Environ Int 74:221–230

Levy JL, Stauber JL, Adams MS, Maher WA, Kirby JK, Jolley DF (2005) Toxicity, biotransformation, and mode of action of arsenic in two freshwater microalgae (*Chlorella* sp. and *Monoraphidium arcuatum*). Environ Toxicol Chem 24:2630–2639

Li N, Wang J, Song WY (2016) Arsenic uptake and translocation in plants. Plant Cell Physiol 57(1):4–13

Li RY, Ago Y, Liu WJ, Mitani N, Feldmann J, McGrath SP, Ma JF, Zhao FJ (2009) The rice aquaporin Lsi1mediates uptake of methylated arsenic species. Plant Physiol 150:2071–2080

Li Y, Dhankher OP, Carreira L, Lee D, Chen A, Schroeder JI, Balish RS, Meagher RB (2005) Overexpression of phytochelatin synthase in Arabidopsis leads to enhanced arsenic tolerance and cadmium hypersensitivity. Plant Cell Physiol 46:387

Li Y, Dhankher OP, Carreira L, Lee D, Chen A, Schroeder JI, Balish RS, Meagher RB (2004) Over expression of phytochelatin synthase in Arabidopsis leads to enhanced arsenic tolerance and cadmium hypersensitivity. Plant Cell Physiol 45:1787–1797

Liu WJ, Wood BA, Raab A, McGrath SP, Zhao FJ, Feldmann J (2010) Complexation of arsenite with phytochelatins reduces arsenite efflux and translocation from roots to shoots in Arabidopsis. Plant Physiol 152:2211–2221

Lloyd JR, Oremland RS (2006) Microbial transformations of arsenic in the environment: from soda lakes to aquifers. Elements 2:85–90

Ma JF, Yamaji N, Mitani N, Xu XY, Su YH, McGrath SP, Zhao FJ (2008) Transporters of arsenite in rice and their role in arsenic accumulation in rice grain. Proc Natl Acad Sci U S A 105:9931–9935

Ma LQ, Komar KM, Tu C, Zhang WH, Cai Y, Kennelley ED (2001) A fern that hyperaccumulates arsenic. Nature 409:579–579

Maeda S, Kusadome K, Arima H, Ohki A, Naka K (1992) Biomethylation of arsenic and its excretion by the alga *Chlorella vulgaris*. Appl Organomet Chem 6:407–413

Maeda S, Fujita S, Ohki A, Yoshifilku I, Higashi S, Takeshita T (1988) Arsenic accumulation by arsenic-tolerant freshwater blue-green algae alga (*Phormidium* sp.) Appl Organomet Chem 2:353–357

Mahmud R, Inoue N, Kasajima SY, Shaheen R (2008) Assessment of potential indigenous plant species for the phytoremediation of arsenic-contaminated areas of Bangladesh. Int J Phytorem 10:117–130

Mandal A, Purakayastha TJ, Patra AK, Sanyal SK (2012) Phytoremediation of arsenic contaminated soil by *Pteris vittata* L. I. Influence of phosphatic fertilizers and repeated harvests. Int J Phytorem 14:978–995

Marin AR, Masscheleyn PH, Patrick WH (1993) Soil redox-pH stability of arsenic species and its influence on arsenic uptake by rice. Plant Soil 152:245–253

Maurel C, Verdoucq L, Luu DT, Santoni V (2008) Plant aquaporins: membrane channels with multiple integrated functions. Annu Rev Plant Biol 59:595–624

Meharg AA (2004) Arsenic in rice –understanding a new disaster for South-East Asia. Trend Plant Sci 9:415–417

Meharg AA, Hartley-Whitaker J (2002) Arsenic uptake and metabolism in arsenic resistant and nonresistant plant species. New Phytol 154:29–43

Meharg AA, Jardine L (2003) Arsenite transport into paddy rice (*Oryza sativa*) roots. New Phytol 157:39–44

Meharg AA, Macnair MR (1992) Suppression of the high-affinity phosphate uptake system: a mechanism of arsenate tolerance in *Holcus lanatus* L. J Exp Bot 43:519–524

Meharg AA, Naylor J, Macnair MR (1994) Phosphorus nutrition of arsenate tolerant and nontolerant phenotypes of velvet grass. J Environ Qual 23:234–238

Mendoza-Cózatl DG, Jobe TO, Hauser F, Schroeder JI (2011) Long-distance transport, vacuolar sequestration, tolerance, and transcriptional responses induced by cadmium and arsenic. Curr Opin Plant Biol 14:554–562

Meng XY, Qin J, Wang LH, Duan GL, Sun GX, Wu HL, Chu CC, Ling HQ, Rosen BP, Zhu YG (2011) Arsenic biotransformation and volatilization in transgenic rice. New Phytol 191:49–56

Mitra A, Chatterjee S, Datta S, Sharma S, Veer V, Razafindrabe BHM, Walther C, Gupta DK (2014) Mechanism of metal transporter in plants. In: Gupta DK, Chatterjee S (eds) Heavy metal remediation transport and accumulation in plants. Nova Science Publishers Inc., New York, pp 1–27

Mitra N, Rezvan Z, Ahmad MS, Hosein MG (2012) Studies of water arsenic and boron pollutants and algae phytoremediation in three springs. Iran Int J Ecos 2:32–37

Miyashita S, Fujiwara S, Tsuzuki M, Kaise T (2011) Rapid biotransformation of arsenate into oxo-arsenosugars by a freshwater unicellular green alga, *Chlamydomonas reinhardtii*. Biosci Biotech Bioch 75:522–530

Miyashita S, Fujiwara S, Tsuzuki M, Kaise T (2012) Cyanobacteria produce arsenosugars. Environ Chem 9:474–484

Mondal P, Majumder CB, Mohanty B (2006) Laboratory-based approaches for arsenic remediation from contaminated water: recent developments. J Hazard Mater 137:464–479

Montes-Bayon M, Meija J, LeDuc DL, Terry N, Caruso JA, Sanz-Medel A (2004) HPLC–ICP-MS and ESI-Q-TOF analysis of biomolecules induced in *Brassica juncea* during arsenic accumulation. J Anal At Spectrom 19:153–158

Morelli E, Mascherpa MC, Scarano G (2005) Biosynthesis of phytochelatins and arsenic accumulation in the marine microalga *Phaeodactylum tricornutum* in response to arsenate exposure. Biometals 18:587–593

Mukhopadhyay R, Shi J, Rosen BP (2000) Purification and characterization of ACR2p, the *Saccharomyces cerevisiae* arsenate reductase. J Biol Chem 275:21149–21157

Munoz LP (2014) The mechanisms of arsenic detoxification by the green microalgae *Chlorella vulgaris*. Middlesex University, Dissertation

Murray LA, Raab A, Marr IL, Feldmann J (2003) Biotransformation of arsenate to arsenosugars by *Chlorella vulgaris*. Appl Organomet Chem 17:669–674

Natarajan S, Stamps RH, Saha UK, Ma LQ (2008) Phytofiltration of arsenic contaminated groundwater using *Pteris vittata* L: effect of plant density and nitrogen and phosphorus levels. Int J Phytorem 10:220–233

Patra M, Bhowmik N, Bandopadhyay B, Sharma A (2004) Comparison of mercury, lead and arsenic with respect to genotoxic effects on plant systems and the development of genetic tolerance. Environ Exp Bot 52:199–223

Peterson PJ, Girling CA (1981) Other trace metals. In: Lepp NW (ed) Effect of heavy metal pollution on plants, Vol 1. Effects of trace metals on plant functions. Applied Science Publishers, London, pp 213–278

Picault N, Cazale AC, Beyly A, Cuine S, Carrier P, Luu DT, Forestier C, Peltier G (2006) Chloroplast targeting of phytochelatin synthase in Arabidopsis: effects on heavy metal tolerance and accumulation. Biochimie 88:1743–1750

Pickering IJ, Gumaelius L, Harris HH, Prince RC, Hirsch G, Banks JA, Salt DE, George GN (2006) Localizing the biochemical transformations of arsenate in a hyperaccumulating fern. Environ Sci Technol 40:5010–5014

Pickering IJ, Prince RC, George MJ, Smith RD, George GN, Salt DE (2000) Reduction and coordination of arsenic in Indian mustard. Plant Physiol 122:1171–1177

Planas D, Healey FP (1978) Effects of arsenate on growth and phosphorus metabolism of phytoplankton. J Phycol 14:337–341

Pohl P (2007) Fractionation analysis of metals in dietary samples using ion-exchange and adsorbing resins. Trend Anal Chem 26:713–726

Poynton CY, Huang JWW, Blaylock MJ, Kochian LV, Elless MP (2004) Mechanisms of arsenic hyperaccumulation in *Pteris* species: root As influx and translocation. Planta 219:1080–1088

Qin J, Lehr CR, Yuan CG, Le XC, Mc Dermott TR, Rosen BP (2009) Biotransformation of arsenic by a Yellowstone thermoacidophilic eukaryotic alga. Proc Natl Acad Sci U S A 106:5213–5217

Qin J, Rosen BP, Zhang Y, Wang GJ, Franke S, Rensing C (2006) Arsenic detoxification and evolution of trimethylarsine gas by a microbial arsenite S-adenosylmethioninemethyltransferase. Proc Natl Acad Sci U S A 103:2075–2080

Raab A, Williams PN, Meharg A, Feldmann J (2007) Uptake and translocation of inorganic and methylated arsenic species by plants. Environ Chem 4:197–203

Raab A, Schat H, Feldmann J, Meharg AA (2005) Uptake, translocation and transformation of arsenate and arsenite in sunflower (*Helianthus annuus*): formation of arsenic–phytochelatin complexes during exposure to high arsenic concentrations. New Phytol 168:551–558

Raab A, Feldmann J, Meharg AA (2004) The nature of arsenic–phytochelatin complexes in *Holcus lanatus* and *Pteris cretica*. Plant Physiol 134:1113–1122

Rahman S, Kim KH, Saha SK, Swaraz AM, Paul DK (2014) Review of remediation techniques for arsenic (As) contamination: a novel approach utilizing bio-organisms. J Environ Manage 134:175–185

Rahman MA, Hasegawa H (2011) Aquatic arsenic: phytoremediation using floating macrophytes. Chemosphere 83:633–646

Roy M, Mukherjee A, Mukherjee S, Biswas J (2014) Arsenic: an alarming global concern. Int J Curr Microbiol App Sci 3:34–47

Salt DE, Smith RD, Raskin I (1998) Phytoremediation. Annu Rev Plant Biol 49:643–668

Sanders JG, Riedel GF (1993) Trace element transformation during the development of an estuarine algal bloom. Estuaries 16:521–532

Sanders JG (1979) Effects of arsenic speciation and phosphate concentration on arsenic inhibition of *Skeletonema costatum* (bacillariophyceae). J Phycol 15:424–428

Schat H, Llugany M, Vooijs R, Hartley-Whitaker J, Bleeker PM (2002) The role of phytochelatins in constitutive and adaptive heavy metal tolerances in hyperaccumulator and nonhyperaccumulator metallophytes. J Exp Bot 53:2381–2392

Schmoger MEV, Oven M, Grill E (2000) Detoxification of arsenic by phytochelatins in plants. Plant Physiol 122:793–801

Shamsuddoha ASM, Bulbul A, Imamul Huq SM (2006) Accumulation of arsenic in green algae and its subsequent transfer to the soil–plant system. Bangladesh J Microbiol 22:148–151

Sharma S, Chatterjee S, Datta S, Mitra A, Vairale MG, Veer V, Chourasia A, Gupta DK (2014) In vitro selection of plants for the removal of toxic metals from contaminated soil: Role of genetic variation in phytoremediation. In: Gupta DK, Chatterjee S (eds) Heavy metal remediation transport and accumulation in plants. Nova Science Publishers Inc, New York, pp 155–177

Sharples JM, Meharg AA, Chambers SM, Cairney JWG (2000) Mechanism of arsenate resistance in the ericoid mycorrhizal fungus *Hymenoscyphus ericae*. Plant Physiol 124:1327–1334

Shin H, Shin HS, Dewbre GR, Harrison MJ (2004) Phosphate transport in *Arabidopsis*: Pht1;1 and Pht1;4 play a major role in phosphate acquisition from both low- and high-phosphate environments. Plant J 39:629–642

Shri M, Dave R, Diwedi S, Shukla D, Kesari R, Tripathi RD, Chakrabarty D (2014) Heterologous expression of *Ceratophyllum demersum* phytochelatin synthase, CdPCS1, in rice leads to lower arsenic accumulation in grain. Sci Rep 4: 5784

Shukla D, Kesari R, Mishra S, Dwivedi S, Tripathi RD, Nath P, Trivedi PK (2012) Expression of phytochelatin synthase from aquatic macrophyte *Ceratophyllum demersum* L. enhances cadmium and arsenic accumulation in tobacco. Plant Cell Rep 31:1687–1699

Shukla D, Kesari R, Tiwari M, Dwivedi S, Tripathi RD, Nath P, Trivedi PK (2013) Expression of *Ceratophyllum demersum* phytochelatin synthase, CdPCS1, in *Escherichia coli* and Arabidopsis enhances heavy metal(loid)s accumulation. Protoplasma 250:1263–1272

Sigg L (1987) Surface chemical aspects of the distribution and fate of metal ions in lakes. In: Stumm W (ed) Aquatic surface chemistry. Wiley Inter science, New York, pp 319–348

Sigg L (1985) Metal transfer mechanisms in lakes; role of settling particles. In: Stumm W (ed) Chemical processes in lakes. Wiley Inter Science, New York, pp 283–310

Singh R, Singh S, Parihar P, Singh VP, Prasad SM (2015) Arsenic contamination, consequences and remediation techniques: a review. Ecotoxicol Environ Saf 112:247–270

Singh S, Juwarkar AA, Kumar S, Meshram J, Fan M (2007) Effect of amendment on phytoextraction of arsenic by *Vetiveria zizanioides* from soil. Int J Enviorn Sci Technol 4:339–344

Singh N, MaL Q, Srivastava M, Rathinasabapathi B (2006) Metabolic adaptations to arsenic induced oxidative stress in *Pteris vittata* L. and *Pteris ensiformis* L. Plant Sci 170:274–282

Song WY, Park J, Mendoza-Cózatl DG, Suter-Grotemeyer M, Shim D, Hortensteiner S, Geisler M, Weder B, Rea PA, Rentsch D, Schroede JI, Lee Y, Martinoia E (2010) Arsenic tolerance in Arabidopsis is mediated by two ABCC- type phytochelatin transporters. Proc Natl Acad Sci U S A 107:21187–21192

Srivastava S, Shrivastava M, Suprasanna P, D'Souza SF (2011) Phytofiltration of arsenic from simulated contaminated water using *Hydrilla verticillata* in field conditions. Ecol Eng 37:1937–1941

Su YH, McGrath SP, Zhu YG, Zhao FJ (2008) Highly efficient xylem transport of arsenite in the arsenic hyperaccumulator *Pteris vittata*. New Phytol 180:434–441

Taboada-de la Calzada A, Villa-Lojo MC, Beceiro-Gonzalez E, Alonso-Rodrıguez E, Prada-Rodrıguez D (1999) Accumulation of arsenic(III) by *Chlorella vulgaris*. Appl Organomet Chem 13:159–162

Takimura O, Fuse H, Murakami K, Kamimura K, Yamaoka Y (1996) Uptake and reduction of arsenate by *Dunaliella* sp. Appl Organomet Chem 10:753–756

Thomas DJ, Li J, Waters SB, Xing W, Adair BM, Drobna Z, Devesa V, Styblo M (2007) Arsenic (+3 oxidation state) methyltransferase and the methylation of arsenicals. Exp Biol Med 232:3–13

Ting YP, Prince IG, Lawson F (1991) Uptake of cadmium and zinc by the alga *Chlorella vulgaris*: II. Multi-ion situation. Biotechnol Bioeng 37:445–455

Tripathi P, Dwivedi S, Mishra A, Kumar A, Dave R, Srivastava S, Shukla MK, Srivastava PK, Chakrabarty D, Trivedi PK (2012) Arsenic accumulation in native plants of West Bengal, India: prospects for phytoremediation but concerns with the use of medicinal plants. Environ Monit Assess 184:2617–2631

Tripathi RD, Srivastava S, Mishra S, Singh N, Tuli R, Gupta DK, Matthuis FJM (2007) Arsenic hazards: strategies for tolerance and remediation by plants. Trend Biotechnol 25:158–165

Tu C, Ma LQ (2003) Interactive effects of pH, arsenic and phosphorus on uptake of As and P and growth of the arsenic hyper accumulator *Pteris vittata* L. under hydroponic conditions. Environ Exp Bot 50:243–251

Tuli R, Chakrabarty D, Trivedi PK, Tripathi RD (2010) Recent advances in arsenic accumulation and metabolism in rice. Mol Breed 26:307–323

Ullrich-Eberius CI, Sanz A, Novacky AJ (1989) Evaluation of arsenate- and vanadate-associated changes of electrical membrane potential and phosphate transport in *Lemna gibba* G1. J Exp Bot 40:119–128

Ultra VU, Tanaka S, Sakurai K, Iwasaki K (2007) Effects of arbuscular mycorrhiza and phosphorus application on arsenic toxicity in sunflower (*Helianthus annuus* L.) and on the transformation of arsenic in the rhizosphere. Plant Soil 290:29–41

Vetterlein D, Szegedi K, Neackermann J, Mattusch J, Neue HU, Tanneberg H, Jahn R (2007) Competitive mobilization of phosphate and arsenate associated with goethite by root activity. J Environ Qual 36:1811–1820

Wallace IS, Choi WG, Roberts DM (2006) The structure, function and regulation of the nodulin 26-like intrinsic protein family of plant aquaglyceroporins. Biochim Biophys Acta Biomem 1758:1165–1175

Wang ZH, Luo ZX, Yan CZ (2013) Accumulation, transformation, and release of inorganic arsenic by the freshwater cyanobacterium *Microcystis aeruginosa*. Environ Sci Pollut Res 20:7286–7295

Wang Y, Zhang CH, Wang S, Shen LY, Ge Y (2013) Accumulation and transformation of different arsenic species in nonaxenic *Dunaliella salina*. Environ Sci 34:4257–4265

Wang Y, Wang S, Xu P, Liu C, Misha L, Wang Y, Wang C, Zhang C, Ge Y (2015) Review of arsenic speciation, toxicity and metabolism in microalgae. Rev Environ Sci Biotechnol 14:427–451

Wang JR, Zhao FJ, Meharg AA, Raab A, Feldmann J, McGrath SP (2002) Mechanisms of arsenic hyperaccumulation in *Pteris vittata*. Uptake kinetics, interactions with phosphate, and arsenic speciation. Plant Physiol 130:1552–1561
Wang S, Mulligan CN (2006) Occurrence of arsenic contamination in Canada: sources, behavior and distribution. Sci Total Environ 366:701–721
Wesenberg D, Krauss GJ, Schaumloffel D (2011) Metallo-thiolomics: investigation of thiol peptide regulated metal homeostasis in plants and fungi by liquid chromatographymass spectrometry. Int J Mass Spectrom 307:46–54
Wild A (1988) Russell's soil conditions and plant growth, 11th edn. Longman, London
Wojas S, Clemens S, Sklodowska A, Antosiewicz D (2010) Arsenic response of AtPCS1- and CePCS-expressing plants— effects of external As(V) concentration on As accumulation pattern and NPT metabolism. J Plant Physiol 167:169–175
Wrench JJ, Addison RF (1981) Reduction, methylation and incorporation of arsenic into lipids by the marine phytoplankton *D. tertiolecta*. Can J Fish Aquat Sci 38:518–523
Wu Z, Ren H, McGrath SP, Wu P, Zhao FJ (2011) Investigating the contribution of the phosphate transport pathway to arsenic accumulation in rice. Plant Physiol 157:498–508
Xu XY, McGrath SP, Zhao FJ (2007) Rapid reduction of arsenate in the medium mediated by plant roots. New Phytol 176:590–599
Xue XM, Raber G, Foster S, Chen SC, Francesconi KA, Zhu YG (2014) Biosynthesis of arsenolipids by the cyanobacterium *Synechocystis* sp. PCC 6803. Environ Chem 11:506–513
Xue HB, Sigg L (1990) Binding of Cu(II) to algae in a metal buffer. Water Res 24:1129–1136
Ye J, Rensing C, Rosen BP, Zhu YG (2012) Arsenic biomethylation by photosynthetic organisms. Trend Plant Sci 17:155–162
Yin XX, Wang LH, Bai R, Huang H, Sun GX (2012) Accumulation and transformation of arsenic in the blue-green alga *Synechocysis* sp. PCC6803. Water Air Soil Pollut 223:1183–1190
Yin XX, Chen J, Qin J, Sun GX, Rosen BP, Zhu YG (2011) Biotransformation and volatilization of arsenic by three photosynthetic cyanobacteria. Plant Physiol 156:1631–1638
Zhang SY, Rensing C, Zhu YG (2014) Cyanobacteria-mediated arsenic redox dynamics is regulated by phosphate in aquatic environments. Environ Sci Technol 489:994–1000
Zhang JY, Ding TD, Zhang CL (2013) Biosorption and toxicity responses to arsenite (As(III)) in Scenedesmus quadricauda. Chemosphere 92:1077–1084
Zhang B, Wang LH, Xu YX (2011) Study on absorption and transformation of arsenic in blue alga (*Synechocystis* sp. PCC6803). Asian J Ecotoxicol 6:629–633
Zhao Y, Shang D, Ning J, Zhai Y (2012) Arsenic and cadmium in the marine macroalgae (*Porphyra yezoensis* and *Laminaria japonica*)—forms and concentrations. Chem Spec Bioavail 24:197–203
Zhao FJ, McGrath SP, Meharg AA (2010) Arsenic as a food chain contaminant: mechanisms of plant uptake and metabolism and mitigation strategies. Annu Rev Plant Biol 61:535–559
Zhao FJ, Ma JF, Meharg AA, Mc Grath SP (2008) Arsenic uptake and metabolism in plants. New Phytol 181:777–794
Zhao R, Zhao M, Wang H, Taneike Y, Zhang X (2006) Arsenic speciation in moso bamboo shoot–a terrestrial plant that contains organoarsenic species. Sci Total Environ 371:293–303
Zhao FJ, Wang JR, Barker JHA, Schat H, Bleeker PM, McGrath SP (2003) The role of phytochelatins in arsenic tolerance in the hyperaccumulator *Pteris vittata*. New Phytol 159:403–410
Zhu YG, Williams PN, Meharg AA (2008) Exposure to inorganic arsenic from rice: a global health issue. Environ Pollut 154:169–171

Genomics and Genetic Engineering in Phytoremediation of Arsenic

Sarma Rajeev Kumar, Gowtham Iyappan, Hema Jagadeesan, and Sathishkumar Ramalingam

Contents

1 Introduction 172
2 Microbial Genomics of Arsenic Resistance 172
3 Genetic Engineering of Arsenic Biotransformation in Microbes 173
3.1 Phytoremediation 174
4 As Uptake and Transport in Plants 177
5 Reduction of As(V) to As(III) Is Essential for Coupling with Phytochelatins 178
6 Sequestration of PC-As(III) Complex to Vacuole 179
7 Concluding Remarks 182
References 182

S.R. Kumar
Molecular Plant Biology and Biotechnology Laboratory, CSIR—Central Institute of Medicinal and Aromatic Plants, Research Centre, Bangalore, India

G. Iyappan
DRDO—BU Centre for Life Sciences, Bharathiar University, Coimbatore, India

H. Jagadeesan
Department of Biotechnology, PSG College of Technology, Coimbatore, India

S. Ramalingam (✉)
DRDO—BU Centre for Life Sciences, Bharathiar University, Coimbatore, India

Plant Genetic Engineering Laboratory, Department of Biotechnology, Bharathiar University, Coimbatore, India
e-mail: rsathish@buc.edu.in

D.K. Gupta, S. Chatterjee (eds.), *Arsenic Contamination in the Environment*,
DOI 10.1007/978-3-319-54356-7_8

1 Introduction

Numerous toxic pollutants including heavy metals and organic compounds like pesticides, herbicides, and insecticides have reached unacceptably high levels in the environment due to continuous use or discharge from industries, agricultural, and municipal processes (Mahar et al. 2016). The concentrations of these contaminants can vary from highly toxic concentrations from an accidental spill to barely detectable amounts that, after long-term exposure, can be detrimental to human health. Arsenic (As), a highly toxic metalloid categorized under group A carcinogen, has influenced human history more than any other toxic element or compound (Nriagu 2001). Ingestion of As can lead to several types of cancers (Bundschuh et al. 2012). Lately, huge populations from different parts of the world have suffered from As poisoning due to consumption of As-contaminated groundwater (Verbruggen and LeDuc 2009). In India, As levels in West Bengal and Madhya Pradesh exceeded the threshold levels (Chaurasia et al. 2012). Use of As-based herbicides, insecticides, and mining activities resulted in As contamination leading to increase in As levels in groundwaters. As a result of consumption of As-contaminated water, several types of cancers have been reported in Taiwan, Argentina, Chile, Bangladesh, and India.

Elemental pollutants like As are particularly difficult to remediate from soil, water, and air as they cannot be degraded to harmless small molecules. In addition, the cost of cleaning up sites contaminated with heavy metals including As is extremely high and also risky due to adverse health effects. Global costs annually ranges from $25 to 50 billion for remediating contaminated sites (Tsao 2003). The high costs involved in remediation have often led to abandoning the polluted sites rather than cleaning up.

2 Microbial Genomics of Arsenic Resistance

Microorganisms have evolved to cope up with As through different cellular mechanisms. Similar to other heavy metals, resistance to As is achieved by one of the several ways including active extrusion of As, extracellular precipitation, and subsequent chelation of As species. Once As enters the cell, they are immediately transformed and compartmentalized in order to reduce the mobility and bioavailability (Slyemi and Bonnefoy 2012; Andres and Bertin 2016). In some bacterial species, As is converted to methylated derivatives like monomethylarsonic acid (MMA) and dimethylarsinic acid (DMA) (Li X et al. 2014). The use of combined biochemical and molecular-genetic approaches has shed light on different physiological mechanisms and process involved in As metabolism in microbes.

An arsenite oxidase operon was first characterized from bacterium *Herminiimonas arsenicoxydans* and complete genome has been sequenced (Muller et al. 2003, 2007). Later, genomes of several As-metabolizing archaea and bacteria were sequenced and genome analyses have led to the identification and characterization of genetic determinants of As-related processes. Further, transcriptome analysis resulted in the identification of As-related genes throughout prokaryotic domain and regulatory networks controlling their expression and As levels (Andres and Bertin 2016). Several As metabolizing gene homologues of Arsenate reductase like *aio* genes involved in transformation of As(III) to As(V) (van Lis et al. 2013) and *arx* involved in As(V) during anaerobic respiration of As(V) were reported (Zargar et al. 2012). Further, using genomic tools, genes involved in As methylation and demethylation have been identified. Products of *arsM* and *arsI* genes encoding an arsenite methyltransferase and a C·As lyase, respectively, are involved in methylation and demethylation.

The availability of genome sequences led to understanding the concept of gene cluster involved in As related metabolic genes. The simple ars cluster with arsRBC operon confers tolerance to As (Kruger et al. 2013). arsRDABC cluster found in a class of bacterial species conferring high level As resistance was identified through a genomic approach (Li X et al. 2014). Presence of several homologues of *arr*, *aio*, and *arx* genes in As metabolic clusters was identified and characterized through functional genomic approaches (Zargar et al. 2012; van Lis et al. 2013). The above-mentioned studies are few among many others, which led to the understanding of As-related genes as operons or metabolic clusters in prokaryotes.

Identification of 71 kb DNA in *Alcaligenes faecalis* containing more than 20 genes involved in As resistance and metabolic processes revealed the concept of Arsenic island (Silver and Phung 2005). A similar kind of genomic island was subsequently identified in *H. arsenicoxydans* (Muller et al. 2007). Freel et al. (2015) identified more than 20 genomic islands in three different *Thiomonas* sp. isolated from acid mine drainage. In addition, genomic studies have led to the identification of several As resistance- and metabolism-related clusters that coexist in a genome of *H. arsenicoxydans* and *Anabaena* sp. (Muller et al. 2007; Pandey et al. 2013). *Achromobacter* genome sequence revealed genomic arsenic island, an intact type III secretion system, and multiple metal (loid) transporters, which together confer resistance to As (Li et al. 2011). Andres et al. (2013) reported the genome of *Rhizobium* with megaplasmid carrying various genes involved in metabolism of arsenite. Hence, it could be possible that multiple copies of As-related genes present in genome, or extrachromosomal regions like plasmids and genomic islands together contribute to As resistance.

3 Genetic Engineering of Arsenic Biotransformation in Microbes

As is a ubiquitous toxic metal pollutant that has influenced the evolution of several As-resistant bacteria and archaea. Either adsorption or transportation confers As resistance through ABC transporters/methylation. Heterologous expression of *arsM* gene from *Rhodopseudomonas palustris* in *E. coli* resulted in the formation of methylated As and exhibited As tolerance (Liu et al. 2011). Similarly, *arsM* when overexpressed in *Sphingomonas desiccabilis* and *Bacillus idriensis* resulted in more than tenfold As bio-volatilization under in vitro conditions (Liu et al. 2011). Integration of S-adenosylmethionine methyltransferase (ArsM) to chromosome of *Pseudomonas putida* KT2440 resulted in a fivefold increase in bio-volatilization as compared to wild-type bacterium (Chen J et al. 2013; Chen Y et al. 2013). Recombinant *P. putida* stably expressing arsM:GFP fusion protein was capable of detoxifying methylated As species (Chen et al. 2014). Genetically engineered *B. subtilis* expressing arsenite S-adenosylmethionine methyltransferase from alga *Cyanidioschyzon merolae* converted As to volatile dimethylarsenate and trimethylarsine (Huang et al. 2015). Although there are several reports of using microbes for removal of As and other heavy metals, practical applications are limited due to several issues. Hence, phytoremediation has been attempted as an alternate system for removing As from contaminated sites.

3.1 Phytoremediation

Phytoremediation is the use of plants to remediate contaminated sites. This phenomenon exploits the innate ability of plants to extract chemicals from water, soil, and air using sunlight as energy source. Another advantage of using plants is that they act as soil stabilizers by reducing the amount of contaminated dust that could leave the site. The main objective of phytoremediation on elemental pollutants is to extract soil borne contaminants, transport, and concentrate them in aboveground tissues by orders of magnitude for reprocessing. Plants can also assist with the cleanup of As-contaminated surface water. The use of microorganisms in engineered bioremediation systems had mixed success. Further, plants can support bacterial remediation schemes for elemental pollutants like As, because of their superior ability to enhance rhizosphere activity . The main advantage is that it is approximately 10–15 times less expensive compared to several conventional strategies (Dhankher et al. 2006). Phytoremediation of different heavy metals has been successfully improved with transgenic plants (Pilon-Smits and Pilon 2002; Cases and De Lorenzo 2005; Eapen and D'Souza 2005; Meagher and Heaton 2005; Doty 2008; Kotrba et al. 2009; Zhao and McGrath 2009; Li G et al. 2014; Li et al. 2016). A straightforward approach for improving the effectiveness of phytoremediation is to overexpress the genes involved in As uptake, transport, or metabolism. The introduction of these genes can be readily achieved in many plant species using *Agrobacterium*-mediated plant transformation. Several genes involved in uptake, translocation, and sequestration of As have been identified from the model plant *Arabidopsis* or other natural hyperaccumulators and their functional role have been validated in heterologous system (Table 1). Nevertheless, their ability is limited owing to their small size, slow growth rates, and restricted growth habitat (Meagher and Heaton 2005).

Table 1 Genetic engineering in phytoremediation of arsenic

Gene source	Transgenic host	Mechanism of tolerance	References
ars C from *E. coli*	*Arabidopsis thaliana*	Arsenate reductase (ArsC) catalyzed glutathione (GSH)-coupled electrochemical reduction of arsenate to the more toxic arsenite. ArsC enhanced arsenic accumulation in shoots; however, transgenic lines did not exhibit any increase in tolerance	Dhankher et al. (2002)
γ-ECS—γ-glutamylcysteine from *E. coli*	*A. thaliana*	γ-ECS mediates biosynthesis of GSH, which is the intermediate for phytochelatin biosynthesis and transgenic lines were moderately tolerant to arsenic compounds	Dhankher et al. (2002)
ars C and γ-ECS	*A. thaliana*	When grown on arsenic, transgenic plants accumulated 4- to 17-fold greater shoot weight and accumulated more than threefold more arsenic compared to WT plants	Dhankher et al. (2002)
Enterobacter cloacae UW4 1-aminocyclopropane-1-carboxylate (ACC) deaminase gene	*Brassica napus*	Tolerance to As was achieved via hyperaccumulation and enhanced biomass. Transgenic plants grew to a significantly greater extent than nontransformed control plants	Nie et al. (2002)
PCS from *A. thaliana*	*A. thaliana*	Overexpression under the constitutive promoter conferred resistance to As. Further, transgenic lines exhibited ~20–100 times higher biomass on 250 and 300 μM arsenate than WT	Li (2004)
ATP sulfurylase from *B. juncea*	*B. juncea*	Transgenic seedlings had up to 2.5-fold higher shoot concentrations of As(III), As(V). Transgenic lines were also tolerant to other heavy metals including Hg, Cd, Cu, and Zn	Wangeline et al. (2004)
Endogenous root *ARS2 (arsenate reductase) of A. thaliana*	*A. thaliana*	RNA knockdown plants reduced ACR2 protein expression in *Arabidopsis* 2% compared to WT. Knockdown lines were more sensitive to high concentrations of arsenate, but not arsenite and accumulated 10- to 16-fold more arsenic in shoots and retained less arsenic in roots than wild type	Dhankher et al. (2006)
PCS from *A. thaliana*	*A. thaliana*	Targeting *AtPCS1* to plastids induced a marked sensitivity to As whereas lines overexpressing same gene in the cytosol were more tolerant to As. This difference in tolerance was implicated to As trafficking pathway in plants	Picault et al. (2006)
PCS from *A. thaliana*	*B. juncea*	Although accumulation of As in transgenic plants was similar to that WT, transgenic plants displayed 1.4 times longer roots on media with 500 μM AsO_4^{3-}	Gasic and Korban (2007)
AsPCS1 of garlic *Allium sativum*—metal chelator	*A. thaliana*	Improved tolerance to arsenate/arsenite and hyperaccumulation of As	Guo et al. (2008)

(continued)

Table 1 (continued)

Gene source	Transgenic host	Mechanism of tolerance	References
GSH1 and *AsPCS1* of *S. cerevisiae* and *A. sativum* respectively	*A. thaliana*	Transgenic plants displayed more than twofold increase in root formation and 3–10 times more accumulation of AsO_2^-	Guo et al. (2008)
glutaredoxin, PvGRX5 from *Pteris vittata*	*A. thaliana*	Transgenic lines were more tolerant to As compared with control lines and contained significantly lower total As compared with control lines following treatment with As (V). Additionally, transgenic lines efficiently reduced arsenate under in vivo conditions	Sundaram et al. (2009)
As(III) methylases (*CmarsM7* and *CmarsM8*) from *Cyanidioschyzon merolae*	*E. coli*	*CmarsM7* and *CmarsM8* conferred resistance to As(III) in *E. coli*. Purified recombinant protein transformed As(III) into monomethylarsenite, DMAs(V), TMAO, and trimethylarsine gas under in vitro conditions	Qin et al. (2009)
PCS1 from *A. thaliana* and *PCS* from *Ceratophyllum demersum*	Tobacco	Upon exposure to less concentration of As(V), both transgenic line accumulated more As in roots and leaves than WT. However, at higher As (V) concentration, reduction in PC levels in both roots and leaves of transgenic lines relative to WT was observed	Wojas et al. (2010)
As *PCS1* of garlic and YCF (ABC transporter) of *S. cerevisiae*	*A. thaliana*	Dual-gene transgenic lines displayed longer roots and the highest accumulation of As. Also double transgenic lines accumulated over two- to tenfold arsenite and two- to threefold arsenate than WT	Guo et al. (2012)
PCS1 from *Ceratophyllum demersum*	Tobacco	Transgenic tobacco plants expressing *PCS1* exhibited severalfold increased PC content and precursor nonprotein thiols. Also, accumulation of As did not compromise the plant growth	Shukla et al. (2012)
PvACR3 from *Pteris vittata* translocate As to shoot	*A. thaliana*	Transgenic lines exhibited less As toxicity in roots and As was translocated to shoots	Chen J et al. 2013; Chen Y et al. 2013
PCS1 from *C. demersum*	*O. sativa*	Transgenic rice lines displayed an increase in PCS activity and accumulation of PCs. Transgenic lines accumulated As in root and shoot; transgenic plants accumulated significantly lower As in grain and husk compared to WT	Shri et al. (2014)
Glutaredoxins (Grx) from rice	*A. thaliana*	Transgenic lines displayed significantly reduced As accumulation in seeds and shoot tissues compared to WT plants during both As(III) and As(V) stress	Verma et al. (2016)
Cytokinin (CK) mutants *A. thaliana*	*A. thaliana*/Tobacco	As(V) causes severe depletion of endogenous CKs in WT *A. thaliana*. *A. thaliana* CK signaling mutants and transgenic lines with reduced endogenous CK levels showed an As(V)-tolerant phenotype. Further, Transgenic CK-deficient *A. thaliana* and tobacco lines exhibited an increase in arsenic accumulation	Mohan et al. (2016)

4 As Uptake and Transport in Plants

Over the past decade, the role of several transporters involved in As uptake in plants has been reported (Shin et al. 2004; Catarecha et al. 2007; Zhu and Rosen 2015). Although different transporters, enzymes have been characterized from different plant species, a general mechanism of As uptake and transport shown in Fig. 1. Under high concentration of heavy metals including As, in addition to transporting essential minerals or nutrients, transporters also mobilize unwanted heavy metals like As, Cadmium (Cd), and Mercury (Hg) (Schroeder et al. 2013). Once taken up, plants translocate different As species through different transporters and distinct pathways (Li et al. 2016). The free As present in soil is converted by microbes to arsenate [As(V)], arsenite [As(III)], and further to methylated As derivates like MMA and DMA. As(V) major form in soil enters plant roots through phosphate (Pi) transporters (since oxyanion form of As(V) is a structural analogue to Pi). Several high affinity Pi transporters involved in As transport have been reported

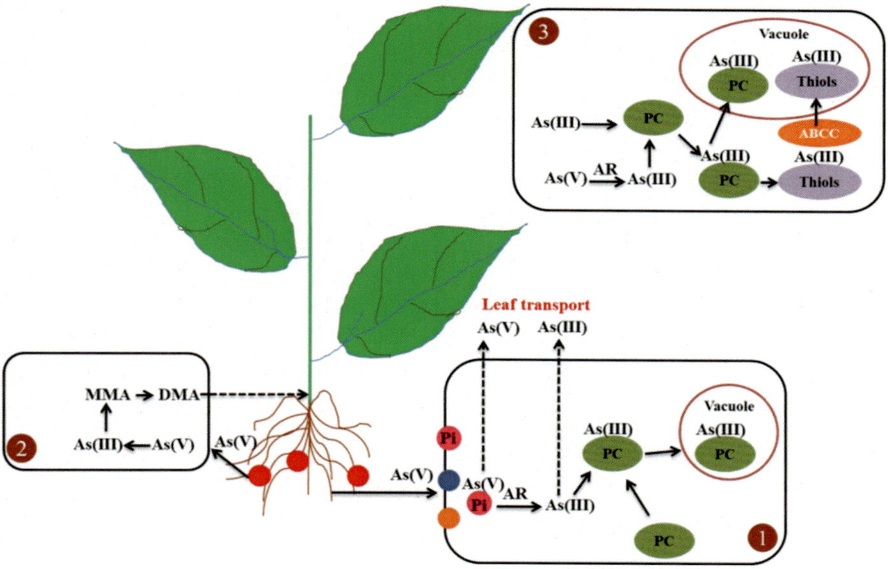

Fig. 1 Arsenic uptake and transport in plants—(*1*) Mechanism in Root-Free As(V) present in soil is transported by different Pi or other transporters (*pink/blue/orange circles*) to root cells and As(V) is reduced to As(III) by Arsenate reductase (AR). As (III) is either conjugated to phytochelatins (PC) (*green circles*) and transported to vacuole as As(III) PC complex or transported to shoots/leaves through xylem or phloem. (*2*) The bacteria present in the rhizosphere also methylates As(III) to DMA or MMA by different methylases and methylated As species are also transported to shoots. (*3*) Leaf metabolism-As(V) is reduced to As(III) by AR or their homologues and also free As(III) transported are conjugated to PC and transported to vacuole. Alternatively, the As(III) thiol derivatives (*violet circles*) are transported to vacuole by various transporters including ABCC transporters (*orange*)

from *A. thaliana* and in other plants (Shin et al. 2004; Catarecha et al. 2007). Overexpression of *AtPht1*:7, a Pi transporter in *Arabidopsis*, enhanced As(V) accumulation in transgenic lines (LeBlanc et al. 2013). Overexpression of high affinity phosphate transporter *OsPht1:8* in rice enhanced As(V) uptake and translocation in transgenic lines (Wu et al. 2011). Uptake of As by roots also depended on the concentration of Pi in soil. Higher concentrations of external Pi reduce As uptake by competitive inhibition and subsequent accumulation in plants (Lou-Hing et al. 2011). In addition, it has also been reported that mutations in Pi transporters reduce As(V) uptake by roots and enhanced As(V) resistance (Remy et al. 2012; Kamiya et al. 2013). Unfortunately, suppression of Pi transporters drastically affects the physiology of plants and hence it is very essential to specifically identify and target particular As/Pi transporters to prevent As uptake without negatively affecting the physiological functions of the plant.

The role of several transcriptional regulators including WRKY family of transcription factors has been implicated in the transport of As. Two classes of WRKY transcription factor, WRKY6 and WRKY45, regulate As(V) uptake by modulating the expression of *AtPht1:1* transporter (Castrillo et al. 2013; Wang et al. 2014). Unlike phosphate transporters, aquaporins, membrane channels that are involved in the transport of water and neutral molecules, are responsible for the transport of As(III), the predominant form of As present in anaerobic environment including submerged soils (Li G et al. 2014). Additionally, the role of Nodulin 26-like intrinsic proteins from *A. thaliana*, *O. sativa*, and *Lotus japonicus* has been associated with the uptake of As(III) (Bienert et al. 2008). *OsNIP3:3* when overexpressed in yeast mutant lacking As(III) efflux transporter were sensitive to As(III), proving its role in NIP transporters for As(III) uptake or transport (Ali et al. 2012). Overexpression of different rice plasma membrane intrinsic proteins OsPIP2;4, OsPIP2;6, and OsPIP2;7 in *A. thaliana* enhanced As(III) tolerance, resulting in active influx and efflux of As(III) in plant roots that are exposed to As(III) (Mosa et al. 2012).

As mentioned above, methylated As species especially MMA and DMA present in the soil is produced as a result of As biomethylation by different soil microorganisms (Li et al. 2016). It has been reported in rice that *OsLsi1* (aquaporin) is critical for uptake of methylated As species and in *OsLsi1* rice mutants the uptake of MMA and DMA was reduced to almost 80% (Li et al. 2009).

5 Reduction of As(V) to As(III) Is Essential for Coupling with Phytochelatins

Inside the plants cells, As species reach different cellular components before undergoing detoxification through xylem or phloem transport. Primarily, As(V) is reduced to As(III) by As(V) reductase (AR) and the As(III) thus generated is coupled with phytochelatins (PCs) and transported to vacuole. To date, several *AR* genes have been isolated and functionally characterized from plant species across different plant families (Zhao et al. 2009). Sanchez-Bermejo et al. (2014) identified AR homologue *ATQ1* from *A. thaliana* and *ATQ1* T-DNA mutants were highly sensitive to As(V). The ratio of As(III)/As(V) was significantly altered in mutant lines. When *ATQ1* was heterologously expressed in *E. coli* mutant ΔarsC, lacking a functional *AR* gene, *ATQ1* enhanced As(III) extrusion in the presence of As(V). *ATQ1* homologue, High Arsenic Content1 (*HAC1*) having AR activity, was functionally characterized in *A. thaliana* by Chao et al. (2014). The coupling of As(III) with PCs is an important mechanism through which plants detoxify As.

With an extensive knowledge generated from the studies using model systems by molecular-genetic and biochemical approaches, several efforts were made to increase the ability of plants to pump out toxic As from soil (Doucleff and Terry 2002). Dhankher et al. (2002) generated transgenic *Arabidopsis* overexpressing *E. coli* arsenate reductase C (*ArsC*) and *E. coli* γ-glutamylcysteine synthetase (*γ-ECS*) genes to enhance mobility and sequestration of As. *ArsC* catalyses the glutathione (GSH) coupled reduction of Arsenate to highly toxic Arsenite and transgenic *A. thaliana* overexpressing *ArsC* was hypersensitive to Arsenate as genes were expressed under light-induced promoter (therefore was not expressed in roots). Subsequently, transgenic plants overexpressing *γ-ECS* (is involved in the formation of glutamylcysteine, a committed step in the synthesis of GSH) were generated that exhibited tolerance to As compared to WT. Interestingly, transgenic lines overexpressing both *ArsC* and *γ-ECS* were able to accumulate threefold more As as compared to WT. Few years later, two independent groups identified As metabolizing genes from *Pteris vittata*, which is an extraordinary arsenic-hyperaccumulating fern. Ellis et al. (2006) identified an arsenate reductase gene (*PvACR2*) that complemented arsenic hyperaccumulation phenotype in *ACR2* yeast mutants. Rathinasabapathi et al. (2006) used a cDNA library-based approach to identify As resistance genes from *P. vittata* and identified a plant cytosolic triosephosphate isomerase exhibiting arsenate reductase activity. Recombinant *E. coli* expressing triose phosphate isomerase displayed more of the free As than the arsenite form confirming the arsenate reductase activity of triosephosphate isomerase.

6 Sequestration of PC-As(III) Complex to Vacuole

Phytochelatin synthase (PCS) catalyzes the last step in the biosynthesis of cysteine-rich thiol reactive peptides phytochelatins that are involved in processing of many thiol-reactive toxicants. Transgenic *Arabidopsis* lines overexpressing *AtPCS* were highly resistant to As 300 μM arsenate as compared to WT plants. Transgenic plants exhibited the expression of many unknown thiol products upon As exposure and also displayed a significant increase in levels of γ-glutamyl cysteine (γ-EC), which is the substrate for phytochelatin synthesis (Li 2004). Another study proved the role of γ-glutamylcysteine synthetase in conferring tolerance to heavy metals including As (Li et al. 2006). A modified bacterial *ECS* when overexpressed in shoots of the ECS-deficient, heavy-metal-sensitive cad2-1 mutant of *Arabidopsis* enhanced arsenate tolerance significantly beyond control plants. Overexpression of *ECS* in mutant lines resulted in an increase in levels of γ-EC, several phytochelatins including PC2, and PC3 compared to mutants. In addition, the shoot and root levels of glutathione were close to fivefold higher in transgenic lines implicating EC and glutathione are efficiently transported from shoots to roots for uptake and transport of metabolites (Li et al. 2006).

Kim et al. (2005) identified a phytochelatin synthase (*PCS1)* from tobacco and its heterologous expression in yeast mutants resulted in enhanced As tolerance. Recently, transgenic tobacco lines overexpressing *NtPCS1* exhibited enhanced tolerance to arsenite; however, there was no change in As accumulation in transgenic lines (Lee and Hwang 2015). Shukla et al. (2012) functionally characterized a *PCS1* isolated from *Ceratophyllum demersum*, a submerged aquatic macrophyte. Transgenic tobacco lines overexpressing *CdPCS1* displayed severalfold increase in PC levels, nonprotein thiols, and enhanced accumulation of As without compromising plant growth or biomass (Shukla et al. 2012). *CdPCS1* overexpressing transgenic rice lines exhibited higher PCS activity, enhanced synthesis of PCs, and higher accumulation of As in root and shoot. Interestingly, transgenic lines accumulated significantly lower As in grain and husk when compared to control plants. The higher level of PCs in transgenic lines presumably improved sequestration and detoxification of As in roots and shoots further restricting its accumulation in grain. This study implicates the role of *CdPCS1* as an inevitable part of metal detoxification mechanism in hyperaccumulator (*C. demersum*). An attempt was made to enhance As accumulation in Indian mustard *Brassica juncea* by overexpressing *AtPCS1*. Although transgenic plants exhibited significantly higher tolerance to As, accumulation of As in transgenic plants was similar to that of WT plants (Gasic and Korban 2007). Guo et al. (2008) overexpressed *PCS1* and *GSH1* derived from *Allium cepa* and yeast respectively in *Arabidopsis*. Transgenic lines overexpressing both genes exhibited improved accumulation of heavy metals including As when compared to transgenic plants expressing either of one.

Song et al. (2010) implicated the crucial role of two ABCC type transporters in As remediation. *Arabidopsis* mutants lacking AtABCC1 and AtABCC2 were extremely sensitive to As and As-based herbicides. Further, heterologous expression of these ABCC transporters in yeast enhanced As tolerance and accumulation. In addition, overexpression of *AtPCS1* and *AtABCC1* resulted in transgenic lines tolerant to As. Transgenic *A. thaliana* expressing a PvACR3, a key arsenite [As(III)] antiporter from *P. vittata,* exhibited enhanced As tolerance. Transgenic lines accumulated approximately 7.5-fold more As in aboveground tissues than wild-type plants when plants were grown in presence of As(V). Subsequent As analysis revealed that PvACR3 significantly reduced As concentrations in roots and at the same time increased shoot As even under 150 μM As(V) (Chen J et al. 2013; Chen Y et al. 2013). Recently, the effect of As alone or in combination with Cd in transgenic tobacco lines overexpressing *A. thaliana PCS1* revealed interesting findings (Zanella et al. 2016). Transgenic lines showed higher PC levels, accumulation of both As and Cd in roots, and detoxification ability. However, metal exposure damaged lateral root apices in transgenic lines. Although As alone or in combination with Cd caused damage to the lateral roots, it did not negatively influence the accumulation and detoxification of either As or Cd (Zanella et al. 2016).

Irrespective of reduction of As(V) to As(III) and coupling with PCs, sequestration of As(III)-PCs to vacuoles is the ultimate step involved in detoxification of As, since the complex is stable only in acidic environment as that of vacuole. A novel vacuolar transporter from rice OsABCC1 involved in transporting As(III)-PCs across tonoplast was recently characterized by Song et al. (2014a, b). OsABCC1 mutants were highly sensitive to As and *A. thaliana* overexpression lines exhibited enhanced tolerance to As. Moreover, when *OsABCC1* was overexpressed in atabcc1/atabcc2 double mutant, As-sensitive phenotype was rescued implicating their role in As detoxification. *P. vittata* PvACR3, an ACR3 As(III) efflux protein, sequesters As into vacuoles and plants devoid of PvACR3 displayed As(III)-sensitive phenotype (Indriolo et al. 2010). Similarly, overexpression of *PvACR3* in *Arabidopsis* enhanced As tolerance (Chen J et al. 2013; Chen Y et al. 2013).

As hyperaccumulating capacity and As tolerance depend on As translocation from the roots to the shoots (Verbruggen et al. 2009; Verbruggen and LeDuc 2009). Complete mechanism through which translocation of As from root to shoot has not been understood yet, which is very essential for future breeding or transgenic approaches for generation of As hyperaccumulators that can be used in phytoremediation. The long distance transport of As in plants depends on the ratio of As present in roots and shoots. Also, in general, the mobility of As from roots to shoots is slow except in hyperaccumulators. This is partly due to the reason that arsenate is reduced to arsenite rapidly in roots and immediately complexes with thiols and transported to root vacuoles. Knockdown of *A. thaliana* arsenate reductase *AtACR2* resulted in As accumulation in the shoots (Dhankher et al. 2006). Silencing of *AtACR2* resulted in more arsenate in the roots that was available for transport to the shoots through xylem mediated by phosphate transporters. Contrarily, T-DNA insertion lines of *AtACR2* displayed reduced As translocation to the shoots (Bleeker et al. 2006). Hence, the underlying mechanism behind the transport of different As species from root to shoot is not yet completely elucidated. The role of *OsLsi2* in xylem loading of As(III) in rice roots is well studied and mutation of *OsLsi2* drastically affected As(III) level in xylem but did not influence uptake of As(III) by roots, implicating *OsLsi2* plays a critical role in phloem loading (Ma et al. 2008). The role of NRAMP (Natural Resistance-Associated Macrophage Protein) in the xylem loading and shoot accumulation of As in plants was reported (Tiwari et al. 2014). Heterologous expression of NRAMP in yeast *fet3:fet4* mutant rescued defective Fe transport and transgenic *A. thaliana* expressing *OsNRAMP1* exhibited enhanced tolerance to As(III) (Sarangi et al. 2009; Tiwari et al. 2014). Despite the fact that methylated As species including DMA is taken up by roots inefficiently, they are translocated more efficiently from roots to shoots than the inorganic As. However, the key regulators of methylated As transport in plants remain unclear (Carey et al. 2010; Ye et al. 2010).

Although there are several studies related to xylem loading and transport of As species, there exists only little knowledge on the mechanism of As transport in the phloem. It has been already reported that PCs and other conjugated thiol peptides are transported through phloem in *A. thaliana* (Chen et al. 2005). However, it is still under investigation whether As(III)-PC complexes can be transported in phloem as these complexes are not stable at neutral or slightly alkaline pH of phloem sap. Knockout of OsABCC1 (involved in the detoxification and reduction of As) in rice was susceptible to As compared to wild-type plants demonstrating the role of ABCC1 transporters in which they act as inhibitors of translocation of As into rice grains by transporting As into the vacuoles (Song et al. 2014a, b).

7 Concluding Remarks

Over the last years, genomics combined with metabolome studies helped to decipher the genetic basis of As resistance. Functional genomic approaches revealed distinct cellular processes of As uptake and metabolism in both prokaryotes and plants. However, several gaps do exist in As-related phenomenon for which molecular and biochemical evidences are lacking. The genes involved in As methylation in plants are not yet identified and characterized. Also, molecular understanding of As is still in its infancy. Future efforts combining biochemical, molecular-genetic, and genomic approaches could fill the above gaps quickly. Generation of transgenic plants that tolerate not only As but also uptake and sequestrate As will be a popular strategy for remediation. Lately, attention has been focused on the role of endophytic bacteria in phytoremediation. These endophytes can assist plants in the phytoremediation process. However, a judicious combination of selection of endophytes and engineering methods only will provide the ultimate solution to cleaning up of heavily contaminated sites. Despite having the potential to remove carcinogens like As and other harmful pollutants from the environment, the use of transgenic plants under open field conditions is limited due to the fear of biosafety issues.

References

Ali W, Isner JC, Isayenkov SV, Liu W, Zhao FJ, Maathuis FJM (2012) Heterologous expression of the yeast arsenite efflux system ACR3 improves *Arabidopsis thaliana* tolerance to arsenic stress. New Phytol 194:716–723

Andres J, Bertin PN (2016) The microbial genomics of arsenic. FEMS Microbiol Rev 40:299–322

Andres J, Arsène-Ploetze F, Barbe V, Brochier-Armanet C, Cleiss-Arnold J, Coppée JY, Dillies MA, Geist L, Joublin A, Koechler S, Lassalle F, Marchal M, Médigue C, Muller D, Nesme X, Plewniak F, Proux C, Ramírez-Bahena MH, Schenowitz C, Sismeiro O, Vallenet D, Santini JM, Bertin PN (2013) Life in an arsenic-containing gold mine: genome and physiology of the autotrophic arsenite-oxidizing bacterium rhizobium sp. NT-26. Genome Biol Evol 5:934–953

Bienert GP, Thorsen M, Schüssler MD, Nilsson HR, Wagner A, Tamás MJ, Jahn TP (2008) A subgroup of plant aquaporins facilitate the bi-directional diffusion of $As(OH)_3$ and $Sb(OH)_3$ across membranes. BMC Biol 6:1–15

Bleeker PM, Hakvoort HWJ, Bliek M, Souer E, Schat H (2006) Enhanced arsenate reduction by a CDC25-like tyrosine phosphatase explains increased phytochelatin accumulation in arsenate-tolerant *Holcus lanatus*. Plant J 45:917–929

Bundschuh J, Nath B, Bhattacharya P, Liu CW, Armienta MA, Moreno Lopez MV (2012) Arsenic in the human food chain: The Latin American perspective. Sci Total Environ 429:92–106

Carey AM, Scheckel KG, Lombi E, Newville M, Choi Y, Norton GJ (2010) Grain unloading of arsenic species in Rice. Plant Physiol 152:309–319

Cases I, De Lorenzo V (2005) Genetically modified organisms for the environment: stories of success and failure and what we have learned from them. Int Microbiol 8:213–222

Castrillo G, Sanchez-Bermejo E, de Lorenzo L, Crevillen P, Fraile-Escanciano A, Tc M (2013) WRKY6 transcription factor restricts arsenate uptake and transposon activation in Arabidopsis. Plant Cell 25:2944–2957

Catarecha P, Segura MD, Franco-Zorrilla JM, Garcia-Ponce B, Lanza M, Solano R (2007) A mutant of the Arabidopsis phosphate transporter PHT1;1 displays enhanced arsenic accumulation. Plant Cell 19:1123–1133

Chao DY, Chen Y, Chen J, Shi S, Chen Z, Wang C (2014) Genome-wide association mapping identifies a new arsenate reductase enzyme critical for limiting arsenic accumulation in plants. PLoS Biol 12:e1002009

Chaurasia N, Mishra A, Pandey SK (2012) Finger print of arsenic contaminated water in India—a review. J Foren Res 3:1–4

Chen Z, Zhu YG, Liu WJ, Meharg AA (2005) Direct evidence showing the effect of root surface iron plaque on arsenite and arsenate uptake in rice (*Oryza sativa*) roots. New Phytol 165:91–97

Chen J, Qin J, Zhu YG, de Lorenzo V, Rosen BP (2013) Engineering the soil bacterium *Pseudomonas putida* for arsenic methylation. Appl Environ Microbiol 79:4493–4495

Chen Y, Xu W, Shen H, Yan H, Xu W, He Z, Ma M (2013) Engineering arsenic tolerance and hyperaccumulation in plants for phytoremediation by a *PvACR3* transgenic approach. Environ Sci Technol 47:9355–9362

Chen J, Sun GX, Wang XX, Lorenzo VD, Rosen BP, Zhu YG (2014) Volatilization of arsenic from polluted soil by *Pseudomonas putida* engineered for expression of the *arsM* arsenic(III) S-adenosine methyltransferase gene. Environ Sci Technol 48:10337–10344

Dhankher OP, Li Y, Rosen BP, Shi J, Salt D, Senecoff JF (2002) Engineering tolerance and hyperaccumulation of arsenic in plants by combining arsenate reductase and gamma-glutamylcysteine synthetase expression. Nat Biotechnol 20:1140–1145

Dhankher OP, Rosen BP, McKinney EC, Meagher RB (2006) Hyperaccumulation of arsenic in the shoots of Arabidopsis silenced for arsenate reductase (ACR2). Proc Natl Acad Sci U S A 103:5413–5418

Doty SL (2008) Enhancing phytoremediation through the use of transgenics and endophytes. New Phytol 179:318–333

Doucleff M, Terry N (2002) Pumping out the arsenic. Nat Biotechnol 20:1094–1095

Eapen S, D'Souza SF (2005) Prospects of genetic engineering of plants for phytoremediation of toxic metals. Biotechnol Adv 23:97–114

Ellis DR, Gumaelius L, Indriolo E, Pickering IJ, Banks JA, Salt DE (2006) A novel arsenate reductase from the arsenic hyperaccumulating fern *Pteris vittata*. Plant Physiol 141:1544–1554

Freel KC, Krueger MC, Farasin J, Brochier-Armanet C, Barbe V, Andrès J, Cholley PE, Dillies MA, Jagla B, Koechler S, Leva Y, Magdelenat G, Plewniak F, Proux C, Coppée JY, Bertin PN, Heipieper HJ, Arsène-Ploetze F (2015) Adaptation in toxic environments: arsenic genomic islands in the bacterial genus *Thiomonas*. PLoS One 10:e0139011

Gasic K, Korban SS (2007) Transgenic Indian mustard (*Brassica juncea*) plants expressing an Arabidopsis phytochelatin synthase (*AtPCS1*) exhibit enhanced As and Cd tolerance. Plant Mol Biol 64:361–369

Guo J, Dai X, Xu W, Ma M (2008) Overexpressing GSH1 and AsPCS1 simultaneously increases the tolerance and accumulation of cadmium and arsenic in *Arabidopsis thaliana*. Chemosphere 72:1020–1026

Guo J, Xu W, Ma M (2012) The assembly of metals chelation by thiols and vacuolar compartmentalization conferred increased tolerance to and accumulation of cadmium and arsenic in transgenic *Arabidopsis thaliana*. J Hazard Mater 199–200:309–313

Huang K, Chen C, Shen Q, Rosen BP, Zhao FJ (2015) Genetically engineering *Bacillus subtilis* with a heat-resistant arsenite methyltransferase for bioremediation of arsenic-contaminated organic waste. Appl Environ Microbiol 81:6718–6724

Indriolo E, Na G, Ellis D, Salt DE, Banks JA (2010) A vacuolar arsenite transporter necessary for arsenic tolerance in the arsenic hyperaccumulating fern *Pteris vittata* is missing in flowering plants. Plant Cell 22:2045–2057

Kamiya T, Islam MR, Duan G, Uraguchi S, Fujiwara T (2013) Phosphate deficiency signaling pathway is a target of arsenate and phosphate transporter OsPT1 is involved in As accumulation in shoots of rice (plant nutrition). Soil Sci Plant Nutr 59:580–590

Kim YJ, Chang KS, Lee MR, Kim JH, Lee CE, Jeon YJ (2005) Expression of tobacco cDNA encoding phytochelatin synthase promotes tolerance to and accumulation of Cd and As in *Saccharomyces cerevisiae*. J Plant Biol 48:440–447

Kotrba P, Najmanova J, Macek T, Ruml T, Mackova M (2009) Genetically modified plants in phytoremediation of heavy metal and metalloid soil and sediment pollution. Biotechnol Adv 27:799–810

Kruger MC, Bertin PN, Heipieper HJ, Arsène-Ploetze F (2013) Bacterial metabolism of environmental arsenic—mechanisms and biotechnological applications. Appl Microbiol Biotechnol 97:3827–3841

LeBlanc MS, McKinney EC, Meagher RB, Smith AP (2013) Hijacking membrane transporters for arsenic phytoextraction. J Biotechnol 163:1–9

Lee BD, Hwang S (2015) Tobacco phytochelatin synthase (*NtPCS1*) plays important roles in cadmium and arsenic tolerance and in early plant development in tobacco. Plant Biotechnol Rep 9:107–114

Li Y (2004) Overexpression of phytochelatin synthase in Arabidopsis leads to enhanced arsenic tolerance and cadmium hypersensitivity. Plant Cell Physiol 45:1787–1797

Li Y, Dankher OP, Carreira L, Smith AP, Meagher RB (2006) The shoot-specific expression of γ-glutamylcysteine synthetase directs the long-distance transport of thiol-peptides to roots conferring tolerance to mercury and arsenic. Plant Physiol 141:288–298

Li RY, Ago Y, Liu WJ, Mitani N, Feldmann J, McGrath SP (2009) The rice aquaporin Lsi1 mediates uptake of methylated arsenic species. Plant Physiol 150:2071–2080

Li X, Hu Y, Gong J, Lin Y, Johnstone L, Rensing C, Wang G (2011) Genome sequence of the highly efficient arsenite-oxidizing bacterium *Achromobacter arsenitoxydans* SY8. J Bacteriol 194:1243–1244

Li G, Santoni V, Maurel C (2014) Plant aquaporins: roles in plant physiology. Biochim Biophys Acta 1840:1574–1582

Li X, Zhang L, Wang G (2014) Genomic evidence reveals the extreme diversity and wide distribution of the arsenic-related genes in Burkholderiales. PLoS One 9:e92236

Li N, Wang J, Song WY (2016) Arsenic uptake and translocation in plants. Plant Cell Physiol 57:4–13

Liu S, Zhang F, Chen J, Sun G (2011) Arsenic removal from contaminated soil via biovolatilization by genetically engineered bacteria under laboratory conditions. J Environ Sci (China) 23:1544–1550

Lou-Hing D, Zhang B, Price AH, Meharg AA (2011) Effects of phosphate on arsenate and arsenite sensitivity in two rice (*Oryza sativa* L.) cultivars of different sensitivity. Environ Exp Bot 72:47–52

Ma JF, Yamaji N, Mitani N, Xu XY, Su YH, McGrath SP, Zhao FJ (2008) Transporters of arsenite in rice and their role in arsenic accumulation in rice grain. Proc Natl Acad Sci U S A 105:9931–9935

Mahar A, Wang P, Ali A, Awasthi MK, Lahori AH, Wang Q (2016) Challenges and opportunities in the phytoremediation of heavy metals contaminated soils: a review. Ecotoxicol Environ Saf 126:111–121

Meagher RB, Heaton ACP (2005) Strategies for the engineered phytoremediation of toxic element pollution: mercury and arsenic. J Ind Microbiol Biotechnol 32:502–513

Mohan TC, Castrillo G, Navarro C, Zarco-Fernández S, Ramireddy E, Mateo C, Zamarreño AM, Paz-Ares J, Muñoz R, García-Mina JM, Hernández LE, Schmülling T, Leyva A (2016) Cytokinin determines thiol-mediated arsenic tolerance and accumulation. Plant Physiol 171:1418–1426

Mosa KA, Kumar K, Chhikara S, Mcdermott J, Liu Z, Musante C (2012) Members of rice plasma membrane intrinsic proteins subfamily are involved in arsenite permeability and tolerance in plants. Transgen Res 21:1265–1277

Muller D, Lievremont D, Simeonova DD, Hubert JC, Lett MC (2003) Arsenite oxidase aox genes from a metal-resistant beta-proteobacterium. J Bacteriol 185:135–141

Muller D, Medigue C, Koechler S, Barbe V, Barakat M, Talla E (2007) A tale of two oxidation states: bacterial colonization of arsenic-rich environments. PLoS Genet 3:53

Nie L, Shah S, Rashid A, Burd GI, Dixon DG, Glick BR (2002) Phytoremediation of arsenate contaminated soil by transgenic canola and the plant growth-promoting bacterium *Enterobacter cloacae* CAL2. Plant Physiol Biochem 40:355–361

Nriagu J (2001) Arsenic poisoning through the ages. In: Frankenburger WT (ed) Environmental chemistry of arsenic. CRC Press, Boca Raton, FL, pp 1–26

Pandey S, Shrivastava AK, Singh VK, Rai R, Singh PK, Rai S, Rai LC (2013) A new arsenate reductase involved in arsenic detoxification in *Anabaena* sp. PCC7120. Funct Integr Genomics 13:43–55

Picault N, Cazalé AC, Beyly A, Cuiné S, Carrier P, Luu DT, Forestier C, Peltier G (2006) Chloroplast targeting of phytochelatin synthase in *Arabidopsis*: effects on heavy metal tolerance and accumulation. Biochimie 88:1743–1750

Pilon-Smits E, Pilon M (2002) Phytoremediation of metals using transgenic plants. Crit Rev Plant Sci 21:439–456

Qin J, Lehr CR, Yuan C, Le XC, McDermott TR, Rosen BP (2009) Biotransformation of arsenic by a Yellowstone thermoacidophilic eukaryotic alga. Proc Natl Acad Sci U S A 106:5213–5217

Rathinasabapathi B, Wu S, Sundaram S, Rivoal J, Srivastava M, Ma LQ (2006) Arsenic resistance in *Pteris vittata* L.: identification of a cytosolic triosephosphate isomerase based on cDNA expression cloning in *Escherichia coli*. Plant Mol Biol 62:845–857

Remy E, Cabrito TR, Batista RA, Teixeira MC, Sa-Correia I, Duque P (2012) The Pht1;9 and Pht1;8 transporters mediate inorganic phosphate acquisition by the *Arabidopsis thaliana* root during phosphorus starvation. New Phytol 195:356–371

Sánchez-Bermejo E, Castrillo G, del Llano B, Navarro C, Zarco-Fernández S, Martinez-Herrera DJ, Leo-del Puerto Y, Muñoz R, Cámara C, Paz-Ares J, Alonso-Blanco C, Leyva A (2014) Natural variation in arsenate tolerance identifies an arsenate reductase in *Arabidopsis thaliana*. Nat Commun 5:4617

Sarangi BK, Kalve SAPR, Chakrabarti T (2009) Transgenic plants for phytoremediation of arsenic and chromium to enhance tolerance and hyperaccumulation. Transgen Plant J 3:57–86

Schroeder JI, Delhaize E, Frommer WB, Guerinot ML, Harrison MJ, Herrera-Estrella L (2013) Using membrane transporters to improve crops for sustainable food production. Nature 497:60–66

Shin H, Shin HS, Dewbre GR, Harrison MJ (2004) Phosphate transport in Arabidopsis: Pht1;1 and Pht1;4 play a major role in phosphate acquisition from both low- and high-phosphate environments. Plant J Cell Mol Biol 39:629–642

Shri M, Dave R, Diwedi S, Shukla D, Kesari R, Tripathi RD, Trivedi PK, Chakrabarty D (2014) Heterologous expression of *Ceratophyllum demersum* phytochelatin synthase, CdPCS1, in rice leads to lower arsenic accumulation in grain. Sci Rep 4:5784

Shukla D, Kesari R, Mishra S, Dwivedi S, Tripathi RD, Nath P, Trivedi PK (2012) Expression of phytochelatin synthase from aquatic macrophyte *Ceratophyllum demersum* L. enhances cadmium and arsenic accumulation in tobacco. Plant Cell Rep 31:1687–1699

Silver S, Phung LT (2005) Genes and enzymes involved in bacterial oxidation and reduction of inorganic arsenic. Appl Environ Microbiol 71:599–608

Slyemi D, Bonnefoy V (2012) How prokaryotes deal with arsenic(dagger). Environ Microbiol Rep 4:571–586
Song WY, Park J, Mendoza-Cózatl DG, Suter-Grotemeyer M, Shim D, Hörtensteiner S (2010) Arsenic tolerance in Arabidopsis is mediated by two ABCC-type phytochelatin transporters. Proc Natl Acad Sci U S A 107:21187–21192
Song WY, Mendoza-Cozatl DG, Lee Y, Schroeder JI, Ahn SN, Lee HS (2014a) Phytochelatin-metal(loid) transport into vacuoles shows different substrate preferences in barley and Arabidopsis. Plant Cell Environ 37:1192–1201
Song WY, Yamaki T, Yamaji N, Ko D, Jung KH, Fujii-Kashino M (2014b) A rice ABC transporter, OsABCC1, reduces arsenic accumulation in the grain. Proc Natl Acad Sci U S A 111:15699–15704
Sundaram S, Wu S, Ma LQ, Rathinasabapathi B (2009) Expression of a *Pteris vittata* glutaredoxin PvGRX5 in transgenic *Arabidopsis thaliana* increases plant arsenic tolerance and decreases arsenic accumulation in the leaves. Plant Cell Environ 32:851–858
Tiwari M, Sharma D, Dwivedi S, Singh M, Tripathi RD, Trivedi PK (2014) Expression in Arabidopsis and cellular localization reveal involvement of rice NRAMP, OsNRAMP1, in arsenic transport and tolerance. Plant Cell Environ 37:140–152
Tsao DT (2003) Overview of phytotechnologies. Adv Biochem Eng/Biotechnol 78:1–50
van Lis R, Nitschke W, Duval S, Schoepp-Cothenet B (2013) Arsenics as bioenergetic substrates. Biochim Biophys Acta 1827:176–188
Verbruggen N, LeDuc DL (2009) Potential of plant genetic engineering for phytoremediation of toxic trace elements. Encyclop Life Supp Syst 24. http://www.eolss.net/Sample-Chapters/C09/E6-199-12-00.pdf
Verbruggen N, Hermans C, Schat H (2009) Mechanisms to cope with arsenic or cadmium excess in plants. Curr Opin Plant Biol 12:364–372
Verma PK, Verma S, Pande V, Mallick S, Deo Tripathi R, Dhankher OP, Chakrabarty D (2016) Overexpression of rice glutaredoxin OsGrx_C7 and OsGrx_C2.1 reduces intracellular arsenic accumulation and increases tolerance in *Arabidopsis thaliana*. Front Plant Sci 7:740
Wang H, Xu Q, Kong YH, Chen Y, Duan JY, Wu WH, Chen YF (2014) Arabidopsis WRKY45 transcription factor activates phosphate Transporter1;1 expression in response to phosphate starvation. Plant Physiol 164:2020–2029
Wangeline AL, Burkhead JL, Hale KL, Lindblom SD, Terry N, Pilon M, Pilon-Smits EAH (2004) Overexpression of ATP sulfurylase in Indian mustard: effects on tolerance and accumulation of twelve metals. J Environ Qual 33:54–60
Wojas S, Clemens S, Skłodowska A, Maria Antosiewicz D (2010) Arsenic response of AtPCS1- and CePCS-expressing plants—effects of external As(V) concentration on As-accumulation pattern and NPT metabolism. J Plant Physiol 167:169–175
Wu Z, Ren H, McGrath SP, Wu P, Zhao FJ (2011) Investigating the contribution of the phosphate transport pathway to arsenic accumulation in rice. Plant Physiol 157:498–508
Ye WL, Wood BA, Stroud JL, Andralojc PJ, Raab A, McGrath SP (2010) Arsenic speciation in phloem and xylem exudates of castor bean. Plant Physiol 154:1505–1513
Zanella L, Fattorini L, Brunetti P, Roccotiello E, Cornara L, D'Angeli S, Della Rovere F, Cardarelli M, Barbieri M, Sanità di Toppi L, Degola F, Lindberg S, Altamura MM, Falasca G (2016) Overexpression of AtPCS1 in tobacco increases arsenic and arsenic plus cadmium accumulation and detoxification. Planta 243:605–622
Zargar K, Conrad A, Bernick DL, Lowe TM, Stolc V, Hoeft S (2012) ArxA, a new clade of arsenite oxidase within the DMSO reductase family of molybdenum oxidoreductases. Environ Microbiol 14:1635–1645
Zhao FJ, McGrath SP (2009) Biofortification and phytoremediation. Curr Opin Plant Biol 12:373–380
Zhao FJ, Ma JF, Meharg AA, McGrath SP (2009) Arsenic uptake and metabolism in plants. New Phytol 181:777–794
Zhu YG, Rosen BP (2015) Perspectives for genetic engineering for the phytoremediation of arsenic-contaminated environments: from imagination to reality. Curr Opin Biotechnol 20:220–224

Potential of Plant Tissue Culture Research Contributing to Combating Arsenic Pollution

David W.M. Leung

Contents

1 Introduction ... 195
2 A Survey of Application of Plant Tissue Culture in Arsenic Toxicity and Phytoremediation Research ... 188
2.1 Effect of Arsenate on Seed Germination ... 189
2.2 Interesting Insights Gained from the Use of In Vitro Seedlings ... 189
3 Development of Low As Accumulating Crop Cultivars ... 190
4 Potential Uses of Low as Accumulating Somaclonal Variants ... 192
5 Conclusion ... 193
References ... 193

1 Introduction

The presence of toxic heavy metals and metalloids such as arsenic (As) in the environment is not only a core concern of the public health domain globally but also poses a grand challenge to scientists and engineers in multiple disciplines for sustainable solutions. The role of plant science stems from the natural ability of plants to take up heavy metals and metalloids even though most of them are not essential for plant growth. The field of phytoremediation research and the aligned basic studies investigating mechanisms of heavy metals and metalloids uptake, resistance/tolerance, accumulation, and transport in the plants is built on the working hypothesis that some plants can be deployed in large scale to assist in situ remediation of contaminated soils and groundwater (Zhao et al. 2010). This is greatly supported with the discovery of hyperaccumulators including *Pteris vittata* (the Chinese brake

D.W.M. Leung (✉)
School of Biological Sciences, University of Canterbury,
Private Bag 4800, Christchurch 8140, New Zealand
e-mail: david.leung@canterbury.ac.nz

D.K. Gupta, S. Chatterjee (eds.), *Arsenic Contamination in the Environment*,
DOI 10.1007/978-3-319-54356-7_9

fern) that grew well in the presence of 1 mM arsenate and accumulated 1250 mg kg^{-1} As in the sporophytes (Yang et al. 2007). By contrast, *Arabidopsis thaliana* did not survive upon exposure to 0.2 mM arsenate.

Ironically, accumulation of toxic metals and metalloids into crop plants, particularly into the edible parts such as seeds and tubers, is a potential food safety issue as this could add to the dietary burden of contaminants in foods. If this food chain contamination is not stopped, then the possibility of large-scale additive poisoning, albeit not immediately or overtly obvious, at the wide population level could not be discounted (Clemens et al. 2013). In this light, food plants must be screened to ascertain the levels of toxic metal/metalloid accumulation under standard agronomic or agricultural production conditions. Low accumulation of toxic metals and metalloids should be considered as a top trait in germplasm improvement of crop plants. It also follows that crop plants should be avoided for phytoremediation purposes (Ashrafzadeh and Leung 2015).

There have been some critical evaluations of biotechnological tools such as plant tissue culture in phytoremediation research (Doran 2009). They are of great utility in assisting the dissection of the biology of uptake, toxicity, detoxification, metabolism and tolerance to environmental contaminants including heavy metals. Clearly, the artificial cultural systems are different from whole plants and have no direct applicability in field remediation of heavy metals or metalloids. There seems to be relatively more research employing plant tissue culture on heavy metals such as lead, cadmium, etc. than arsenic. Here, the objective was to discuss the plant tissue culture tools developed that could enable scientists to contribute to the challenge of dealing with the arsenic (As) contamination in agricultural soil and groundwater.

2 A Survey of Application of Plant Tissue Culture in Arsenic Toxicity and Phytoremediation Research

Plant tissue cultures including callus, cell suspension and hairy root cultures are composed of cells cloned from a piece of plant material (explant) and generally exhibit basic plant cell properties similar to many of those in the whole plants. Plant tissue culture is useful in laboratory studies of metal toxicity but had no direct use in remediation of polluted soil and water (Doran 2009; Ibanez et al. 2016). Use of plant tissue culture in metal toxicity research in general has several merits. In particular, compared to use of whole plants for investigations into the toxic effects of certain concentrations of metals, plant tissue culture can be studied without the possible interference from the activities of microbes that might be associated with the surface of the whole plant body.

2.1 Effect of Arsenate on Seed Germination

Germination of tobacco (*Nicotiana tabacum* cv. Wisconsin) was not affected negatively by 10–200 μM arsenate (As^{v}) (Talano et al. 2014). However, there are studies showing that increasing concentrations of arsenate could be inhibitory to seed germination of rice (Abedin and Meharg 2002; Shri et al. 2009) and wheat (Liu et al. 2005; Li et al. 2007). It is possible that different types of seeds might differ in their relative sensitivity to exogenous arsenate, but it is also possible that the microbes associated with seed surface could influence the response of seeds to toxic levels of exogenous arsenate. However, the latter could be discounted as all the seeds used were surface-sterilised before arsenic exposure.

Tobacco seedlings grown on a modified Murashige and Skoog medium (Murashige and Skoog 1962) from surface-sterilised seeds were found to be useful for As toxicity testing as they can respond to increasing concentrations of As with a reduction in growth (Talano et al. 2014). Conceivably, the findings from the use of in vitro seedlings in jars or petri dishes on racks of shelves in a growth room could be used in a more economic space-saving way to conduct screening of variation of seedlings of different genotypes in As tolerance compared to glasshouse or field evaluation of the seedlings.

2.2 Interesting Insights Gained from the Use of In Vitro Seedlings

The response of seedlings to many stressors including many heavy metals (e.g. cadmium and lead) seems to share a common morphogenic basis in that a reduction in primary root length is accompanied by a stimulation of lateral root formation (Potters et al. 2009). However, As inhibited both primary lateral growth in in vitro tobacco seedlings (Talano et al. 2014). At As levels (10 or 50 μM) not inhibitory to root growth, it was then found that hairy root (HR) cultures were more efficient than seedling roots to remove As from liquid medium (Talano et al. 2014). At least the rhizofiltration potential of As by hairy root cultures has been demonstrated. In addition, future studies could uncover the differences in gene expression profiles (e.g. using transcriptome analysis) and different metabolic pathways (Ibanez et al. 2016) in relation to As uptake, accumulation and tolerance between seedling roots and HR.

The perennial sporophyte of the Chinese brake fern (*P. vittata*) is now a well-known As hyperaccumulator (Yang et al. 2007). Insights gained from this plant should better inform the practical requirements of phytoremediation. Basic tissue culture tools including spore germination and gametophyte development under in vitro conditions as well as callus induction from gametophyte explants and sporophyte regeneration from the callus have been developed (Antonio et al. 2007; Zheng et al. 2008). These tissue culture tools can be used to aid screening and mass clonal propagation of selected plants for phytoremediation. They could also be used to aid genetic dissection and improvement of the As hyperaccumulation mechanism in *P. vittata* based on in vitro cell mutant cell line selection, plant transformation (Muthukumar et al. 2013), omics analysis (Kumar et al. 2015) and the emerging gene editing tools applicable to plants such as CRISPR (Ren et al. 2016). This powerful genome modification tool involves selection of a genome target such as a candidate gene involved in detoxification of arsenic in plant cells in vitro. Following the designed modifications, the mutated gene construct must be delivered into isolated plant protoplasts. Putative mutant plants must then be regenerated and screened for the phenotype and genotype of interest. Therefore, a good knowledge of plant protoplast isolation, culture and transformation would help the application of this new genome editing technology.

3 Development of Low As Accumulating Crop Cultivars

The concept that it is possible to reduce accumulation of As in plants through a genetic modification has been proved though, ironically, with the help of a cDNA cloned from the As hyperaccumulator fern, *P. vittata*. The cDNA was a glutaredoxin gene (PvGRX5) which codes for a protein known to have glutathione-dependent oxidoreductase activity involved in protecting some proteins from oxidative stress-induced damages. PvGRX5 cDNA was transferred to *A. thaliana* following *Agrobacterium*-mediated stable transformation and it was expressed in the transgenic plants under a constitutive promoter (Sundaram et al. 2009). All the transgenic lines expressing the PvGRX5 CDNA had a higher glutaredoxin activity than the wild type or vector control (transformation minus the fern cDNA). These lines also exhibited improved root growth (As tolerance) when they were grown in the presence of 1 mM arsenate compared to the wild type and the vector control. The PvGRX5-transgenic plant lines had also significantly lower As content in their leaves. The PvGRX5 was the first plant glutaredoxin gene implicated in arsenic tolerance and used for construction of plant lines with reduced As accumulation. The mechanisms for a lower As content in the transgenic leaves could be due to increased As efflux or there was a reduced transport of As within the plant (from root to the shoot). Recently, overexpression of two rice glutaredoxin genes has been linked to reduce As content in transgenic *A. thaliana* seeds (Verma et al. 2016). This further suggests that it would be possible to minimise As content in seeds of food plants. In another recent study, the transfer of a phytochelatin synthase gene (*CdPCS1* of *Ceratophyllum demersum*) also led to a significant reduction of As content in rice grains.

It is most desirable to have As-free rice grains (Kumar et al. 2015), but there is no published report of any plants that can completely prevent the uptake and accumulation of As. Phytoremediation of As contaminated soil has not been put to practical field application and there does not seem to be any timeframe or stage gate for this green technology to have any significant practical impact on the removal of environmental As in the near future. It is therefore proposed here that the least plant

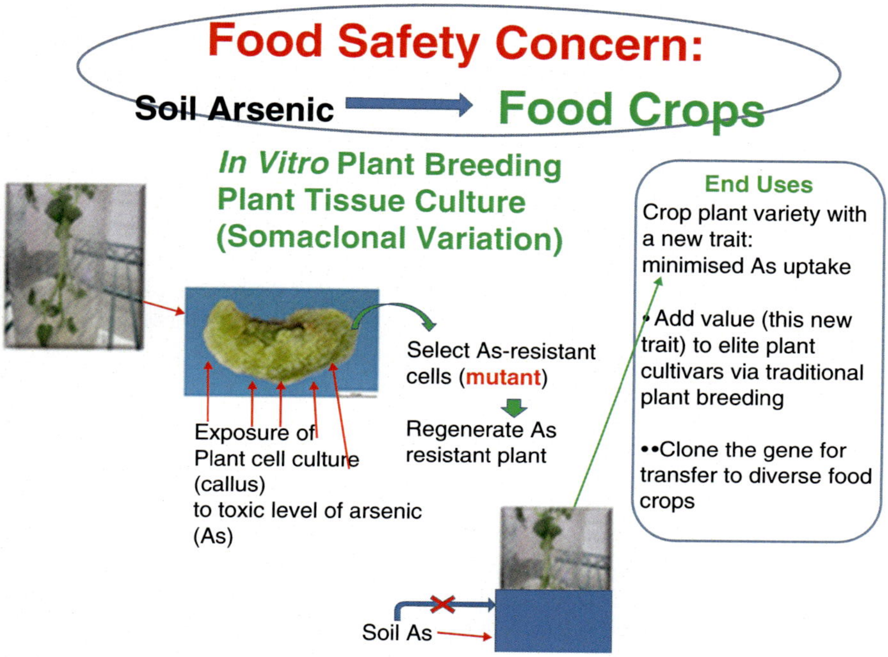

Fig. 1 An outline of in vitro cell selection (somaclonal variant selection) to obtain novel low arsenic accumulating crop plant and its potential uses

science can offer as a practical solution to help alleviate the toxicity arising from As exposure of the human populations is to develop low bioaccumulating As crop cultivars while scientists and engineers work on the best large-scale and practical strategy(ies) to remediate As contaminated soils and groundwater.

The transgenic plants engineered for low bioaccumulation of As may be controversial from the perspective of public perception and other food safety regulatory issues in some countries, for example, New Zealand and Germany. In vitro plant breeding based on somaclonal variation occurring during plant tissue culture is an attractive practical alternative (Ashrafzadeh and Leung 2015). The general practical strategy that may be used to obtain low As accumulating plants starting from plant material accumulating more As is outlined in Fig. 1. There are no in vitro recombinant DNA manipulations involved in generating the putative novel plant variants with the altered As accumulation trait. Therefore, they are generally not regarded as genetically modified organisms and can be field trialled for agronomic performance without the substantial regulations associated with field trials or field release of transgenic plants, for example, in New Zealand.

The strategy outlined for in vitro breeding of low bioaccumulating As plants is adapted from that has been used in obtaining many somaclonal variants with improved resistance to heavy metals such as Al, Cd, Pb, and Zn (Ashrafzadeh and Leung 2015). A key requirement is to expose callus of a plant of interest to a toxic but sub-lethal level of a selective agent, namely As. In this way, As-resistant callus cells could be selected against the non-resistant cells which can be placed on a plant regeneration medium. Then another key requirement is to validate that the plants are more resistant to As than the mother plant used for initiation of callus for As selection. The third requirement is to determine the As bioaccumulation potential of the As-resistant plants (somaclonal variants). Conceivably, there are two possible mechanisms that could underpin As resistance in the variant plants: (1) the variants could exclude or minimise As uptake and accumulation, and (2) the variants could accumulate toxic levels of As but have improved tolerance mechanisms. If the growth of these two types of variants are compared with the mother plants (control) for in vitro As-resistant plant selection, they would appear both to outperform the control or exhibit improved As resistance based on two widely opposite mechanisms. For example, in an in vitro cell line selection of *Brassica juncea* using Cd as the selection agent, some somaclonal variant plants were capable of accumulating Cd, some were capable of excluding Cd, and some were not different from the mother plants (Nehnevajova et al. 2007). Once the low As accumulating plant lines of interest have been identified, they could be evaluated further to see in what other ways they might vary from the mother plant material under field conditions. Any unacceptable agronomic performance, for example, with significant yield reduction, would preclude them as outright candidates to be developed as new cultivars without further breeding manipulations (Ashrafzadeh and Leung 2015). The stability of the low As accumulating character of the plant lines must at least be validated in different soil types.

4 Potential Uses of Low as Accumulating Somaclonal Variants

A somaclonal variant plant line with confirmed low As accumulation potential might or might not have satisfactory agronomic performance to be qualified as a potentially new cultivar but it could still be a valuable bioresource material. The line could be bred with other elite breeding lines for the transfer and addition of the low As accumulating trait into the elite breeding materials. The somaclonal line can also be studied in comparison with the mother plant material to gain deeper insights about any morphological, physiological, and biochemical mechanisms associated with the low As accumulating trait. Omics studies (Kumar et al. 2015) can be carried out to gain insights at the system biology level of low As accumulating trait. The genes that have relatively large contributions to promoting As efflux, restricting uptake, transport and accumulation of As particularly in the edible parts of the variant crop line, could be isolated and cloned using molecular biology techniques. The cloned genes could be transferred to other plants using plant transformation to add the low As enabling genes to other crop plant germplasm. A comparison of the low As accumulating somaclonal variants with high As accumulating plant lines at transcriptome analysis level might lead to the identification of potential candidate genes for gene editing (correction) with the CRISPR technique (Ren et al. 2016) in plants accumulating more As than the low As accumulating somaclonal variant lines.

5 Conclusion

Plant science research has two broad foci in relation to As contamination of agricultural land and groundwater. The discovery of As hyperaccumulator such as the Chinese brake fern has fuelled the hope for the deployment of phytoremediation as a more attractive alternative to civil engineering approaches to soil remediation measures (Doran 2009). Phytoremediation has yet to make a large-scale practical impact on reducing the exposure of people to As. It seems that the least plant scientists can do to protect people from exposure through dietary intake of As is to develop low As accumulating crop cultivars for human consumption. Plant tissue culture technology is relatively mature to aid these two broad fronts of tackling the As pollution problems, particularly large-scale propagation of useful plants/variants for phytoremediation or producing safer crop plants from in vitro selection of somaclonal variants exhibiting low As accumulation (Shri et al. 2014).

References

Abedin MJ, Meharg AA (2002) Relative toxicity of arsenite and arsenate on germination and early seedling growth of rice (*Oryza sativa* L.) Plant Soil 243:57–66

Antonio T, Mantovani M, Fusconi A, Gallo C (2007) In vitro culture of *Pteris vittata*, an arsenic hyperaccumulating fern, for screening and propagating strains useful for phytoremediation. Caryologia 60:160–164

Ashrafzadeh S, Leung DWM (2015) In vitro breeding of heavy-metal resistant plants: a review. Hortic Environ Biotechnol 56:131–136

Clemens S, Aarts MGM, Thomine S, Verbruggen N (2013) Plant science: the key to preventing slow cadmium poisoning. Trend Plant Sci 18:92–99

Doran PM (2009) Application of plant tissue cultures in phytoremediation research: incentives and limitations. Biotechnol Bioeng 103:60–76

Ibanez S, Talano M, Ontanon O, Suman J, Medina MI, Macek T, Agostini E (2016) Transgenic plants and hairy roots: exploiting the potential of plant species to remediate contaminants. New Biotechnol 33:625–635

Kumar S, Dubey RS, Tripathi RD, Charkrabarty D, Trivedi PK (2015) Omics and biotechnology of arsenic stress and detoxification in plants: current updates and perspective. Environ Int 74:221–230

Li CX, Feng SL, Shao Y, Jiang LN, Lu XY, Hou XL (2007) Effects of arsenic on seed germination and physiological activities of wheat seedlings. J Environ Sci 19:725–732

Liu X, Zhang S, Shan X, Zhu YG (2005) Toxicity of arsenate and arsenite on germination, seedling growth and amylolytic activity of wheat. Chemosphere 61:293–301

Murashige T, Skoog F (1962) A revised medium for rapid growth and bioassays with tobacco tissue cultures. Physiol Plant 15:473–497

Muthukumar B, Joyce BL, Elless MP, Stewart CN (2013) Stable transformation of ferns using spores as targets: *Pteris vittata* and *Ceratopteris thalictroides*. Plant Physiol 163:648–658

Nehnevajova E, Herzig R, Erismann KH, Schwitzguebel JP (2007) In vitro breeding of *Brassica juncea* L. to enhance metal accumulation and extraction properties. Plant Cell Rep 26:429–437

Potters G, Pasternak TP, Guisez Y, Jansen MAK (2009) Different stresses, similar morphogenic responses: integrating a plethora of pathways. Plant Cell Environ 32:158–169

Ren C, Liu XJ, Zhang Z, Wang Y, Duan W, Li SH, Liang ZC (2016) CRISPR/Cas9-mediated efficient targeted mutagenesis in Chardonnay (*Vitis vinifera* L). Sci Rep 6:32289

Shri M, Kumar S, Chakrabarty D, Trivedi PK, Mallick S, Misra P, Shukla D, Mishra S, Srivastava S, Tripathi RD, Tuli R (2009) Effect of arsenic on growth, oxidative stress, and antioxidant system in rice seedlings. Ecotoxicol Environ Saf 72:1102–1110

Shri M, Dave R, Dwivedi S, Shukla D, Kesari R, Tripathi RD, Trivedi PK, Chakrabarty D (2014) Heterologous expression of *Ceratophyllum demersum* phytochelatin synthase, CdPCS1, in rice leads to lower arsenic accumulation in grain. Sci Rep 4:5784

Sundaram S, Wu S, Ma LQ, Rathinasabpathi B (2009) Expression of a *Pteris vittata* glutaredoxin PvGRX5 in transgenic *Arabidopsis thaliana* increases plant arsenic tolerance and decreases arsenic accumulation in the leaves. Plant Cell Enviorn 32:851–858

Talano MA, Oller WAL, Gonzalez P, Gonzalez SO, Agostini E (2014) Effects of arsenate on tobacco hairy root and seedling growth, and its removal. In Vitro Cell Dev Biol-Plant 50:217–225

Verma PK, Verma S, Pande V, Mallick S, Tripathi RD, Dhankher OP, Chakrabarty D (2016) Overexpression of rice glutaredoxin OsGrx_C7 and OsGrx_C2.1 reduces intracellular arsenic accumulation and increases tolerance in *Arabidopsis thaliana*. Front. Plant Sci 7:740

Yang X, Chen H, Xu W, He Z, Ma M (2007) Hyperaccumulation of arsenic by callus, sporophytes, and gametophytes of *Pteris vittata* cultured *in vitro*. Plant Cell Rep 26:1889–1897

Zhao FJ, McGrath PS, Meharg AA (2010) Arsenic as a food chain contaminant: mechanism of plant uptake and metabolism and mitigation strategies. Annl Rev Plant Biol 61:535–559

Zheng Y, Xu W, He Z, Ma M (2008) Plant regeneration of the arsenic hyperaccumulator *Pteris vittata* L. from spores and identification of its tolerance and accumulation of arsenic and copper. Acta Physiol Plant 30:249–255

Potential Role of Microbes in Bioremediation of Arsenic

Anindita Mitra, Soumya Chatterjee, and Dharmendra K. Gupta

Contents

1 Introduction ... 196
2 Mechanism of Uptake of As by Microbes ... 197
 2.1 As Uptake by Bacteria ... 197
 2.2 As Uptake by Fungus ... 198
3 Biochemical Fate of As in Microbes ... 199
 3.1 As Metabolism in Bacteria ... 199
  3.1.1 Arsenite Oxidation ... 199
  3.1.2 Arsenate Reduction ... 200
  3.1.3 Methylation of As Species ... 200
  3.1.4 Arsenite Efflux in Microbes ... 201
 3.2 As Metabolism in Fungus ... 201
4 Bioremediation of As by Microbes ... 202
 4.1 Bioremediation of As by Using Biological Sulfate Reduction ... 203
 4.2 As Removal from Contaminated Groundwater by Using Ferrous Oxide and Microorganism ... 203
 4.3 Bioremediation Through Arsenic Biomineralization ... 204
 4.4 Arsenic Bioremediation Through Production of Hyperaccumulating Bacterial Strains ... 204
 4.5 Removal of As by Using Fixed Base Bioreactor System ... 205
 4.6 Bioremediation Through Genetically Engineered Bacterial Strain ... 205
 4.7 Bioremediation by Arsenite Oxidizing Bacteria ... 206

A. Mitra
Bankura Christian College, Bankura, West Bengal, India

S. Chatterjee (✉)
Defence Research Laboratory, Defence Research and Development Organization (DRDO), Ministry of Defence, Post Bag No. 2, Tezpur 784001, Assam, India
e-mail: drlsoumya@gmail.com

D.K. Gupta
Institut für Radioökologie und Strahlenschutz (IRS), Gottfried Wilhelm Leibniz Universität Hannover, Herrenhäuser Str. 2, Hannover 30419, Germany

D.K. Gupta, S. Chatterjee (eds.), *Arsenic Contamination in the Environment*,
DOI 10.1007/978-3-319-54356-7_10

4.8 Bioremediation by Arsenate-Reducing Bacteria 206
4.9 Mycoremediation of As 206
4.10 Bioremediation of As Through Constructed Wetlands 207
5 Conclusion 208
References 208

1 Introduction

Arsenic (As), an extremely toxic metalloid, derived from natural, geothermal and anthropogenic sources is widely distributed in soil and water. Arsenic accumulated in the edible parts of plants such as seeds, grain or leaves get into the human body through food chain and causes severe health problems (Abedin et al. 2002). Besides causing various health hazards, arsenic also known to have mutagenic and genotoxic effects on human increasing the risk of developing skin, kidney, lung, and bladder cancer (Karagas et al. 1998). Arsenic contamination has become a major public health concern in countries like India, China, Argentina, and the USA (Gehle 2009; WHO 2001). Therefore, efficient removal of arsenic from contaminated water and soil is urgently needed.

In general, metals or metalloid are removed by methods including chemical precipitation, ion-exchange, adsorption and membrane separation. Major disadvantages of this method include high cost, technologically complex operation, maintenance and generation of toxic sludge (Chatterjee et al. 2013a). Bioremediation of arsenic by microorganism provides a cost-effective technology and environmental friendly way and therefore has been widely preferred for heavy metal removal (Valls and Lorenzo 2002). Bacterial and fungal species capable of removing arsenic from the contaminated water and soil may be ideal candidates for bioremediation and therefore, may be an alternative for exiting physicochemical methods of arsenic removal (Takeuchi et al. 2007).

Fungi are dominating living biomass of soil (Hoshino and Morimoto 2008) and are involved in arsenic biogeochemical cycle through redox and methylation of arsenic (Singh et al. 2015). Some fungal species have been isolated from arsenic-contaminated soil and their interaction with arsenic has been studied by Srivastava et al. (2011). Though the fungi have an advantage over bacteria for bioremediation of heavy metals owing to their biomass, hyphal network, and longer life cycle, still now they are not exploited for remediation of soil contaminated with pollutants (Singh et al. 2015).

In nature, depending on the species diversity, bacteria and fungi respond to arsenic in various ways like chelation, compartmentalization, exclusion, and immobilization (Di Toppi and Gabbrielli 1999). Exploration of the molecular and genetic basis of arsenic metabolism in microbes will therefore provide an extensive knowledge base for developing proficient and selective arsenic bioremediation tools. In this chapter, we will highlight the uptake and metabolism of arsenic in bacteria and fungi and their potential utility on environmental arsenic remediation.

2 Mechanism of Uptake of As by Microbes

2.1 As Uptake by Bacteria

In bacterial cell, arsenate (As^{V}) and arsenite (As^{III}) enter through phosphate (Pi) transporter and aquaglyceroporins respectively (Fig. 1) (Tripathi et al. 2007). Resistance to both arsenite and arsenate are found widely in a number of Gram positive and Gram negative bacteria, develops under the control of genes contained within *ars* operon (Tripathi et al. 2007). The operon consists of three genes *arsR* (encodes a trans-acting repressor) *arsB* (encodes the membrane localized arsenite permease pump) and *arsC* (encodes an intracellular arsenate reductase). The repressor senses the presence of As^{III} and controls the expression of *ArsB* and *ArsC* proteins. The arsenate reductase reduces As^{V} to As^{III} using glutathione (GSH) as reductant and *ArsB* outflows As^{III} from the cells by functioning as an $As(OH)_3$-H^+antiporter (Tripathi et al. 2007). In *E. coli*, another arsenic transporter has been identified as glycerol uptake facilitator is named as GlpF (Sanders et al. 1997).

In some Gram negative bacteria two additional genes *arsA* and *arsD* are found in the *ars* operon (in the order RDABC) (Lin et al. 2006). *ArsA* is an intracellular ATPase that binds to *ArsB* and converts the As^{III} carrier proteins into an ATP driven As^{III} efflux pump. *ArsD* is a weak As^{III} sensitive repressor of the operon with a func-

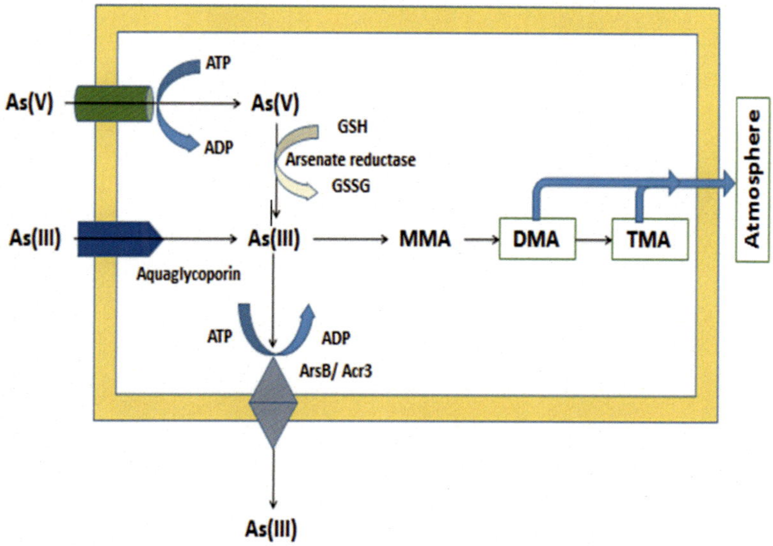

Fig. 1 General mechanism of arsenic uptake and detoxification in bacteria (*ArsB/ACR3* arsenic resistance membrane protein, *GSH* glutathione, *GSSG* glutathione disulfide, *DMA* dimethylarsinic acid, *MMA* monomethylarsonic acid, *TMA* trimethylarsine)

tional similarity with *ArsR* (Rosen 2002). Very recently, Rahman et al. (2015) reported that the bacteria *Lysinibacillus sphaericus* B1-CDAstrain, collected from highly arsenic-contaminated region of Bangladesh, harbors the arsenic marker genes *acr*3, *ars*R, *ars*B and *ars*C. The strain is highly resistant to arsenic and accumulates arsenic inside the cell. It is hypothesized that *ars*B/*acr*3 genes are the primary determinants in arsenic resistance (Achour et al. 2007).

2.2 As Uptake by Fungus

The fungus *Saccharomyces cerevisiae* absorbs arsenic from the environment through three different transporter proteins (Fig. 2) The arsenate (As^{V}) is taken up through a phosphate transporter Pho87p (Persson et al. 1999) as in bacteria. For arsenite (As^{III}) transport two different transporter systems have been identified. One is aquaglyceroporin Fps1p, a glycerol transporter encoded by *FPS1* gene (Wysocki et al. 1997, 2001) similar to bacterial systems and the other is hexose transporter (Hxt1p to Hxt1 plus Gal2p) (Liu et al. 2004). Disruption of *FPS1* gene and in presence of glucose arsenite uptake was reduced by 80%, confirming that the hexose transporters are mainly responsible for arsenite uptake (Tsai et al. 2009) (Fig. 2).

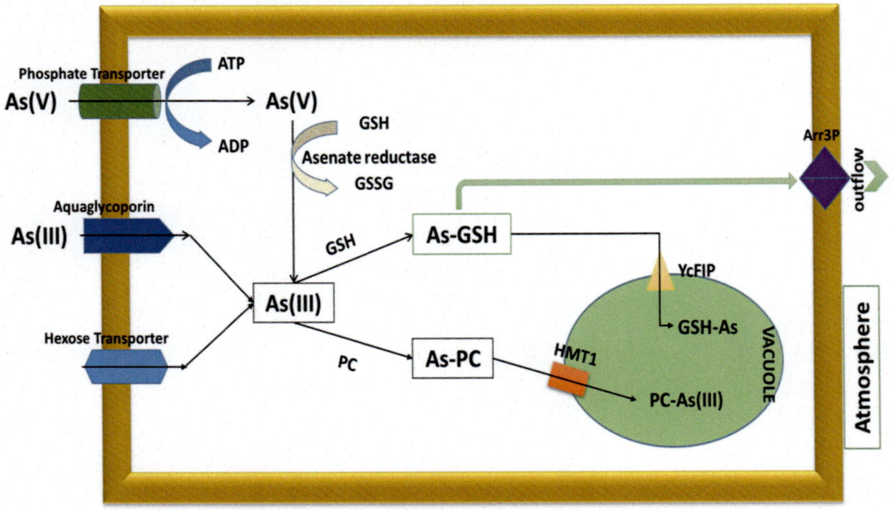

Fig. 2 General mechanism of arsenic uptake and detoxification in fungi (*As* Arsenic, *ATP* adenosine tri-phosphate, *ADP* adenosine di-phosphate, *GSH* glutathione, *GSSG* glutathione disulfide, *PC* phytochelatin)

3 Biochemical Fate of As in Microbes

3.1 As Metabolism in Bacteria

3.1.1 Arsenite Oxidation

Arsenite (As^{III}) oxidation to generate arsenate (As^{V}) was viewed as arsenic detoxification strategy in some microorganism as it may lead to the production of less toxic As^{V} species than the toxic arsenite which has high affinity for peptide-thiol groups. A diverse, As^{III} oxidizer includes both hetrotrophic As^{III} oxidizer (HAO) and chemolithoautotrophic As^{III} oxidizer (CAO) (Oremland and Stolz 2003; Stolz et al. 2010). In case of heterotrophic oxidation As^{III} was converted to less toxic As^{IV} and may be used as supplementary energy source (vanden Hoven and Santini 2004) whereas in CAO As^{III} is used as an electron donor and CO_2 fixation is coupled with reduction of oxygen (Santini et al. 2000; Lievremont et al. 2009). Arsenite oxidase encoded by the genes, *aoxA* and *aoxB* played an essential role in the oxidation process (Muller et al. 2003) and cloned from *Cenibacterium arsenoxidans* and *Alcaligenes faecalis* (Muller et al. 2003; Silver and Phung 2005). Other bacterial strain such as *Sulfolobus acidocaldarius* strain BC, *Comamonas terrae* sp., *Herminiimonas arsenicoxydans* are reported to have the ability to oxidize arsenite to a less toxic arsenate (Mahimairaja et al. 2005; Chipirom et al. 2012). About 25 bacterial and archaeal genera carrying *aox* gene were identified and isolated from arsenic rich environment (Heinrich-Salmeron et al. 2011). Arsenite oxidase enzyme contains two subunits and different names were assigned for the homologous genes encoding these two subunits like *aox*A-*aox*B/*aro*A-*aro*B/*aso*A-*aso*B) (Yamamura and Amachi 2014). However, the nomenclature for such genes recently unified and the name *aio* was newly assigned, and therefore, the two subunits now denoted as *aio*A-*aio*B (Lett et al. 2012; Sanyal et al. 2016). Phylogenetic diversity of the *aio*A gene has been detected in a number of bacterial species *Achromobacter* sp., *Pseudomonas* sp., *Alcaligenes faecalis*, *Thiobacillus ferrooxydans*, and *T. acidophilus* from various soil and aquatic environments (Inskeep et al. 2007). Arsenite can be oxidized also in anaerobic condition. A chemolithoautotrophic bacteria *Alkalilimnicola ehrlichii* MLHE-1 couples As^{III} oxidation to nitrate reduction (Zargar et al. 2010) and the purple sulfur bacterium *Ectothiorhodospira* sp. PHS-1 can use As^{III} as the electron donor for anoxygenic photosynthesis and produces As^{V} anaerobically (Kulp et al. 2008). But both these bacterial strains were found to lack *aio* gene and instead possess *arx*A genes closely related to *arr*A (Zargar et al. 2010, 2012). Another interesting group of prokaryotic microorganisms Thermus strain HR13 oxidizes As^{III} to As^{V} under aerobic condition and uses arsenate as the terminal electron acceptor under anaerobic condition (Gihring and Banfield 2001).

3.1.2 Arsenate Reduction

Some bacteria like *Corynebacterium glutamicum* can survive under arsenic stress by reducing arsenate and two different classes of arsenate reductase enzymes were identified in this strain (Lim et al. 2014). One is single-cysteine monomeric enzymes named as Cg-ArsC1 and Cg-ArsC2 catalyzing the reduction process which is coupled to the redox pathway using amycothiol transferase mechanism. The second type is three-cysteine containing homodimer Cg-ArsC′ that uses a reduction mechanism linked to the thioredoxin pathway (Mateos et al. 2012). In microbes the mechanism of arsenate reduction has been studied extensively (Messens and Silver 2006; Bhattacharjee and Rosen 2007). Different types of cytosolic arsenate reductase using different reductants during their catalytic process have been isolated from a variety of microorganisms. For example, *ArsC* protein from of *E. coli* uses glutaredoxin (Grx) and glutathione (GSH) as reductants, whereas a second type *ArsC* from *Bacillus subtilis* uses thioredoxin as reductants (Zhao et al. 2008). Liao et al. (2011) isolated 11 arsenic reducing bacterial strain from seven different genera belong to *Pseudomonas*, *Psychrobacter*, *Citrobacter*, *Bacillus*, *Bosea*, *Vibrio,* and *Enterobacter* from groundwater samples in west central Taiwan. These communities of bacteria can survive in high arsenic concentrations in water. Other bacterial strains able to reduce arsenate to arsenite are *Sulfurospirillum barnesii* and *Sulfurospirillum arsenophilum*, *Pyrobaculum arsenaticum* and *Chrysiogenes arsenatis* (Afkar 2012; Huber et al. 2000).

In anaerobic condition another pathway of arsenate reduction occurs in microorganisms via dissimilatory reduction mechanism where arsenite acts as terminal electron acceptor (Switzer Blum et al. 1998). In this process, the reduction of arsenate to arsenite is coupled with the oxidation of organic (acetate, lactate, butyrate, pyruvate, malate, glycerol, and ethanol) and inorganic (sulfides and hydrogen) substrates and ATP is synthesized (Newman et al. 1998). The strain *Alkaliphilus oremlandii* can use arsenate and thiosulfate as terminal electron acceptors while electron donors are acetate, pyruvate, formate, lactate, glycerol, or fructose (Fisher et al. 2008). Another strain *Desulfuribacillus alkaliarsenatis* can also grow by elemental sulfur or thiosulfate reduction in which the electron donors are H_2, formate, pyruvate, or lactate (Sorokin et al. 2012).

3.1.3 Methylation of As Species

Methylation of arsenic is an important method of detoxification in microbes because methylated compounds are highly volatile. Acidophilic iron and sulfur-oxidizing bacteria are able to leach high concentration of arsenic from contaminated soils (Ramasamy et al. 2006). Biomethylation of arsenic can form volatile methylated compounds (mono-, di-, and trimethyl arsenic) and non-volatile methylated compounds, e.g., methyl arsonate and dimethyl arsinate (Bentley and Chasteen 2002). In microbes, arsenite is methylated by S-adenosylmethyltransferase (SAM) enzyme (Zhao et al. 2008). In bacteria, the methyl donor may be S-adenosylmethionine (Qin et al. 2006) or alternatively methyl-cobalamin (Bentley and Chasteen 2002). The reducing agents used by methyltransferase are thioredoxin (Silver and Phung 2005). Genes encoding the SAM are found throughout the microbial genomes. The products of arsenite methylation are mono- and di-methylarsine compounds (MMA^{V}, MMA^{II}, DMA^{V}, DMA^{III}); the final product being trimethylarsine (TMA), a volatile gas (Bentley and Chasteen 2002). In a soil bacteria *Rhodopseudomonas palustris* the *arsM* gene has been reported to encode the enzyme SAM (Qin et al. 2006).

3.1.4 Arsenite Efflux in Microbes

Another common mechanism of arsenic detoxification in microbes is arsenite efflux followed after conversion of arsenate to arsenite (Bhattacharjee and Rosen 2007). It has been reported that in *E. coli* two efflux proteins namely *ArsB* and *ArsAB* mediate the arsenite efflux from the cell (Zhao et al. 2008). *ArsB* was found to function efficiently in association with *ArsA*, an ATPase, to form a pump than functioning alone in extruding arsenite from the cells (Dey et al. 1994). The mechanism of function of *ArsB* depends on proton motive force (Zhao et al. 2008). High altitude Andean lakes (HAAL) in the central Andes region display unexplored ecosystems rich in arsenic due to natural geochemical phenomena and active volcanism (Ordonez et al. 2009) and also have a rich source of microbes showing high resistance to arsenic (Albarracin et al. 2015). Genome sequencing of the strain *Sphingomonas* sp. S17 identified in HAAL ecosystem revealed the presence of both arsenic resistance gene *acr3*, involved in the cell detoxification against arsenite and *arsB* (Ordonez et al. 2015).

3.2 As Metabolism in Fungus

To reduce the cytotoxic effect of arsenic following uptake, three contiguous gene clusters ARR1, ARR2, and ARR3 play important role in the detoxification process in *S. cerevisiae* (Tsai et al. 2009). Inside the yeast cells, As^{V} is reduced to As^{III} by *Arr*2p, an arsenate reductase (Mukhopadhyay et al. 2000) like bacterial arsenate reductase *Ars*C. However, the bacterial arsenate reductase is a 141-residue monomer, whereas *Arr*2p is a homodimer of two 130-residue monomers (Tsai et al. 2009). *Arr*2p uses Grx and GSH as reductants (Bhattacharjee and Rosen 2007). ARR1 gene of Yeast encodes a transcription factor that regulates the transcription of *Arr*2p and the *Arr*3p. Arsenite extrusion pump *Arr*3p transports $As^{(III)}$-GSH complexes out of the membrane (Ghosh et al. 1999). Mutation in ARR2 gene in *S. cerevisiae* eliminated arsenate resistance (Mukhopadhyay and Rosen 1998). Although arsenite is more toxic than arsenate, it has higher affinity to bind with many intracellular chelating proteins or peptides containing thiol ligands, such as glutathione (GSH), phytochelatin (PCs) to form inactive complexes (Singhal et al. 1987; Cobbett and Goldsbrough 2002), thus ultimately eliciting less toxicity. The arsenite-peptide complexes are finally sequestered into the vacuoles. Overexpression of the *S. cerevisiae* GSH1 gene, encoding a γ-glutamylcysteine synthetase (γ-ECS) required for GSH biosynthesis, elevated the tolerance and accumulation of arsenic in transgenic *Arabidopsis thaliana* (Guo et al. 2008). PCs are cysteine-rich peptides bind arsenite efficiently and widely found in yeasts and plants (Schmoger et al. 2000; Wunschmann et al. 2007). Overexpression of the tobacco PC synthase in *S. cerevisiae* showed enhancement in Cd and As tolerance (Kim et al. 2005) but no effect was found in Cd or As accumulation. Thus, overexpression of gene related to GSH and PC synthesis in *S. cerevisiae* is responsible for higher arsenic resistance. Another pathway of As detoxification is the formation of metal sulfide particles as found in *Schizosaccharomyces pombe* and *Candida glabrata* has also been reported (Dameron and Winge 1990; Krumov et al. 2007).

The As tolerance of *S. cerevisiae* is achieved by two different pathways which suppress the cytotoxicity induced by arsenic species. One mechanism transport As^{III}-GSH complexes outside the cell through arsenite extrusion pump *Arr3p*, located in cell membrane (Ghosh et al. 1999). Another pathway of arsenic tolerance is achieved by overexpression of *Arr3p* in yeast (Bobrowicz et al. 1997) or by transporting As^{III}-GSH complex into the vacuole through the *Ycf1p*, a membrane efflux pump (Ghosh et al. 1999). The *Ycf1p* pump is a member of the ABC transporter super family that facilitates the transport of a wide range of GSH-conjugated substrates into the vacuole with the expenditure of ATP (Tsai et al. 2009). Deletion of the *Ycf1* gene results in *S. cerevisiae* becoming hypersensitive to arsenic, supporting the notion that both of these mechanisms are indispensable for survival at high arsenic concentrations and these two pathways function in a synergistic fashion as reported by genetic analysis (Tsai et al. 2009). Many species of fungi have the ability to produce methylated arsenic compounds including *Scopulariopsis brevicaulis* (Challenger 1945), *Candida humicola* (Cox and Alexander 1973), *Penicillium* sp. (Huysmans and Frankenberger 1991). In fungal cells the methyl group donor is S-adenosylmethionine and the methylation product is dimethylarsinic acid (DMA) (Drewniak and Sklodowska 2013). Like bacteria in fungus too, the methylation process is catalyzed by methyltransferase, an essential cytoplasmic protein which uses thioredoxin as a reducing agent during its catalysis (Silver and Phung 1996).

4 Bioremediation of As by Microbes

Microbes can thrive adverse condition such as high and low pH, high temperature, and higher concentration of toxic chemicals and sometimes can develop resistance against particular toxic chemicals in the environment because of their higher rate of genetic mutation. Under favorable condition some microbes can biotransform the hazardous toxic compounds into simpler nontoxic ones. Therefore, use of microbes for cleanup of contaminated lands and water are now common practice. The microbial process that influences the bioremediation of metals is as follows;

1. Efflux of metal ions outside cell.
2. Biosorption of the positively charged heavy metal ions (cation) to the negatively charged microbial cell membrane and from there inside the cell through specific transporter and subsequent bioaccumulation.
3. Immobilization of metal ions by microbial enzymes and biotransformation to less toxic form.

A number of gram-negative and gram-positive bacteria having *ars* genetic system use arsenite and arsenate in their metabolism. These bacteria can represent potential candidates for removal of As available in industrial areas or other arsenic-laden environments. The expression of *arsM* gene in bacteria significantly increases the efficiency of bacteria to transform arsenic into volatile methylated arsenic in aqueous and soil system. Therefore, genetically engineered bacteria harboring *arsM* gene could be an inexpensive and efficient strategy for bioremediation of arsenic-contaminated environment (Liu et al. 2011).

Takeuchi et al. (2007) demonstrated that naturally occurring marine bacteria *Marinomonas communis,* not genetically engineered, have the potentiality to remove arsenic from the culture medium amended with arsenate. Therefore, the bacteria may be a powerful candidate for bioremediation of arsenic-contaminated aquatic environment. *Lactobacillus acidophilus* are also reported to have the ability to fix and remove arsenic from water contaminated with high concentration of arsenic (50–1000 ppb) and maximum removal was found after 4 h of exposure (Singh and Sarma 2010).

4.1 Bioremediation of As by Using Biological Sulfate Reduction

Biological sulfate reduction (BSR) is the process which targets the reduction of sulfate to sulfide by using sulfate-reducing bacteria where sulfate is an electron acceptor and finally precipitating arsenic from reaction with sulfide (Gibson 1990). Kirk et al. (2004) showed that the presence of sulfate-reducing bacteria in groundwater efficiently reduced the level of arsenic. Sulfate-reducing bacteria generate sulfide which react with arsenic to form a low solubility complex and thus can precipitate arsenic.

4.2 As Removal from Contaminated Groundwater by Using Ferrous Oxide and Microorganism

Gallionella ferruginea and *Leptothrix ochracea* were found to support biotic oxidation of iron. This biotic oxidation when applied in laboratory condition; it was found that microorganisms oxidize As^{III} to As^{V} which then became adsorbed in Fe^{III}

resulting overall arsenic removal from contaminated water (Katsoyiannis and Zouboulis 2004). Sengupta et al. (2009) successfully applied this principle in field by using calculated volume of aerated groundwater to boost As^{III} oxidizing microbe population in the aquifer and to promote oxidation of Fe^{II} to Fe^{III} and thereby trapping As^{V}. This method was proved to be effective in reducing As concentration even below the regulatory standard of 10 μg L^{-1} from initial concentration of 250 μg L^{-1} and was tested extensively in field condition. In another report by Sun et al. (2010) simultaneous oxidation of As^{III} and Fe^{III} linked to denitrification was performed in continuous flow sand columns inoculated with As^{III} oxidizing densifying sludge. The soluble Fe^{II} oxidized into a mixture of insoluble Fe^{III} oxides dominated by hematite, resulting in enhanced immobilization of As in the column.

4.3 Bioremediation Through Arsenic Biomineralization

According to the report of Rodriguez-Freire et al. (2016) arsenic-containing water can be remediated through the formation of less soluble arsenic sulfide minerals by promoting concurrent biological reduction of As^{V} and SO_4^{2-} in a continuous bioreactor with a controlled pH ranging from 6.25 to 6.50,where ethanol serve as electron donor. The experiment resulted in 91.2% removal of total soluble As from the reactor in comparison with the control reactor lacking SO_4^{2-}. Among all the As^{V} and SO_4^{2-} reducing microorganisms, only a few such as *Sulfurospirillum arsenophilum* NIT-13 (Ahmann et al. 1994), *Desulfomirobium* sp. str. Ben-RB, and *Desulfovibrio* sp. str. Ben-RA are able to mineralize As (Macy et al. 2000). This study confirms that As biomineralization can be used as a remediation technique which is sustainable for long-term operation in a continuous bioreactor (Rodriguez-Freire et al. 2016).

4.4 Arsenic Bioremediation Through Production of Hyperaccumulating Bacterial Strains

The development of bio tools to remove As from contaminated environments has been gaining notice because of their potential in providing an effective technology for arsenic remediation. *Ochrobacterum tritici* SCII24^{T} is a highly As-resistant bacterium, possess two arsenic resistance operon *ars*1 (composed of five genes *ars*R, *ars*D, *ars*A, cystathione β synthase domain, *ars*B) and operon *ars*2 (composed of six genes encoding two additional *Ars*R, two *Ars*C, one *ACR*3, and an *Ars*H like protein) as previously reported by Branco et al. (2008). For the development of new bio tool a mutant *O. tritici* was constructed in which the genes coding for the arsenite efflux pumps were inactivated to achieve a strain able to accumulate arsenic (Sousa et al. 2015). The result of the experiment showed that double mutant *O. tritici* for *ars*B and *acr*3_1 was able to accumulate highest concentration of arsenic at moderate arsenite concentration and single *acr*3 mutant showed highest resistance to As^{III}, whereas only *ars*B mutants were found to be highly As^{III} sensitive. Therefore, *ars*B, *acr*31 double mutant *O. tritici* can be performed as a promising bio tool for environmental arsenic detoxification (Sousa et al. 2015).

4.5 Removal of As by Using Fixed Base Bioreactor System

Recently a method has been applied for simultaneous removal of nitrate and arsenic from a synthetic groundwater contaminated with nitrate, sulfate, and As^{V}, using a fixed-based bioreactor system. The bioreactor consisted of two consecutive columns; nitrate was removed from the first column and arsenic was mainly immobilized in the subsequent column through As-sulfide precipitation and surface precipitation on iron sulfides (Upadhyaya et al. 2010).

4.6 Bioremediation Through Genetically Engineered Bacterial Strain

In aqueous and soil systems the presence of bacteria expressing *ars*M significantly increased the methylation of arsenic from the medium. Thus volatilization approach using GE bacteria has been suggested as a potential bio tool for removal of As (Liu et al. 2011). Ability of biovolatilization by microbes varies genetically and a number of factors such as As species, soil moisture, soil temperature, organic materials are responsible for As volatilization (Edvantoro et al. 2004). Therefore, genetically engineered (GE) organisms with improved methylating capacity are more preferred for bioremediation of As (Rahman et al. 2014). For example, *Staphylococcus* sp. NBRIEAG-8 expressing *ars* gene and related upregulated proteins are capable of taking up and volatilizing As more efficiently (Srivastava et al. 2012).

The arsenic sensitive *E. coli* strain expressing the *ars*M gene cloned from *R. palustris* became As tolerant and catalyzed the formation of a number of methylated intermediates from arsenite including the volatile TMA, thus ultimately extruding the arsenic from cell and medium (Qin et al. 2006; Yuan et al. 2008). Two indigenous bacterial strain *Sphingomonas desiccabilis* and *Bacillus idriensis* genetically engineered for *ars*M generated tenfold higher levels of volatile arsenic than the wild type and were able to remove 2.2–4.5% arsenic from soil within 30 days (Liu et al. 2011). Another example is genetically engineered *Bacillus subtilis* 168 expressing the gene *CmarsM* (encodes the enzyme adenosyl methionine methyltransferase) from the thermophilic algae *Cyanidioschyzon merolae* are efficiently able to remove As from the organic manure (Huang et al. 2015). The altered *B. subtilis* 168 converted most of the inorganic As in the medium into methylated organic form of arsenic within 48 h and volatilized substantial amounts of dimethylarsine and trimethylarsine. The rate of As methylation and volatilization increased with temperature and at 50 °C modified strain significantly enhanced As volatilization when inoculated with As-contaminated organic manure (Huang et al. 2015).

4.7 Bioremediation by Arsenite Oxidizing Bacteria

Oxidation of arsenite (As^{III}) is an important remediation technique for As because As^{V} is less soluble and can be removed more effectively (Rahman et al. 2014). As^{III} can be oxidized to As^{V} by different prokaryote such as aerobic (*Herminiimonas arsenicoxydans*, *Thiomonas* sp., Rhizobium sp. strain NT26) or anaerobic (*Alkalilimnicola ehrlichii*) bacterial strain (Richey et al. 2009). Simeonova et al. (2005) proposed a method for treating arsenic-contaminated wastewater by using arsenite oxidizing bacteria. They applied the bacteria *H. arsenicoxydans* immobilized on Ca-alginate beads which oxidize As^{III}. Another process described by Lievremont et al. (2003) in which arsenite oxidation was performed by *H. arsenicoxydans* immobilized on cabazite. The arsenate produced was then adsorbed by a low-cost adsorbent, kutnahorite. Similar process was also described by Mokashi and Paknikar (2002) where, *Microbacterium lacticum* immobilized on brick pieces packed in a glass column mediated the oxidation process, followed by the removal of As^{V} with zero valent iron or activated charcoal.

4.8 Bioremediation by Arsenate-Reducing Bacteria

Arsenic-contaminated water can be treated by arsenate-reducing bacteria (Drewniak and Sklodowska (2013). Microbial reduction of As^{V} can lead to the mobilization of arsenite As^{III} which is then removed by precipitation or complexation with sulfide. Microorganisms that can promote coupled reduction of arsenate and sulfate are mostly selected in this remedial process, e.g., *Desulfomicrobium* Ben-RB and *Desulfotomaculum auripigmentum* lead to formation of arsenic trisulfide precipitates (Drewniak and Sklodowska 2013). The study of Chen et al. (2016) showed that biochar amendment in arsenic-contaminated mine tailings stimulated biological reduction of As^{V} by microbes and increase the relative abundance of As^{V} and Fe^{III} reducing bacteria mostly Geobacter, Anaeromyxobacter, Desulfosporosinus, and Pedobacter.

4.9 Mycoremediation of As

Fungi can be used as a potential agent for bioremediation of as they have high growth capacity due to mycelia branching. They are good accumulators of various heavy metals and thus are promising bioremediation agents.

1. From the study of Vala (2010) it was found that a facultative marine fungus *Aspergillus candidus* can grow luxuriantly at different concentrations (25 and 50 mg L^{-1}) of As^{III} and AS^{V} contaminated medium and maximum arsenic removal was recorded on the third day of treatment. Therefore, the fungus *A. candidus* could be a promising candidate for arsenic remediation.
2. Another study of Vala and Sutariya (2012) revealed that two marine fungi *A. flavus* and *Rhizopus* sp. when exposed to 25 and 50 mg L^{-1} concentration of sodium arsenate showed tolerance and also accumulated arsenic. These two species can be good agents for arsenic bioremediation, and their potential should be explored further.
3. In a study carried out by Singh et al. (2015), a total of 54 fungal strains were isolated from the arsenic-contaminated soil samples to test the As tolerance of fungal species. Higher rate of arsenic bioaccumulation and biovolatilization were observed in seven fungal strains namely *A. oryzae* FNBR_L35; *Fusarium* sp. FNBR_B7, FNBR_LK5, and FNBR_B3; *A. nidulans* FNBR_LK1; *Rhizomucor variabilis* sp. FNBR_B9; and *Emericella* sp. FNBR_BA5. These strains have been proposed for use in As remediation of As-contaminated soil and for promotion of plant growth (Singh et al. 2015).
4. The use of heavy metal solubilizing or bioleaching microorganisms in the contaminated soil to facilitate phytoextraction and enhancement of plant biomass production is now gaining interest to the scientist due to its cost-effectiveness. Some fungi like *Trichoderma atroviride*, *T. harzianum*, and *T. pseudokoningii* made a direct linkage between soil to plants, increasing heavy metal availability via solubilization while promoting plant growth and reducing metal toxicity (Adams et al. 2007; Shafiq and Jamil 2012; Cao et al. 2008). The fungal strain *T. virens* PDR-28 is a promising candidate for mycoremediation as it has significantly increased maize biomass and heavy metal uptake through ACC deaminase activity, siderophore production, and solubilization at abandoned mine sites as reported by Babu et al. (2014).

4.10 Bioremediation of As Through Constructed Wetlands

Constructed wetlands represent another strategy for As removal from ground and surface water as they are cost-effective natural systems that successfully remove various pollutants (Marchand et al. 2010; Chatterjee et al. 2013b). The removal process in constructed wetlands includes precipitation, co-precipitation, and sorption often mediated by microorganism. Arsenic can be precipitated as arsenosulfides, arsenates, co-precipitated with other sulfides or metal oxides or can be adsorbed onto the wetland matter (Lizama et al. 2011). Manipulation of the important factors for As removal like pH, presence of Fe and S, temperature, available sources of carbon and dissolved O_2 may be used to enhance the particular removal process (Drewniak and Sklodowska 2013).

5 Conclusion

Arsenic contamination is a major global problem and incautious industrial and mining activities have increased the mobilization of arsenic into the biosphere in addition to natural processes. Bioremediation process provides an effective and innovative measure for treatment of arsenic-contaminated lands and aquatic systems. Several microorganisms in nature exist which have evolved mechanisms to encounter this environmental challenge. The most important parameters for bioremediation by microbes are nature of pollutants, soil texture, pH, moisture content and hydrogeology, nutritional status, microbial diversity, temperature, and redox potential (Dua et al. 2002). The understanding of the metabolic and biochemical pathways involved in arsenic tolerance and accumulation is indispensable for engineering microbes for effective arsenic remediation. Although the initial reports are promising, microbial remediation of arsenic-contaminated areas is sometimes disappointing mainly because of the re-liberation of immobilized or adsorbed metalloids by some bacteria back into the environment, responsiveness of microbes to redox potential change and fluctuation in the valence state of the concerned toxic metal (Singh et al. 2015). Recently, scientific attention has been focused on the application of endophytic bacteria and fungi for their potential to enhance phytoremediation of heavy metals and pollutants (Stepniewska and Kuzniar 2013). Different genera of arsenic-resistant endophytic bacteria isolated from *Pteris vittata* and *Pteris multifida* have been found to possess the ability of both As^{V} reduction and As^{III} oxidation as reported by various researchers. These endophytes can be utilized for upgradation of phytoextraction processes (Zhu Ling et al. 2014; Tiwari et al. 2016). Thus, new discoveries and tools will certainly empower the scientists in applying bioremediation practices for mitigating arsenic contamination.

Acknowledgements The authors sincerely acknowledge Directors DRL, IRS, and Principal BCC for their kind support. Further, a sincere apology is rendered to the many colleagues whose works could not be referred to in this chapter due to space limitations.

References

Abedin MJ, Cresser MS, Meharg AA, Feldmann J, Cotter-Howells J (2002) Arsenic accumulation and metabolism in rice (*Oryza sativa* L.) Environ Sci Technol 36:962–968

Achour AR, Bauda P, Billard P (2007) Diversity of arsenite transporter genes from arsenic-resistant soil bacteria. Res Microbiol 158:128–137

Adams P, De-Leij FA, Lynch JM (2007) *Trichoderma harzianum* Rifai 1295-22 mediates growth promotion of crack willow (*Salix fragilis*) saplings in both clean and metal-contaminated soil. Microb Ecol 54:306–313

Afkar E (2012) Localization of the dissimilatory arsenate reductase in *Sulfurospirillum barnesii*s-train SeS-3. Amer J Agric Biol Sci 7:97–105

Ahmann D, Roberts AL, Krumholz LR, Morel FMM (1994) Microbe grows by reducing arsenic. Nature 371:750

Albarracin VH, Kurth D, Ordonez OF, Belfiore C, Luccini E, Salum GM, Piacentini RD, Farias ME (2015) High-up: a remote reservoir of microbial extremophiles in central Andean wetlands. Front Microbiol 6:1404

Babu AG, Shim J, Bang KS, Shea PJ, Oh BT (2014) *Trichoderma virens* PDR-28: a heavy metal-tolerant and plant growth-promoting fungus for remediation and bioenergy crop production on mine tailing soil. J Environ Manag 132:129–134

Bentley R, Chasteen TG (2002) Microbial methylation of metalloids: arsenic, antimony, and bismuth. Microbiol Mol Biol Rev 66:250–271

Bhattacharjee H, Rosen BP (2007) Arsenic metabolism in prokaryotic and eukaryotic microbes. In: Nies DH, Silver S (eds) Molecular microbiology of heavy metals. Springer-Verlag, Berlin, pp 371–406

Bobrowicz P, Wysocki R, Owsianik G, Goffeau A, Ulaszewski S (1997) Isolation of three contiguous genes, ACR1, ACR2 and ACR3, involved in resistance to arsenic compounds in the yeast *Saccharomyces cerevisiae*. Yeast 13:819–828

Branco R, Chung AP, Morais PV (2008) Sequencing and expression of two arsenic resistance operons with different functions in the highly arsenic-resistant strain *Ochrobactrum tritici* SCII24T. BMC Microbiol 8:95

Cao LX, Jiang M, Zeng ZR, Du AX, Tan HM, Liu YH (2008) *Trichoderma atroviride* F6 improves phytoextraction efficiency of mustard (*Brassica juncea* (L.) Coss. Var. Foliosa bailey) in Cd, Ni contaminated soils. Chemosphere 71:1769–1773

Challenger F (1945) Biological methylation. Chem Rev 36:315–361

Chatterjee S, Mitra A, Datta S, Veer V (2013a) Phytoremediation protocol: an overview. In: Gupta DK (ed) Plant based remediation process. Springer-Verlag, Berlin, pp 1–18

Chatterjee S, Mitra A, Datta S, Veer V (2013b) Use of wetland plants in bioaccumulation of heavy metals. In: Gupta DK (ed) Plant based remediation process. Springer-Verlag, Berlin, pp 117–139

Chen Z, Wang Y, Xia D, Jiang X, Fu D, Shen L, Wang H, Li QB (2016) Enhanced bioreduction of iron and arsenic in sediment by biochar amendment influencing microbial community composition and dissolved organic matter content and composition. J Hazard Mater 311:20–29

Chipirom K, Tanasupawat S, Akaracharanya A, Leepepatpiboon N, Prange A, Kim KW, Chul LK, Lee JS (2012) *Comamonas terrae* sp. nov., an arsenite-oxidizing bacterium isolated from agricultural soil in Thailand. J Gen Appl Microbiol 58:245–251

Cobbett C, Goldsbrough P (2002) Phytochelatins and metallothioneins: roles in heavy metal detoxification and homeostasis. Annu Rev Plant Biol 53:159–182

Cox DP, Alexander M (1973) Production of trimethylarsine gas from various arsenic compounds by three sewage fungi. Bull Environ Contam Toxicol 9:84–88

Dameron CT, Winge DR (1990) Peptide-mediated formation of quantum semiconductors. Trend Biotechnol 8:3–6

Dey S, Dou DX, Rosen BP (1994) ATP-dependent arsenite transport in everted membrane vesicles of *Escherichia coli*. J Biol Chem 269:25442–25446

Di Toppi LS, Gabbrielli R (1999) Response to cadmium in higher plants. Environ Exp Bot 41:105–130

Drewniak L, Sklodowska A (2013) Arsenic-transforming microbes and their role in biomining processes. Environ Sci Pollut Res 20:7728–7739

Dua M, Sethunathan N, Johri AK (2002) Biotechnology bioremediation success and limitations. Appl Microbiol Biotechnol 59:143–152

Edvantoro BB, Naidu R, Meghraj M, Merrington G, Singleton I (2004) Microbial formation of volatile arsenic in cattle dip site soils contaminated with arsenic and DDT. Appl Soil Ecol 25:207–217

Fisher E, Dawson AM, Polshyna G, Lisak J, Crable B, Perera E, Ranganathan M, Thangavelu M, Basu P, Stolz JF (2008) Transformation of inorganic and organic arsenic by *Alkaliphilus oremlandii* sp. nov. strain OhILAs. Annu Rev N Y Acad Sci 1125:230–241

Gehle K (2009) Substances USAfT, Registry D. Arsenic toxicity. Agency for Toxic Substances and Disease Registry, Atlanta, GA

Ghosh M, Shen J, Rosen BP (1999) Pathways of As (III) detoxification in *Saccharomyces cerevisiae*. Proc Natl Acad Sci U S A 96:5001–5006

Gibson GR (1990) Physiology and ecology of the sulphate-reducing bacteria. J Appl Microbiol 69:769–797

Gihring TM, Banfield JF (2001) Arsenite oxidation and arsenate respiration by a new Thermus isolate. FEMS Microbiol Lett 204:335–340

Guo JB, Dai XJ, Xu WZ, Ma M (2008) Overexpressing *GSH1* and *AsPCS1* simultaneously increases the tolerance and accumulation of cadmium and arsenic in *Arabidopsis thaliana*. Chemosphere 72:1020–1026

Heinrich-Salmeron A, Cordi A, Brochier-Armanet C, Halter D, Pagnout C, Abbaszadeh Fard E, Montaut D, Seby F, Bertin PN, Bauda P, Arsene Ploetze F (2011) Unsuspected diversity of arsenite-oxidizing bacteria as revealed by widespread distribution of the aoxB gene in prokaryotes. Appl Environ Microbiol 77:4685–4692

Hoshino YT, Morimoto S (2008) Comparison of 18SrDNA primers for estimating fungal diversity in agricultural soils using polymerase chain reaction denaturing gradient gel electrophoresis. Soil Sci Plant Nutr 54:701–710

Huang K, Chen C, Shen Q, Rosen BP, Zhao FJ (2015) Genetically engineering *Bacillus subtilis* with a heat-resistant arsenite methyltransferase for bioremediation of arsenic-contaminated organic waste. Appl Environ Microbiol 81:6718–6724

Huber R, Sacher M, Vollmann A, Huber H, Rose D (2000) Respiration of arsenate and selenate by hyperthermophilic archaea. Syst Appl Microbiol 23:305–314

Huysmans KD, Frankenberger WT (1991) Evolution of trimethylarsine by a *Penicillium* sp. isolated from agricultural evaporation pond water. Sci Total Environ 105:13–28

Inskeep WP, Macur RE, Hamamura N, Warelow TP, Ward SA, Santini JM (2007) Detection, diversity and expression of aerobic bacterial arsenite oxidase genes. Environ Microbiol 9:934–943

Karagas MR, Tosteson TD, Blum J, Morris JS, Baron JA, Klaue B (1998) Design of an epidemiologic study of drinking water arsenic exposure and skin and bladder cancer risk in a U.S. population. Environ Health Perspect 106:1047–1050

Katsoyiannis IA, Zouboulis AI (2004) Application of biological processes for theremoval of arsenic from ground waters. Water Res 38:17–26

Kim YJ, Chang KS, Lee MR, Kim JH, Lee CE, Jeon YJ, Choi JS, Shin HS, Hwang SB (2005) Expression of tobacco cDNA encoding phytochelatin synthase promotes tolerance to and accumulation of Cd and As in *Saccharomyces cerevisiae*. J Plant Biol 48:440–447

Kirk MF, Holm TR, Park J, Jin Q, Sanford RA, Fouke BW, Bethke CM (2004) Bacterial sulfate reduction limits natural arsenic contamination in groundwater. Geology 32:953–956

Krumov N, Oder S, Perner-Nochta I, Angelov A, Posten C (2007) Accumulation of CdS nanoparticles by yeasts in a fed-batch bioprocess. J Biotechnol 132:481–486

Kulp TR, Hoeft SE, Asao M, Madigan MT, Hollibaugh JT, Fisher JC, Stolz JF, Culbertson CW, Miller LG, Oremland RS (2008) Arsenic(III) fuels an oxygenic photosynthesis in hot spring biofilms from Mono Lake, California. Science 321:967–970

Lett MC, Muller D, Lievremont D, Silver S, Santini J (2012) Unified nomenclature for genes involved in prokaryotic aerobic arsenite oxidation. J Bacteriol 194:207–208

Liao VHC, Chu YJ, Su YC, Hsiao SY, Wei CC, Liu CW, Liao CM, Shen WC, Chang FJ (2011) Arsenite-oxidizing and arsenate-reducing bacteria associated with arsenic-rich groundwater in Taiwan. J Contam Hydrol 123:20–29

Lievremont D, N'Negue MA, Behra P, Lett MC (2003) Biological oxidation of arsenite: batch reactor experiments in presence of kutnahorite and chabazite. Chemosphere 51:419–428

Lievremont D, Bertin PN, Lett MC (2009) Arsenic in contaminated waters: biogeochemical cycle, microbial metabolism and bio treatment processes. Biochimie 91:1229–1237

Lim KT, Shukor MY, Wasoh H (2014) Physical, chemical, and biological methods for the removal of arsenic compounds. Biomed Res Int Article ID: 503784

Lin YF, Walmsley AR, Rosen BP (2006) An arsenic metallochaperone for an arsenic detoxification pump. Proc Natl Acad Sci U S A 103:15617–15622

Liu ZJ, Boles E, Rosen BP (2004) Arsenic trioxide uptake by hexose permeases in *Saccharomyces cerevisiae*. J Biol Chem 279:17312–17318

Liu S, Zhang F, Chen J, Sun G (2011) Arsenic removal from contaminated soil via biovolatilization by genetically engineered bacteria under laboratory conditions. J Environ Sci 23:1544–1550

Lizama AK, Fletcher TD, Sun G (2011) Removal processes for arsenic in constructed wetlands. Chemosphere 84:1032–1043

Macy JM, Santini JM, Pauling BV, O'Neill AH, Sly LI (2000) Two new arsenate/sulfate-reducing bacteria: mechanisms of arsenate reduction. Arch Microbiol 173:49–57

Mahimairaja S, Bolan NS, Adriano DC, Robinson B (2005) Arsenic contamination and its risk management in complex environmental settings. Adv Agron 86:1–82

Marchand L, Mench M, Jacob DL, Otte ML (2010) Metal and metalloid removal in constructed wetlands, with emphasis on the importance of plants and standardized measurements: a review. Environ Pollut 158:3447–3461

Mateos LM, Rosen BP, Messens J (2012) The arsenic stress defense mechanism of *Corynebaterium glutamicum* revealed. Understanding the Geological and Medical Interface of Arsenic-AS 2012. In: Proceeding of the fourth international congress on arsenic in the environment, Cairns, Australia, pp 209–210

Messens J, Silver S (2006) Arsenate reduction: thiol cascade chemistry with convergent evolution. J Mol Biol 362:1–17

Mokashi SA, Paknikar KM (2002) Arsenic (III) oxidizing *Microbacterium lacticum* and its use in the treatment of arsenic contaminated groundwater. Lett Appl Microbiol 34:258–262

Mukhopadhyay R, Rosen BP (1998) Saccharomyces cerevisiae ACR2 gene encodes an arsenate reductase. FEMS Microbiol Lett 168:127–136

Mukhopadhyay R, Shi J, Rosen BP (2000) Purification and characterization of Acr2p, the *Saccharomyces cerevisiae* arsenate reductase. J Biol Chem 275:21149–21157

Muller D, Lievremont D, Simeonova DD, Hubert JC, Lett MC (2003) Arsenite oxidase aox genes from a metal-resistant b-proteobacterium. J Bacteriol 185:135–141

Newman DK, Ahmann D, Morel FM (1998) A brief review of microbial arsenate respiration. Geomicrobiol J 15:255–268

Ordonez OF, Flores MR, Dib JR, Paz A, Farías ME (2009) Extremophile culture collection from Andean lakes: extreme pristine environments that host a wide diversity of microorganisms with tolerance to UV radiation. Microb Ecol 58:461–473

Ordonez OF, Lanzarotti EO, Kurth DG, Cortez N, Farías ME, Turjanski AG (2015) Genome comparison of two Exiguo bacterium strains from high altitude Andean lakes with different arsenic resistance: identification and 3D modeling of the Acr3 efflux pump. Front Environ Sci 3:50

Oremland RS, Stolz JF (2003) The ecology of arsenic. Science 300:939–944

Persson BL, Petersson J, Fristedt U, Weinander R, Berhe A, Pattison J (1999) Phosphate permeases of *Saccharomyces cerevisiae*: structure, function and regulation. Biochim Biophys Acta Rev Biomembr 1422:255–272

Qin J, Rosen BP, Zhang Y, Wang GJ, Franke S, Rensing C (2006) Arsenic detoxification and evolution of trimethylarsine gas by a microbial arsenite S-adenosylmethioninemethyltransferase. Proc Natl Acad Sci U S A 103:2075–2080

Rahman S, Ki-Hyun K, Saha SK, Swaraz AM, Paul DK (2014) Review of remediation techniques for arsenic (As) contamination: a novel approach utilizing bio-organisms. J Environ Manag 134:175–185

Rahman A, Nahar N, Nawani NN, Jana J, Ghosh SD, Olsson B, Mandal A (2015) Data in support of the comparative genome analysis of *Lysinibacillus* B1-CDA, a bacterium that accumulates arsenics. Data Brief 5:579–585

Ramasamy K, Kamaludeen, Banu SP (2006) Bioremediation of metals: microbial processes and techniques. In: Singh SN, Tripathi RD (eds) Environmental bioremediation technologies. Springer-Verlag, Berlin, pp 173–187

Richey C, Chovanec P, Hoeft SE, Oremland RS, Basu P, Stolz JF (2009) Respiratory arsenate reductase as a bidirectional enzyme. Biochem Biophys Res Commun 382:298–302

Rodriguez-Freire L, Moore SE, Sierra-Alvarez R, Root RA, Jon C, James A (2016) Field arsenic remediation by formation of arsenic sulfide minerals in a continuous anaerobic bioreactor. Biotechnol Bioeng 113:522–530

Rosen BP (2002) Biochemistry of arsenic detoxification. FEBS Lett 5:86–92

Sanders OI, Rensing C, Kuroda M, Mitra B, Rosen BP (1997) Antimonite is accumulated by the glycerol facilitator GlpF in *Escherichia coli*. J Bacteriol 179:3365–3367

Santini JM, Sly LI, Schnagl RD, Macy JM (2000) A new chemolithoautotrophic arsenite-oxidizing bacterium isolated from a gold mine: phylogenetic, physiological, and preliminary biochemical studies. Appl Environ Microbiol 66:92–97

Sanyal SK, Jahan MT, Chakrabarty RP, Hoque S, Anwar MH, Sultana M (2016) Diversity of arsenite oxidase gene and arsenotrophic bacteria in arsenic affected Bangladesh soils. AMB Express 6:21

Schmoger MEV, Oven M, Grill E (2000) Detoxification of arsenic by phytochelatins in plants. Plant Physiol 122:793–801

SenGupta B, Chatterjee S, Rott U, Kauffman H, Bandopadhyay A, DeGroot W, Nag NK, Carbonell-Barrachina AA, Mukherjee S (2009) A simple chemical-free arsenic removal method for community water supply—a case study from West Bengal, India. Environ Pollut 157:3351–3353

Shafiq M, Jamil S (2012) Role of plant growth regulators and a saprobic fungus in enhancement of metal phytoextraction potential and stress alleviation in pearl millet. J Hazard Mater 237:186–193

Silver S, Phung LT (1996) Bacterial heavy metal resistance: new surprises. Annu Rev Microbiol 50:753–789

Silver S, Phung LT (2005) Genes and enzymes involved in bacterial oxidation and reduction of inorganic arsenic. Appl Environ Microbiol 71:599–608

Simeonova DD, Micheva K, Muller DAE, Lagarde F, Lett MC, Groudeva VI, Lievremont D (2005) Arsenite oxidation in batch reactors with alginate-immobilized ULPAs1 strain. Biotechnol Bioeng 91:441–446

Singh AL, Sarma PN (2010) Removal of arsenic(III) from waste water using *Lactobacillus acidophilus*. Biorem J 14:92–97

Singh M, Srivastava PK, Verma PC, Kharwar RN, Singh N, Tripathi RD (2015) Soil fungi for mycoremediation of arsenic pollution in agriculture soils. J Appl Microbiol 119:1278–1290

Singhal RK, Anderson ME, Meister A (1987) Glutathione, a 1st line of defense against cadmium toxicity. FEBS Lett 1:220–223

Sorokin DY, Tourova TP, Sukhacheva MV, Muyzer G (2012) *Desulfuribacillus alkaliarsenatis* gen. nov. sp. nov., a deep-lineage, obligatory anaerobic, dissimilatory sulfur and arsenate-reducing, haloalkaliphilic representative of the order Bacillales from soda lakes. Extremophiles 16:597–605

Sousa T, Branco R, Piedade AP, Morais PV (2015) Hyper accumulation of arsenic in mutants of Ochrobactrum tritici silenced for arsenite efflux pumps. PLoS One 10:e0131317

Srivastava PK, Vaish A, Dwivedi S, Chakrabarty D, Singh N, Tripathi RD (2011) Biological removal of arsenic pollution by soil fungi. Sci Total Environ 409:2430–2442

Srivastava S, Verma PC, Singh A, Mishra M, Singh N, Sharma N, Singh N (2012) Isolation and characterization of *Staphylococcus* sp. strain NBRIEAG-8 from arsenic contaminated site of West Bengal. Appl Microbiol Biotechnol 95:1275–1291

Stepniewska Z, Kuzniar A (2013) Endophytic microorganisms—promising applications in bioremediation of greenhouse gases. Appl Microbiol Biotechnol 97:9589–9596

Stolz JF, Basu P, Oremland RS (2010) Microbial arsenic metabolism: new twists on an old poison. Microbe 5:53–59

Sun W, Sierra-Alvarez R, Hsu I, Rowlette P, Field JA (2010) Anoxic oxidation of arsenite linked to chemolithotrophic denitrification in continuous bioreactors. Biotechnol Bioeng 105:909–917

Switzer Blum J, Burns Bindi A, Buzzelli J, Stolz JF, Oremland RS (1998) *Bacillus arsenicoselenatis*, sp. nov., and *Bacillus selenitireducens*, sp. nov.: two haloalkaliphiles from Mono Lake, California that respire oxyanions of selenium and arsenic. Arch Microbiol 171:19–30

Takeuchi M, Kawahata H, Gupta LP, Kita N, Morishita Y, Yoshiro O, Komai T (2007) Arsenic resistance and removal by marine and non-marine bacteria. J Biotechnol 127:434–442

Tiwari S, Sarangi BK, Thul ST (2016) Identification of arsenic resistant endophytic bacteria from *Pteris vittata* roots and characterization for arsenic remediation application. J Environ Manage 180:359–365

Tripathi RD, Srivastava S, Mishra S, Singh N, Tuli R, Gupta DK, Matthuis FJM (2007) Arsenic hazards: strategies for tolerance and remediation by plants. Trend Biotechnol 25:158–165

Tsai SL, Singh S, Chen W (2009) Arsenic metabolism by microbes in nature and the impact on arsenic remediation. Curr Opin Biotechnol 20:659–667

Upadhyaya G, Jackson J, Clancy TM, Hyun SP, Brown J, Hayes KF, Raskin L (2010) Simultaneous removal of nitrate and arsenic from drinking water sources utilizing a fixed-bed bioreactor system. Water Res 44:4958–4969

Vala AK (2010) Tolerance and removal of arsenic by a facultative marine fungus *Aspergillus candidus*. Bioresour Technol 101:2565–2567

Vala AK, Sutariya V (2012) Trivalent arsenic tolerance and accumulation in two facultative marine fungi. Jundishapur J Microbiol 5:542–545

Valls M, Lorenzo VD (2002) Exploiting the genetic and biochemical capacities of bacteria for the remediation of heavy metal pollution. FEMS Microbiol Rev 26:327–338

vanden Hoven RN, Santini JM (2004) Arsenite oxidation by the heterotroph *Hydrogenophaga* sp. str. NT-14: the arsenite oxidase and its physiological electron acceptor. Biochim Biophys Acta 1656:148–155

World Health Organization (2001) Environmental health criteria 224: arsenic and arsenic compounds. WHO, Geneva, pp 1–108

Wunschmann J, Beck A, Meyer L, Letzel T, Grill E, Lendzian KJ (2007) Phytochelatins are synthesized by two vacuolar serine carboxypeptidases in *Saccharomyces cerevisiae*. FEBS Lett 581:1681–1687

Wysocki R, Bobrowicz P, Ulaszewski S (1997) The *Saccharomyces cerevisiae* ACR3 gene encodes a putative membrane protein involved in arsenite transport. J Biol Chem 272:30061–30066

Wysocki R, Chery CC, Wawrzycka D, Van Hulle M, Cornelis R, Thevelein JM, Tamas MJ (2001) The glycerol channel *Fps1p* mediates the uptake of arsenite and antimonite in *Saccharomyces cerevisiae*. Mol Microbiol 40:1391–1401

Yamamura S, Amachi S (2014) Microbiology of inorganic arsenic: from metabolism to bioremediation. J Biosci Bioeng 118:1–9

Yuan CG, Lu XF, Qin J, Rosen BP, Le XC (2008) Volatile arsenic species released from *Escherichia coli* expressing the AsIII S-adenosylmethioninemethyltransferase gene. Environ Sci Technol 42:3201–3206

Zargar K, Hoeft S, Oremland R, Saltikov CW (2010) Identification of a novel arsenite oxidase gene, arxA, in the haloalkaliphilic, arsenite oxidizing bacterium *Alkalilimnicola ehrlichii*strain MLHE-1. J Bacteriol 192:3755–3762

Zarger K, Conrad A, Bernick DL, Lowe TM, Stolc V, Hoeft S, Oremland RS, Stolz J, Saltikov CW (2012) ArxA, a new clade of arsenite oxidase within the DMSO reductase family of molybdenum oxidoreductase. Environ Microbiol 14:1635–1645

Zhao FJ, Ma JF, Meharg AA, McGrath SP (2008) Arsenic uptake and metabolism in plants. New Phytol 181:777–794

Zhu Ling J, Guan DX, Luo J, Rathinasabapathi B, Ma LQ (2014) Characterization of arsenic resistant endophytic bacteria from hyperaccumulators *Pteris vittata* and *Pteris multifida*. Chemosphere 113:9–16

Index

A
Accessibility, 128
Acute poisoning
  psychosis development, 40
Acute promyelocytic leukemia (APL), 54
Algae, 148
  Arsenosugars and Arsenolipids formation, 152
  As^{III} Oxidation, 151
  As^{V} Reduction, 151–152
  phytoremediation, AS, 156
Animal and human systems
  histone acetylation, 76–78
  histone markers, 83
  histone phosphorylation, 79
  miRNAs, 80, 81
  PEMT gene, 82
Antioxidant response elements (AREs), 53
APL. *See* Acute promyelocytic leukemia (APL)
ARGONAUTE (AGO) proteins, 80
Arsenic (As), 48–52, 123–124, 148–153, 174, 197–199
  AdoMet methyl donor, 47
  affecting factors, 155
  Algal Cells (*see* Algae)
  APL, 54
  AS3MT catalyzation, 45
  biogeochemical cycle, 19
  bioremediation, 202–207
  and cancer (*see* Cancer)
  CML, 54
  contamination (*see* Contamination in groundwater)
  detection
    atomic spectrometry-based methods, 27
    electrochemical detection, 28
    radiochemical methods, 28–29
    X-Ray spectroscopy, 29
  environmental perspectives and human health, 5, 38
  gene expression, role in, 69
  genetic susceptibility, 55
  IARC, 44
  iAs bio-transform, 45
  iAs, MMA and DMA, 45
  identification of elements, 16
  -induced carcinogenesis, 45
  inorganic forms, 38
  kinetics and metabolism, 22
  malaria treatment, 53
  methylation and toxicity interlinkage, 46–47
  microbial genomics, 172, 173
  MMA5, MMA3 and DMA3, 45
  non-hyperaccumulator and hyperaccumulator plants, 153, 154
  organic forms, 38 (*see also* Chemical and physicochemical process) (*see also* Small scale arsenic removal units)
  properties, 15
  reduction, 178–179
  sequestration of PC-As(III) Complex, 180–182
  toxic effects of crop plants (*see* Crop plants and As effects)
  toxicity to humans, 43–44
  uptake in plants, 178
  uptake mechanism, bacteria (*see* Bacteria)
  uptake, algae and Plants (*see* Uptake and metabolism, AS)

D.K. Gupta, S. Chatterjee (eds.), *Arsenic Contamination in the Environment*,
DOI 10.1007/978-3-319-54356-7

Arsenic (As) contamination, 6–9
exposure
in 1990s, 6
groundwater
remediation measures and recommendations, 9
urinary arsenic species, 8
high-profile poison, 3
mid-1800s to the mid-1900s, 3, 4
toxic effects (*see* Toxic effects of arsenic)
Arsenic (As) poisoning
asymptomatic/preclinical stage, 41
fertilizers, production of, 40
malignancy, 42
symptomatic/clinical stage, 41–42
Arsenic (As) sources
exposure, 14

B
Bacteria, 199–201
uptake and metabolism
Arsenate reduction, 200
Arsenite oxidation, 199
methylation of As species, 201
uptake mechanism, AS, 198
Bacteria, volatile arsenic, 23
Base-excision repair (BER) mechanism, 49
BATs. *See* Best available technologies (BATs)
BER. *See* Base-excision repair (BER) mechanism
Best available technologies (BATs), 105
Biogeochemical cycle, Arsenic, 19
Biological sulfate reduction (BSR), 203
Bioremediation of Arsenic (AS)
biological sulfate reduction (BSR), 203
genetically engineered bacterial strain, 205
production of hyperaccumulating bacterial strains, 205
Biosand filter (BSF), 112
Bucket Treatment Unit (BTU), 114

C
Cancer
carcinogens, 48
chromosomal and genomic instability, 51
EGFR and RTK signalling pathway, 48
functions, arsenic exposure, 50
JAK and STAT3 activation, 52
and micro-RNA expression, 52
MMAIII-mediated inhibition, 48
NER and BER mechanisms, 49
SAM, methly donor, 51
Chemical and physicochemical process
arsenic oxidizing agents, 105
arsenic removal techniques, 106, 107
BATs, 105
RO/NF method, 109
sorption, 109
Chinese brake fern, 187
Chronic myelogenous leukemia (CML), 54
Chronic poisoning, 40–41
CIM. *See* Composite iron matrix (CIM)
CML. *See* Chronic myelogenous leukemia (CML)
Coagulation/filtration method, 106
Composite iron matrix (CIM), 114
Cortex, 129
Coupled restriction enzyme digestion-random amplification (CRED-RA) technique, 85
Crop plants and AS effects
bioaccumulation, 127
bioavailability in rice rhizosphere, 134
ecosystems, 123
transportation and removal, 129, 130
uptake and transport, 126
water, 124

D
Decontamination, AS, 8–9
Dimethylarsinous acid (DMA3), 45
DNA methylation
footprint, 73
gene promoters, 74, 75
plants toxicity, 83–85
tumorigenesis/oncogenesis, 74

E
EGFR. *See* Epidermal growth factor receptor (EGFR)
Electrochemical detection, 28
Electrophile response elements (EREs), 53
Environmental transport and distribution, arsenic, 19–21
Epidermal growth factor receptor (EGFR), 52
Epigenetic regulation. *See* Animal and human systems
EREs. *See* Electrophile response elements (EREs)

F
Fixed bed adsorption, 108
Fungus, 202
uptake and metabolism
arsenic, 202

G
Gene promoters
  hypermethylation, 75
  hypomethylation, 74
Genetic engineering in phytoremediation of arsenic, 175–176
Genetic susceptibility, 55

H
Histone acetylation, 76–78
Histone methylation, 78
Histone phosphorylation, 79
Human health. *See* Arsenic (As)

I
Indigenous methods, 110, 111
International Agency for Research on Cancer (IARC), 44
International Development Enterprises (IDE), 113

K
Kanchan Arsenic Filter (KAF), 116
3-Kolshi filter, 114

L
Layered double hydroxide (LDH), 108
Lime softening, 106
Long noncoding RNA (lncRNA) genes, 85
Low-as accumulating crop cultivars, 191, 192

M
Magc-Alcan filter, 113
Mean total arsenic concentrations, environment, 21
Melanosis
  melano-keratosis, 41
Metal ions, 128
Methanoarchaea, 23, 25
Methylation-sensitive amplification polymorphism (MSAP), 84
Micro-RNA expression, 52
miRNAs (miR)
  AGO proteins, 80
  PHLPP gene, 80
  plants toxicity, 86, 88, 89
  RNA polymerase II and pre-miRNAs, 80
  RSIC guidelines, 80
Monomethylarsonic acid (MMA5), 45
Monomethylarsonous acid (MMA3), 45
MSAP. *See* Methylation-sensitive amplification polymorphism (MSAP)

N
Nano-filtration (NF) technology, 109
Nirmal filter, 113
Nrf2-KEAP1 signaling pathway, 53
Nucleotide excision repair (NER), 49

O
Organic arsenic (oAs) compounds, 16

P
Phytoremediation, AS, 174
  algae, 155–156
  bacterial arsenate reductase and γ-ECS genes, 159
  synthesis of chelators, 157–158
PI3K/AKT signaling pathway, 52–53
Plant tissue culture
  development of low-as accumulating crop cultivars, 190–192
Plants toxicity
  induced DNA methylation, 83–85
  miRNAs, 86–89
  stress response, histone modification, 86
PMI. *See* Primary methylation index (PMI)
Point-Of-Use (POU) treatment, 110
Prevention of AS uptake, food crops
  As^{III} Efflux Rate, 160
  limitation, 160
  manipulation, Lsi1 and Lsi2 Genes, 160
Primary methylation index (PMI), 45

R
Radiochemical methods, arsenic detection, 28
Remediation, AS, 9
Reverse osmosis (RO), 109
RNA-induced Silencing Complex (RSIC), 80

S
S-Adenosyl methionine (SAM), 51
SDMRs. *See* Stress-specific differentially methylated region (SDMRs)
Secondary methylation index (SMI), 45
SFFV. *See* Spleen focus forming virus (SFFV)
Shapla and Surokka filter, 113–114

Small scale arsenic removal units
BSF, 112–113
BTU, 114
KAF, 115, 116
3-Kolshi filter, 114
SAR, 116–117
SONO filter, 114–115
SMI. *See* Secondary methylation index (SMI)
Somaclonal variation, plant, 192, 193
SONO filter, 114, 115
Sorption method
fixed bed adsorption, 108
LDH, 108
Spleen focus forming virus (SFFV), 51
Spleen proviral integration oncogene B (SPIB), 51
Stress-specific differentially methylated region (SDMRs), 85
Subterranean Arsenic Removal (SAR) technology, 116

T
Toxic effects of arsenic, 7, 8
Tumorigenesis/oncogenesis, 74

U
Uptake and metabolism, As
algal cells, 148
arsenate, plants, 148–150
arsenite, plants, 150
bacteria, 198
fungus, 198
organic species, 151

V
Volatile arsenicals, 18
Volatilization, arsenic
bacteria, 23
fungi, 22
methanoarchaea, 23–25

X
X-ray cross complementing protein 1 (XRCC1), 49
X-ray spectroscopy, arsenic detection, 29

Printed in the United States
By Bookmasters